AWS Certified Solutions Architect – Professional Exam Guide (SAP-C02)

Gain the practical skills, knowledge, and confidence to ace the AWS (SAP-C02) exam on your first attempt

Patrick Sard

Yohan Wadia

AWS Certified Solutions Architect – Professional Exam Guide (SAP-C02)

Authors: Patrick Sard and Yohan Wadia

Reviewers: Vineethkumar Marpadge, Vishal Munguskar, and Subhajit Bhattacharya

Publishing Product Manager: Sneha Shinde

Editorial Director: Alex Mazonowicz

Development Editor: Shubhra Mayuri

Presentation Designer: Salma Patel

Editorial Board: Vijin Boricha, Megan Carlisle, Wilson D'souza, Ketan Giri, Saurabh Kadave, Alex Mazonowicz, Gandhali Raut, and Ankita Thakur

First Published: February 2024

Production Reference: 3171024

Published by Packt Publishing Ltd.
Grosvenor House
11 St Paul's Square
Birmingham
B3 1RB

ISBN: 978-1-80181-313-6

www.packtpub.com

Contributors

About the Authors

Patrick Sard, as an AWS Solutions Architect, specializes in capturing and conveying best practices, offering prescriptive guidance on application and systems design across Amazon platforms and technologies. His role involves direct interaction with customers and partners, creating technical content, conducting events, evangelism, training, and providing operational event support. Additionally, Patrick contributes to the evolution of Amazon platforms and technologies by providing direct input and feedback from the field to the engineering teams developing these offerings.

With more than 20 years experience in IT, in his current and past roles, Patrick has been responsible for establishing and maintaining strategic vision, driving and influencing innovation, and delivering integrated and persuasive architectural thought leadership. With end-to-end project lifecycle experience, encompassing sales, inception, design, implementation, and post-sales support, he combines acute capabilities and insights into technology and business. This enables him to analyze and address client needs from various perspectives. In close collaboration with technical leaders, he drives the creation and execution of strategies promoting growth and innovation.

Yohan Wadia is a seasoned Cloud Solutions Architect with expertise in designing, implementing, and optimizing cloud-based solutions to meet the diverse needs of modern businesses. With 14 years of hands-on experience in Cloud Computing, AWS, and other cloud technologies, he has demonstrated a strong proficiency in architecting robust, scalable, and cost-effective solutions tailored to his clients' specific requirements. Yohan is based out of Brussels, Belgium together with his wife, Anahita.

About the Reviewers

Vineethkumar Marpadage has more than 5 years of exceptional work experience in various cloud technologies. He is an active Subject Matter Expert in AWS CodeSuite and Elastic Beanstalk Services who holds an AWS Solutions Architect Professional Certification alongside 5 other AWS Certifications, including Security Specialty. With vast knowledge on AWS Technologies, he designed complex architectures and managed multiple projects on AWS that spans across Applications, DevOps, and infrastructure monitoring. His unwavering commitment to staying abreast of emerging technologies and opportunities underscores his dedication to continuous growth and innovation.

Vishal Munguskar is an AWS DevOps engineer with a passion for cloud-based solutions. He has spent the last three years honing his skills in design, implementation, and management. Vishal proudly holds two AWS certifications: AWS Certified Solutions Architect - Associate and AWS Certified DevOps Engineer – Professional. He has a rich background with experience in AWS DevOps tools and practices, including CI/CD, automation, monitoring, and security. Currently, he is engrossed in the DSA project (Healthpharma sector) and is responsible for maintaining the security and data integrity of the organization's AWS cloud infrastructure. Vishal strives to ensure a smooth and reliable data flow and to work towards optimizing his organization's AWS cloud resources for scalability. Collaborating with other engineers and stakeholders to deliver high-quality products and services is part of his daily routine.

Subhajit Bhattacharya is an AWS Solutions Architect with over 10 years of experience in creating secure, scalable, and resilient cloud solutions. He has a proven track record across IoT, insurance, financial, and banking sectors, with expertise in modernizing and migrating applications to the cloud. Skilled in a wide range of AWS services, Subhajit holds multiple certifications, including AWS Solutions Architect Professional, and is recognized for his innovative approaches and leadership in cloud architecture.

Table of Contents

2

3

6

Meeting Reliability Requirements   111

7

Ensuring Business Continuity   139

8

Meeting Performance Objectives   153

9

Establishing a Deployment Strategy   175

10

Designing for Cost Efficiency   199

11

Improving Operational Excellence   217

12

Improving Reliability 237

13

Improving Performance   255

14

Improving Security   265

19

Determining a New Architecture for Existing Workloads   343

20

Determining Opportunities for Modernization and Enhancements   361

21

Preface

Amazon Web Services (**AWS**) has been the leading cloud service provider for over 15 years. Millions of companies across the globe—including the fastest-growing start-ups, largest enterprises, and leading government agencies—are using AWS to lower costs, become more agile, and innovate faster.

This book covers all four domains of the **AWS Solutions Architect Professional** (**SA Pro**) certification, an advanced certification from AWS that aims to give certified individuals credibility on the market and trust in their ability to design enterprise-grade solutions with AWS.

This book will help you reinforce your skills in delivering scalable, highly available, fault-tolerant, secure, performant, and cost-optimized solutions that keep up with the most demanding requirements and constraints. It will also stimulate you to think from multiple different perspectives before jumping into solution design.

By the end of this book, you will have validated your advanced AWS technical skills and expertise and, most likely, built enough confidence to take the AWS SA Pro certification exam with success.

Who This Book Is for

This book is for you if you are an experienced IT professional with a good understanding of designing cloud architectures on AWS. Prior knowledge of the AWS platform and services is expected.

Although not required for the exam, hands-on experience in working with AWS-based applications for two or more years is recommended.

What This Book Covers

This book is aligned with the AWS SA Pro certification contents outline updated in 2022 and covers the following topics.

Chapter 1, Determining an Authentication and Access Control Strategy for Complex Organizations, explains the concepts supporting **Identity and Access Management** (**IAM**) on AWS. It covers aspects such as cross-account access control and user federation, along with the multiple ways an organization can provide their users with access to AWS by leveraging their existing directory service.

Chapter 2, Designing Networks for Complex Organizations, covers the AWS services that can be used to design hybrid networks, allowing organizations to access AWS resources from their on-premises environments and vice versa and communicate across multiple AWS accounts.

Chapter 3, Designing a Multi-Account AWS Environment for Complex Organizations, explains how to organize resources across multiple AWS accounts for an organization. It discusses how to approach billing and resource isolation and how to increase security across your entire organization as well as for individual business units.

Chapter 4, Ensuring Cost Optimization, focuses on the various mechanisms and services available to keep your AWS bill under control.

Chapter 5, Determining Security Requirements and Controls, examines access control aspects for resources spread across your organization's AWS accounts. It takes you through the relevant services and patterns to apply security and compliance controls.

Chapter 6, Meeting Reliability Requirements, explores several architectural patterns and relevant AWS services to help you choose a design and implementation strategy for your reliability requirements.

Chapter 7, Ensuring Business Continuity, walks you through different strategies to protect your critical workloads on AWS in case of a disaster.

Chapter 8, Meeting Performance Objectives, puts the focus on finding a solution design that meets your performance objectives. It covers the best practices and strategies to implement when designing for performance on AWS.

Chapter 9, Establishing a Deployment Strategy, explores the various options offered on AWS for deploying and updating workloads.

Chapter 10, Designing for Cost Efficiency, discusses the various pricing models offered by AWS and how to select the optimal one for your requirements and constraints.

Chapter 11, Improving Operational Excellence, discusses the importance of reviewing your existing operational strategy through AWS best practices to identify areas of improvement.

Chapter 12, Improving Reliability, guides you in assessing your workload design through the lens of AWS reliability best practices.

Chapter 13, Improving Performance, covers the specifics of performance engineering to help you improve your workload's performance efficiency by following AWS best practices.

Chapter 14, Improving Security, focuses on AWS security practices to help you reinforce the security of your workloads.

Chapter 15, Improving Deployment, takes you through the deployment strategies and AWS capabilities that can help you improve deployment for an existing solution.

Chapter 16, Exploring Opportunities for Cost Optimization, discusses the aspects that can help you optimize your costs further on AWS.

Chapter 17, Selecting Existing Workloads and Processes to Migrate, dives into migration readiness, application discovery, application portfolio analysis, and how to select and prioritize workloads for migration.

Chapter 18, Selecting Migration Tools and Services, presents an overview of the tools and AWS services that you can leverage to prepare for a migration.

Chapter 19, Determining a New Architecture for Existing Workloads, guides you through the vast number of options available for compute, storage, and databases when migrating a workload.

Chapter 20, Determining Opportunities for Modernization and Enhancements, explores serverless and container options, as well as purpose-built databases and new cloud-native integration patterns.

SA Pro Certification – November 2022 Release

A new version of AWS' SA Pro certification was released at the end of 2022. There are some small yet important differences compared to the previous version of the exam. The following table illustrates the changes made to the exam content outline.

AWS SA Pro certification v1 (SAP-C01)		AWS SA Pro certification v2 (SAP-C02)	
Domain	% of exam	Domain	% of exam
Design for organizational complexity	12.5%	Design solutions for organizational complexity	26%
Design for new solutions	31%	Design for new solutions	29%
Migration planning	15%	Continuous improvement for existing solutions	25%
Cost control	12.5%	Accelerate workload migration and modernization	20%
Continuous improvement	29%		

Table 0.1: The differences between both versions of the AWS SA Pro certification

In the new exam version, the domains have been re-balanced and, as you will note from the exam description, AWS underlines that they are now putting more emphasis on architecting solutions aligned with the AWS Well-Architected Framework. Also, notably, cost optimization has become an integral part of the solution design process, instead of being treated separately.

How to Get the Most Out of This Book

This book is directly aligned with the SA Pro certification from AWS. It is advisable to stick to the following steps when preparing for your SA Pro exam:

1. Read this book from end to end.

2. Go through the AWS SA Pro certification guidelines.

3. Refer to the AWS Well-Architected Framework and AWS whitepapers and documentation, many of which are highlighted in the *Further Reading* sections throughout this book.

4. Review the practice exam questions in the book and in the AWS SA Pro certification guide.

5. Attempt the online practice exam. Make a note of the concepts you are weak in, revisit them in the book, and re-attempt the practice questions.

6. Keep attempting the practice exams until you are able to score at least 80% within the time limit.

7. Review exam tips on the AWS website.

SA Pro certification candidates will gain a lot of confidence if they approach their SA Pro certification preparation using these steps.

Online Practice Resources

With this book, you will unlock unlimited access to our online exam-prep platform (*Figure 0.1*). This is your place to practice everything you learn in the book.

> **How to access the resources**
>
> To learn how to access the online resources, refer to *Chapter 21, Accessing the Online Practice Resources* at the end of this book.

Figure 0.1: Online exam-prep platform on a desktop device

Sharpen your knowledge of AWS SAP_C02 concepts with multiple sets of practice questions, interactive flashcards, and exam tips accessible from all modern web browsers.

Download the Color Images

We also provide a PDF file that has color images of the screenshots/diagrams used in this book.

You can download it here: `https://packt.link/euXEF`.

Conventions Used

There are a number of text conventions used throughout this book.

`Code in text`: Indicates code words in text, database table names, folder names, filenames, file extensions, pathnames, dummy URLs, user input, and Twitter handles. Here is an example: "You will use the `detect_labels` API from Amazon Rekognition in the code."

Any block of code, command-line input, or output is written as follows:

```
class MyFirstCdkStack extends Stack {
  constructor(scope: App, id: string, props?: StackProps) {
  super(scope, id, props);
```

Bold: Indicates a new term, an important word, or words that you see onscreen. For example, words in menus or dialog boxes appear in the text like this. Here is an example: "In **CloudWatch**, each **Lambda** function will have a **log group** and, inside that log group, many **log streams**."

> **Tips or important notes**
> Appear like this.

Get in Touch

Feedback from our readers is always welcome.

General feedback: If you have questions about any aspect of this book, mention the book title in the subject of your message and email us at customercare@packt.com.

Errata: Although we have taken every care to ensure the accuracy of our content, mistakes do happen. If you have found a mistake in this book, we would be grateful if you would report this to us. Please visit www.packtpub.com/support/errata, selecting your book, clicking on the Errata Submission Form link, and entering the details. We ensure that all valid errata are promptly updated in the GitHub repository, with the relevant information available in the Readme.md file. You can access the GitHub repository: https://packt.link/WDFVz.

Piracy: If you come across any illegal copies of our works in any form on the Internet, we would be grateful if you would provide us with the location address or website name. Please contact us at copyright@packt.com with a link to the material.

If you are interested in becoming an author: If there is a topic that you have expertise in and you are interested in either writing or contributing to a book, please visit authors.packtpub.com.

Share Your Thoughts

Once you've read *AWS Certified Solutions Architect – Professional Exam Guide (SAP-C02)*, we'd love to hear your thoughts! Scan the QR code below to go straight to the Amazon review page for this book and share your feedback.

https://packt.link/r/1801813132

Your review is important to us and the tech community and will help us make sure we're delivering excellent quality content.

Download a Free PDF Copy of This Book

Thanks for purchasing this book!

Do you like to read on the go but are unable to carry your print books everywhere?

Is your eBook purchase not compatible with the device of your choice?

Don't worry, now with every Packt book you get a DRM-free PDF version of that book at no cost.

Read anywhere, any place, on any device. Search, copy, and paste code from your favorite technical books directly into your application.

The perks don't stop there, you can get exclusive access to discounts, newsletters, and great free content in your inbox daily.

Follow these simple steps to get the benefits:

1. Scan the QR code or visit the link below:

https://packt.link/free-ebook/9781801813136

2. Submit your proof of purchase.
3. That's it! We'll send your free PDF and other benefits to your email directly.

1

Determining an Authentication and Access Control Strategy for Complex Organizations

This chapter introduces the first objective of this book, that is, determining an authentication and access control strategy to address the requirements of complex organizations.

To pass your **Amazon Web Services (AWS)** Solutions Architect Professional certification, you will start by revisiting the key concepts and mechanisms supporting **Identity and Access Management (IAM)** on AWS. You will then investigate cross-account access control and user federation, which are essential support for complex organizations. Finally, you will cover the multiple ways an organization can provide its users access to AWS by leveraging its existing directory service.

The following topics will be covered in this chapter:

- Identity and Access Management
- Examining access control
- Leveraging access delegation
- Considering user federation
- Reviewing AWS Directory Service

Since you are preparing for **AWS Solutions Architect Professional** certification, you should have already been exposed to AWS environments and services. You may already be familiar with most of the concepts covered in this chapter, but it's worth revisiting them as to ensure you have the core knowledge needed to pass the certification.

Making the Most Out of this Book – Your Certification and Beyond

This book and its accompanying online resources are designed to be a complete preparation tool for your AWS SAP-C02 Exam.

The book is written in a way that you can apply everything you've learned here even after your certification. The online practice resources that come with this book (*Figure 1.1*) are designed to improve your test-taking skills. They are loaded with practice questions, interactive flashcards, and exam tips to help you work on your exam readiness from now till your test day.

> **Before You Proceed**
>
> To learn how to access these resources, head over to *Chapter 21, Accessing the Online Practice Resources*, at the end of the book.

Figure 1.1: Dashboard interface of the online practice resources

Here are some tips on how to make the most out of this book so that you can clear your certification and retain your knowledge beyond your exam:

1. Read each section thoroughly.

2. **Make ample notes**: You can use your favorite online note-taking tool or use a physical notebook. The free online resources also give you access to an online version of this book. Click the BACK TO THE BOOK link from the Dashboard to access the book in Packt Reader. You can highlight specific sections of the book there.

3. **Practice Questions**: Go through the practice questions provided online with this book. Use them to test yourself on the concepts learned. If you get some answers wrong, go back to the book and revisit the concepts you're weak in.

4. **Flashcards**: After you've gone through the book and start reviewing the online flashcards. They will help you memorize key concepts.

5. **Exam Tips**: Review these from time to time to improve your exam readiness even further.

Now that you have gone through the preceding tips to help you maximize the benefits of this book and the online resources provided with it, you can proceed to the first main topic of this chapter.

Diving into Identity and Access Management

AWS Identity and Access Management (IAM) is used to define and control who can access which resources in an AWS environment. IAM concepts and how they provide security controls are a key part of the exam. Here are some key concepts:

Every new AWS account comes with a root user that has full access to all AWS services and all the resources in the account. As a best practice, it is recommended to do the following:

- Immediately protect that root user with **multi-factor authentication (MFA)**.

- Secure the root user credentials and only use them if you need to perform specific service and account management tasks that only the root user can perform.

> **Note**
> See `https://packt.link/eVR8z` for more details on tasks that only the root user can perform.

IAM users

An IAM user is an entity designed to be associated with a single individual or application. It is used to allow access to AWS resources either through the AWS Management Console (providing a username and password) and/or programmatically (using an access key and a secret access key) from the **command-line interface (CLI)** or one of the AWS **software development kits (SDKs)**. IAM users are given permissions either by being directly assigned IAM policies or by being assigned to an IAM user group.

MFA

The security of IAM users can be enhanced by enabling MFA. Users then must provide two forms of authentication. The first is identity credentials such as username/password or access key/secret access key. The second form takes the shape of a temporary six-digit numeric code. This can be provided by a hardware device, an application on a mobile device such as a smartphone or tablet, or sent by AWS to a mobile device as an SMS.

IAM User Groups

An IAM user group is a collection of IAM users. It cannot be used to access AWS services directly. Its main purpose—other than grouping related users together—is to assign the same permissions to all the users in the group.

Instead of granting permissions individually to users, it is recommended that you give permissions to a group, and then you add the users who need these permissions to the group. When a user should no longer have the permissions granted to the group, you simply remove them from the group. Managing permissions for users then becomes a lot easier.

As an example, think of a group representing a company's software developers and another group representing its system administrators. Because each user in a group automatically inherits the permissions assigned to the group, it then becomes easier for an AWS administrator to maintain the permissions required by each group member (software developers or system admins, in the given example) at a group level rather than individually at a user level.

An IAM user can be assigned to multiple IAM user groups, in which case it inherits the permissions of all the user groups it is a member of.

IAM Roles

An IAM role is an identity that possesses specific permissions. It is like an IAM user in which it provides access to AWS resources and defines what the user or application assuming that role can do on AWS. It is different from an IAM user in that a role is not associated with a single individual or application but can be assumed by multiple entities. IAM roles are used to provide temporary credentials to entities (individuals, AWS services, or applications).

> **Important Note**
>
> An IAM user or role cannot span multiple AWS accounts. The *Leveraging Access Delegation* section will cover how cross-account access (accessing resources in AWS account A from AWS account B) can be granted.

IAM Policies

An IAM policy is an object that allows access control on AWS. It can be assigned either to an IAM identity (user, user group, or role) or to an AWS resource. When access to an AWS resource is requested, IAM will evaluate the permissions defined by all the policies entering the scope of that request and based on their intersection, decide whether access is allowed or denied. IAM supports multiple types of policies: **identity-based policies**, **resource-based policies**, **permissions boundaries**, **organizations' service control policies (SCPs)**, **access control lists (ACLs)**, and session policies. You will explore these in the following sections.

Identity-Based Policies

Identity-based policies are **JavaScript Object Notation (JSON)** policy documents that are attached to IAM identities (users, user groups, or roles). They control what actions an IAM identity (user, user group, or role) can perform on which AWS resources and under which conditions. They are further subdivided into the following categories:

1. **Managed policies**: Managed policies are named policies and can be assigned to any number of IAM identities. They can be of two sorts, as outlined here:

   I. AWS-managed policies: Policies created and managed by AWS

   II. Customer-managed policies: Policies created and managed by you

2. **Inline policies**: Inline policies are permissions directly attached to a single IAM identity. Their life cycle is the same as that identity's life cycle.

Here is an example of an identity-based policy that gives read-only access to all **Simple Storage Service (S3)** objects on its AWS account:

```
{
  "Version": "2012-10-17",
  "Statement": [
    {
      "Effect": "Allow",
      "Action": [
        "s3:Get*",
        "s3:List*"
      ],
      "Resource": "*"
    }
  ]
}
```

> **Note**
>
> Remember that an IAM identity has by default no permissions at all on AWS. You must assign one or more identity-based policies for it to be able to do something on AWS.

Resource-Based Policies

Resource-based policies are JSON policy documents that are attached to AWS resources. They control what actions can be performed on the attached resource(s) by which principal (user or role) under which conditions. As opposed to identity-based policies, resource-based policies are always inline policies.

Here is an example of a resource-based policy providing permissions to any principal (user or role) in the account to get any object from the S3 bucket identified by the `Resource` attribute:

```
{
  "Version": "2012-10-17",
  "Id": "Policy123456789",
  "Statement": [
    {
      "Sid": "",
      "Action": [
        "s3:GetObject"
      ],
      "Effect": "Allow",
      "Resource": "arn:aws:s3:::my-bucket/*",
      "Principal": "*"
    }
  ]
}
```

> **Note**
>
> Remember that resource-based policies provide an opportunity to further protect your AWS resources by limiting not just what actions can be performed but also the IAM entities allowed to perform them.

Permissions Boundaries

Permissions boundaries allow us to define the maximum permissions that identity-based policies can give to IAM entities (user or role). An entity can then only perform actions allowed by both its identity-based policies and its permissions boundaries. Setting a permissions boundary does not give permissions on its own but it limits what the entity can do. It is also worth noting that permissions boundaries do not affect the permissions provided by resource-based policies. A resource-based policy can provide permissions to an IAM entity beyond the scope defined by permissions boundaries.

Look at an example to learn how this all works.

Suppose that you have an IAM user with the following identity-based policy:

```
{
    "Version": "2012-10-17",
    "Statement": {
        "Effect": "Allow",
        "Action": "iam:ChangePassword",
        "Resource": "*"
    }
}
```

The policy gives them the ability to change their user's password.

Now, imagine that you set for the same user the following permissions boundary:

```
{
    "Version": "2012-10-17",
    "Statement": [
        {
            "Effect": "Allow",
            "Action": [
                "lambda:*",
                "ec2:*"
            ],
            "Resource": "*"
        }
    ]
}
```

The permissions boundary policy limits the user to any action on both AWS Lambda and **Amazon Elastic Compute Cloud (EC2)** resources in their AWS account.

Now, if the user tries to change their password (after all, they were given the permissions to do so) the operation will fail. Why? The user was given permissions in their identity-based policy to change their password, but their permissions boundary only allowed actions to be performed on AWS Lambda and Amazon EC2, not on AWS IAM. For the `iam:ChangePassword` action to work, the user's permissions boundary would need to be expanded to include at least that action on AWS IAM.

On a second note, the user was not given permission to perform any other action than `iam:ChangePassword`. So, even though their permissions boundary would authorize them to perform any action on AWS Lambda and Amazon EC2, they simply cannot do so because their identity-based policy is too restrictive.

Additionally, imagine that you have defined on the same account the previous sample resource-based policy providing permissions to any principal (user or role) in the account to get any object from the S3 bucket identified by the Resource attribute. Even though the permissions boundary policy limits your user's actions to AWS Lambda and Amazon EC2 resources, your user will nevertheless be able to get any object from the S3 bucket specified in the resource-based policy. Why is that? Because permissions boundaries only affect the scope of permissions defined by identity-based policies, not by resource-based policies.

That said, permissions boundaries are an efficient mechanism to thwart privilege escalation by limiting what IAM entities (user or role) can do independently of the identity-based policies that are attached to them. Diving deeper into this goes beyond the scope of this chapter.

> **Note**
>
> Make sure to review `https://packt.link/4xWr4` to clearly understand how this can be achieved in practice.

Remember that permissions boundaries do not add permissions to what an IAM entity can do; they only limit what it can do.

Organizations SCPs

AWS Organizations is a service that allows us to centrally manage multiple AWS accounts belonging to the same organization. It provides the ability to structure them according to a hierarchy of **organizational units (OUs)**. It also provides a feature called SCPs that allows us to limit permissions for all member accounts in either an entire organization or a single OU. It allows an AWS administrator to enforce those controls from a central place and easily adapt them to the evolution and needs of your organization over time.

You will cover SCPs in more detail while learning AWS Organizations in a later chapter.

> **Note**
>
> Remember that SCPs are one efficient mechanism to enforce security controls, following your organization's security policies systematically and repeatedly without having to duplicate the same policies in each individual AWS account.

ACLs

ACLs are service policies that let you control which principals in another account are allowed to access a resource in the current account. They are somewhat like resource-based policies but present clear differences:

- They are expressed using Extensible Markup Language (XML) and not JSON

- They cannot be used to control access within the same account as the principal requesting access.

ACLs can help address some very specific use cases in which resource-based policies may not be your best option. Such use cases include, for instance, controlling access to S3 objects that do not belong to the S3 bucket owner, or setting different access permissions for individual objects inside the same folder within a given S3 bucket.

Amazon S3 but also services such as AWS **Web Application Firewall (WAF)**, and Amazon **Virtual Private Cloud (VPC)** support ACLs.

> **Note**
>
> To dive deeper into ACLs and specific use cases where they can prove useful, consult the ACL overview page in the Amazon S3 documentation at `https://packt.link/bd4MI`.

Session Policies

Session policies are policies passed as a parameter when programmatically creating a temporary session for a role or a federated user. They are meant to limit the permissions from the role's or user's identity-based policy that are allowed during the session. Like permissions boundaries, they cannot be used to grant more permissions than those already allowed by the identity-based policy.

To create a temporary session for a role, you use either the `AssumeRole`, the `AssumeRoleWithSAML`, or the `AssumeRoleWithWebIdentity` application programming interface (API) operation from the AWS **Security Token Service (STS)**. For federated users, temporary sessions are created using the `GetFederationToken` API operation from AWS STS.

You will review the importance of sessions when going through further details about access control later in this chapter.

Identity-Based Versus Resource-Based Policies

So, what most AWS administrators end up wondering is: *Should I rather use identity-based policies or resource-based policies?* Well, it depends—it really does.

It is not a binary decision. Very likely, you will end up using a combination of both plus several of the other types of policies we've previously discussed.

A first observation is that not all AWS resources support resource-based policies, so identity-based policies remain the only way of giving access to the resources that do not. A second key aspect is that identity-based policies provide a means to manage access to AWS resources independently of these AWS resources and their life cycle. So, it makes sense for an AWS administrator who wants to centralize access control as much as possible in a single place to rely on identity-based policies to control access to AWS resources.

Does this mean that resource-based policies would not be useful? No—they will prove very useful for providing additional control to the security-savvy resource owner who wants to further ensure that only specific entities within their organization are allowed to access their resources.

For instance, consider a situation where multiple teams have access to resources in your account. As the owner of sensitive data sitting in a specific S3 bucket, you want to restrict access to people who you know have been approved to access that data. You happen to know that these people are all part of a specific OU inside your organization. You could then make sure to restrict access, independently of the permissions that anyone is assigned in your organization, by enforcing specific conditions for accessing your sensitive data. For that, you would typically use resource-based policies. In this example, you could add a condition such as the following to the resource-based policy assigned to your resources sitting, for instance, in an S3 bucket:

```
"Condition" : { "ForAnyValue:StringLike" : {
"aws:PrincipalOrgPaths":[" o-alphab2avo/r-abcd/ou-wxyz- hal45678/*"]
}}
```

And does this mean that other policies, such as SCPs and session policies, are not so important? No—it most certainly does not. Complex organizations with multiple AWS accounts will also, at least, leverage SCPs on top of identity-based and resource-based policies. You will cover SCPs in more detail when you learn AWS Organizations.

> **Note**
>
> **Identity-Based Policies and Resource-Based Policies Are Permissions Policies**
>
> Provided that there are no further constraints (SCPs, permissions boundaries, session policies, or ACLs), an entity will have access to a resource in the same account if access is not explicitly denied and if at least one policy statement grants access, whether it is within an identity-based policy or a resource-based policy.

Now examine how these different functionalities offered by IAM can be leveraged to control access to AWS resources.

Examining Access Control

In this section, you will investigate two different approaches organizations can take to control access, either based on a principal's role or based on specific properties, also known as attributes, characterizing a principal.

Role-Based Access Control (RBAC)

This is the traditional access control approach where the permissions defining the actions that a principal (user or role) can perform are based on the function that the person has in their job. You typically define different policies for the roles you need in your organization and then assign these policies to IAM identities (users, user groups, or roles). Note that AWS already includes some managed policies for job functions.

Since granting the least privilege is a best practice, you should restrict the permissions that you grant to the various job functions to the strict minimum each of them needs to perform its job. Typically, you do that by explicitly listing the AWS resources each job function has access to, including the relevant actions that can be performed.

The main issue in doing that effectively is that the AWS resources a job function needs access to are likely to be quite variable over time. Every time a new AWS resource that one or more job functions need access to get added (for instance, a new S3 bucket, a new DynamoDB table, or a new **Relational Database Service (RDS)** database), you must update the policies for these job functions. Every time a specific job function changes team or project, you must update its permissions. The more granular control you want to have, the more policies you will have, and the more often you will need to update them.

So, the temptation rapidly grows for an AWS administrator to provide slightly too wide permissions by using *wildcards* ('*') in job functions' policies instead of actual resources' **Amazon Resource Names (ARNs)**.

Therefore, a lack of flexibility and management overhead is often the main drawback of a traditional RBAC approach.

Attribute-Based Access Control (ABAC)

This is where an ABAC approach can help. With ABAC, you can create fewer and more compact policies.

You can, for instance, give an identity access to all resources of a certain type (for example, all S3 buckets) within the account but then specify a condition filtering on the value of a certain tag (or attribute), which would only allow access to the resource if the principal tag matched the resource tag. That tag could represent anything meaningful to your organization and context (for example, **business unit (BU)** or project), and you could filter on multiple tags (or attributes).

ABAC allows much more flexibility and is far less painful in terms of administration while still providing fine-grained access control to implement the least-privilege best practice.

ABAC really shines in helping you simplify permissions management at scale; therefore, it comes in handy for complex organizations. These organizations are likely to use identity federation for their workforce to access AWS resources leveraging their already existing **identity provider (IdP)**, such as Microsoft **Active Directory (AD)**. What you will see later when discussing user federation is that you can pass session tags when you create a temporary session for a role or a federated user, and when integrating your IdP with AWS, you can specify which user attributes you would like to pass as session tags. So, imagine that you set up attributes such as job function, department, and team as session tags. These session tags are then assigned to the principal and can be used in your identity-based or resource-based policies.

As an example, to illustrate the preceding information, the following identity-based policy would give, to the identity it is assigned to, full access to EC2 resources in the account, with the condition that the principal is in a **systems administrator (SysAdmin)** job function, provided that the job function is being passed as a session tag through identity federation:

```
{
  "Version": "2012-10-17",
  "Statement": [
    {
      "Effect": "Allow",
      "Action": "ec2:*",
      "Resource": "*",
      "Condition": {"StringEquals": {"aws:PrincipalTag/
        jobfunction": "SysAdmin"}}
    }
  ]
}
```

So, now that you have covered IAM and how to leverage it to establish granular access control on AWS resources, which other security mechanisms do you need to investigate? Well, to start with, you need to consider cases where the permissions provided to a principal are simply not enough for what they need to achieve. What can you do in such cases while still adhering to your least-privilege best practice? This is where access delegation can help, and this is what you are going to cover next.

Leveraging Access Delegation

You are now going to investigate **access delegation**. Access delegation is essentially used for the following reasons:

- Providing an entity temporary access to resources that they do not have access to with their current privileges. This could be one of the following:

  - A user that needs temporarily elevated privileges to perform a specific task

  - An application or AWS service that requires specific privileges

- Providing an entity access to resources located in another AWS account.

Now, start by examining these cases.

Temporary Access Delegation

Take for instance, the first use case where you need to provide trusted users, applications, or AWS services with temporary security credentials so that they can access your AWS resources. As the name implies, the security credentials that will be provided are temporary, which has the following benefits:

- The access provided is limited to a short period of time, typically ranging from a few minutes to a few hours.

- No need to store or manage (for example, rotate) temporary credentials like you would have to with permanent credentials because they are short-lived.

- No need to define an AWS identity for each user that requires access. Instead, you can leverage identity federation, which relies on temporary security credentials.

AWS STS is the central AWS service for requesting temporary security credentials on AWS. In a nutshell, AWS STS creates a new session on AWS with temporary security credentials that include an access key pair (access key, secret access key) and a session token. These credentials can then be used by end users or applications to access your resources. You can additionally programmatically pass session policies and session tags using AWS STS. As you noticed earlier, the resulting permissions are then the intersection of the temporarily assumed IAM role's identity-based policies and the session policies.

Accessing Resources from One Account to Another

Complex organizations, as with most enterprises, will maintain multiple AWS accounts to host their workloads on AWS. In some circumstances, entities in one account need to access resources in another account. Access across accounts is not permitted by default on AWS; resources in one account are fully isolated within the account and cannot be accessed from other AWS accounts unless specific permissions are explicitly given.

IAM roles' Trust Policies

Cross-account access is made possible because of IAM roles. IAM roles have a distinct capacity to act both as an identity and as a resource, and as such, you can associate both identity-based policies and resource-based policies with IAM roles. In the case of IAM roles, resource-based policies are also called **trust policies**.

To illustrate how this works, consider that an organization stores some data on S3 in a central account and makes it available to other accounts in the organization. By default, none of the other accounts has permission to perform any action on the S3 bucket belonging to the central account.

One solution consists of using resource-based policies and giving read or write access, as needed, to that central S3 bucket and to all the other accounts. But because not all AWS services support resource-based policies, look at a more generic solution that could also work with other AWS services and not just Amazon S3.

In this case, the generic approach consists of leveraging IAM roles and establishing trust between the central account and the other accounts. An IAM role can be assumed by another principal (user or role), including from another account, if they have the right permissions. For this, you create an IAM role in the account where the resources need to be accessed from other accounts, where you will establish trust. The central account is the trusting account, while all the other accounts that need to access resources in that account are the trusted accounts.

In the central account (call it `AccountA`), you define a role (call it `roleA`), and you assign that role a trust policy in which you specify that it can be assumed by other principals from any other accounts that need to. Suppose that you need to get it accessed by `userB` from another account, `AccountB`. The trust policy would look something like this:

```
{
  "Version": "2012-10-17",
  "Statement": [
    {
      "Effect": "Allow",
      "Principal": {
        "AWS": "arn:aws:iam::AccountB:user/userB"
      },
      "Action": "sts:AssumeRole"
    }
  ]
}
```

This trust policy instructs IAM that `roleA` in `accountA` can be assumed by `userB` from `accountB`.

So, you need to provide roleA enough privileges to be able to read or write to S3 on the account. You do this by either assigning an identity-based policy to roleA or by attaching resource-based policies to the relevant S3 buckets, to give the necessary permissions to roleA. See the previous sections on identity-based and resource-based policies for more details.

Now, most importantly, in accountB, you must also give userB permissions to assume roleA in accountA. You set such permissions in an identity-based policy that you assign to userB, such as this:

```
{
  "Version": "2012-10-17",
  "Statement": {
    "Effect": "Allow",
    "Action": "sts:AssumeRole",
    "Resource": "arn:aws:iam::AccountA:role/roleA"
  }
}
```

When userB in accountB assumes roleA in accountA, they inherit the permissions assigned to roleA and can then perform any action roleA is entitled to on resources in accountA.

AWS Resource Access Manager (RAM)

There is also an alternative for sharing resources across multiple accounts.

AWS RAM is a central service that allows you to share resources you own in one account with multiple accounts either within your own AWS OU or beyond. There is one caveat, though: you cannot share all types of AWS resources with RAM as only a limited set is supported.

> **Note**
>
> For a list of shareable resources, check the documentation at https://packt.link/qEeQE.

That said, for those supported resources it is a convenient and easy way to share resources across your entire organization or with specific AWS accounts. It works quite simply. AWS RAM allows you to share resources in one account with a set of principals. Such principals can be individual AWS accounts, the member accounts of an organization, an OU in AWS Organizations, or even individual IAM entities (users or roles). It is worth noting that when you share resources within an organization that has resource sharing enabled, principals in your organization immediately get access to the shared resources. AWS RAM does not send them an invitation in this case. However, when you share resources in a standalone AWS account or in an organization without resource sharing enabled, AWS RAM would send the principals an invitation that they must accept to gain access to the shared resources.

This is a widespread approach among complex organizations since they often need to share resources located in and managed from a central AWS account with multiple AWS accounts (for instance, EC2 images, Route 53 **Domain Name System (DNS)** forwarding rules, AWS Transit Gateway, and more).

So far, you have covered IAM, how to leverage it to establish granular access control on AWS resources, and how to delegate access when you need either additional permissions on a temporary basis or access resources in a different AWS account. You will now investigate a security mechanism that enables organizations to reuse their existing IdP(s) to access AWS—namely, the concept of user federation.

Considering User Federation

It is only natural for organizations to want to reuse their existing IdPs to give their workforce, customers, or partners access to AWS without having to create and manage a separate set of identities on AWS. This avoids multiplying long-lived security credentials unnecessarily and, as such, limits the security risks. You can leverage either **AWS Single Sign-On (AWS SSO)** or AWS IAM to enable user federation depending on the use case.

AWS SSO is well suited for cases where you want to establish user federation across multiple AWS accounts and leverage your existing corporate or a third-party IdP. You can then assign permissions to your users based on their group membership in your IdP's directory and control access by modifying users and groups on your IdP. You can also implement ABAC, whether via the user information synchronized with your IdP via **System for Cross-domain Identity Management (SCIM)** or by passing user attributes in **Security Assertion Markup Language 2.0 (SAML 2.0)** assertions.

You can also use AWS IAM for user federation with individual accounts. IAM is well suited to cases where, for instance, you want to use separate IdPs for different accounts—in some cases, your corporate IdP, and in others, a third-party **OpenID Connect (OIDC)** IdP. You can also implement ABAC by passing user attributes in SAML 2.0 assertions.

In both cases, if you use an ABAC approach you can easily change, add, or revoke user permissions by changing the user attributes centrally in your IdP. Whether these attributes get synchronized through SCIM or are passed in SAML 2.0 assertions, AWS SSO will pass them as session tags, which are temporary tags that remain valid for the duration of the federated session you establish. As you also saw previously, you can then use these session tags to filter access to AWS resources by using conditions within IAM policies.

So far, you have covered IAM, how to leverage it to establish granular access control on AWS resources, and how to delegate access when you need either additional permissions on a temporary basis or to access resources in a different AWS account and user federation to leverage your existing IdPs for authentication. The last part of your journey through authentication and authorization mechanisms on AWS will take you to the directory service provided by AWS: AWS Directory Service.

Reviewing AWS Directory Service

AWS Directory Service offers several choices for organizations to deploy existing applications on AWS that rely on Microsoft AD or Lightweight **Directory Access Protocol (LDAP)**. This is the native AWS service to use when you need a directory to manage users, groups, devices, and access.

AWS Directory Service proposes different options to use Microsoft AD with AWS services, as follows:

- **Simple AD**: A low-scale and low-cost directory with basic Microsoft AD compatibility
- **AD Connector**: A proxy service to connect to a remote Microsoft AD on-premises
- **Managed Microsoft AD**: A Microsoft AD environment managed by AWS

The following sections will discuss the main differences between these three options and when to use one or the other.

Simple AD

Simple AD is a Microsoft AD-compatible directory that provides basic AD features such as managing user accounts, group memberships, and group policies, joining a (Linux or Windows) EC2 instance to your directory, and Kerberos-based SSO.

Simple AD is a standalone directory running on AWS to create and manage user identities and control access to applications. It is compatible with various applications and tools that only require basic AD features. For example, Simple AD supports Amazon WorkSpaces and Amazon QuickSight, among others. You can also use it to log in to the AWS Management Console.

Limitations

Simple AD does not support MFA or trust relationships with other domains and many other advanced AD features. Simple AD is not compatible with Amazon RDS for SQL Server, and it can support applications or services on AWS only.

When to Use It

Use Simple AD if you need to support Windows workloads that need basic AD features, to work with compatible AWS applications, or to support Linux workloads that need an LDAP service.

AD Connector

AD Connector is a scalable proxy service that forwards requests to your on-premises AD. It offers an easy way to connect compatible AWS applications—for instance, Amazon WorkSpaces, Amazon QuickSight, or Amazon EC2 for Microsoft Windows Server instances—to your existing on-premises Microsoft AD. It does not require you to synchronize your directory and does not add extra cost or complexity—there's no need, for instance, to set up a federation infrastructure.

AD Connector supports numerous AWS applications and services such as Amazon WorkSpaces, Amazon WorkDocs, Amazon QuickSight, or Amazon Connect. It also lets you join your EC2 Windows instances to your on-premises AD domain seamlessly. Users can also leverage it to sign in to the AWS Management Console and manage AWS resources using their existing AD credentials.

AD Connector does not cache any information on AWS, which has both benefits (your users' information is never stored on AWS) and drawbacks (every sign-in request is a request back to your on-premises directory).

There is a one-to-one relationship between AD connectors and your AD domains; each on-premises domain requires a separate AD connector before it can be used for authentication.

Limitations

AD Connector is not compatible with Amazon RDS for SQL Server or with Amazon FSx for Windows File Server.

When to Use It

AD Connector is recommended when you want to use your existing on-premises directory with compatible AWS services.

Managed Microsoft AD

Managed Microsoft AD essentially lets you run Microsoft AD as a managed service on AWS. A fully featured AD, it supports Microsoft SharePoint, Microsoft SQL Server, Always On availability groups, and numerous .NET applications. Some AWS-managed applications and services are also supported, among which are Amazon WorkSpaces, Amazon QuickSight, Amazon Connect, Amazon RDS for Microsoft SQL Server, and Amazon FSx for Windows File Server.

Managed Microsoft AD is provided as a single-tenant solution; you do not share it or its components with any other AWS customer. It comes by default with two domain controllers deployed in separate **Availability Zones (AZs)** for increased resiliency.

You can leverage user credentials stored in AWS Managed Microsoft AD to work with compatible applications. Alternatively, you can also establish trust with your existing AD infrastructure and use credentials whether they are stored on an AD running on-premises or on Amazon EC2. By joining your EC2 instances to AWS Managed Microsoft AD, users can access Windows workloads on AWS with the same Windows SSO experience as on-premises.

AWS Managed Microsoft AD lets you sign in to the AWS Management Console. Leveraging AWS SSO, you can further obtain short-term credentials that you can use with the AWS SDK and CLI. It also gives you access, through integration with SAML, to sign in to various cloud applications. It also offers built-in integrations with Microsoft Office 365 and other business applications (for example, Salesforce) with credentials stored in AWS Managed Microsoft AD. You can further reinforce security by enabling MFA.

When to Use It

AWS Managed Microsoft AD is the recommended option when you need genuine AD features to support AWS applications or Windows workloads. That includes Amazon RDS for Microsoft SQL Server and Amazon FSx for Windows File Server. Similarly, it remains your best option if you require access to business applications such as Office 365, Salesforce, and more.

Summary

In this first chapter, you have reviewed the core IAM concepts of AWS. You then investigated cross-account access control and user federation, which are essential elements for supporting complex organizations. Finally, you looked at the various flavors offered by AWS Directory Service. All these functionalities are core for securing access to AWS resources for complex organizations. So, do make sure these elements are crystal clear in your mind before moving on and, especially if that is not the case, have a look at the additional resources provided in the next section.

The next chapter of this book will take you through the AWS networking capabilities you need to know about to select and configure the optimal network topology for your organization.

Further Reading

For more information, please refer the following resources:

- Getting started with IAM: `https://packt.link/jmFBw`

- RBAC versus ABAC: `https://packt.link/DxXF8`

- Best Practices for Security, Identity, & Compliance: `https://packt.link/IfCum`

- Security Pillar – AWS Well-Architected Framework: `https://packt.link/Y3Lez`

- AD services on AWS: `https://packt.link/P0Ys5`

2
Designing Networks for Complex Organizations

Networking is a key aspect in meeting the security and compliance requirements of an organization. It determines whether and how resources in your Amazon Web Services (AWS) environment can be accessed from anywhere in your organization and beyond.

This chapter will cover the services on AWS that can be used to design hybrid networks, allowing an organization to reach AWS resources from its on-premises environments and vice versa. You will learn how to connect to AWS services without going through the internet and will also look at network communication across multiple AWS accounts.

The following topics will be covered in this chapter:

- Establishing **virtual private network (VPN)** connections
- Introducing **AWS Direct Connect (DX)**
- Introducing AWS Storage Gateway
- Leveraging **virtual private cloud (VPC)** endpoints
- Introducing AWS Transit Gateway

Establishing VPN Connections

The first option when it comes to protecting connectivity between an enterprise's on-premises infrastructure and its AWS environment is to establish a VPN connection. AWS offers several alternatives to achieve that. The following section details each of them.

AWS Managed VPN

The first one is **AWS Managed VPN**, or **Site-to-Site VPN**. This is a fully managed service that provides an **Internet Protocol Security (IPsec)** VPN connection over the internet from your on-premises network equipment to AWS-managed network equipment attached to your AWS VPC.

The VPN concentrator end on the AWS side can be either a **virtual private gateway (VGW)** attached to a single VPC, as illustrated in the following diagram, or **a transit gateway (TGW)** attached to multiple VPCs (see *Figure 2.2*). The other end connecting to your on-premises equipment is called a **customer gateway (CGW)**:

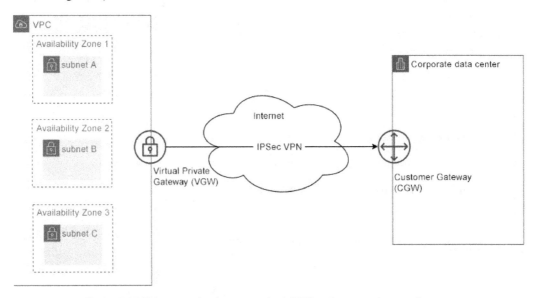

Figure 2.1: VPN connection between single VPC and on-premises equipment

The architecture you choose depends on your AWS environment network topology. *Figure 2.2* shows the TGW option:

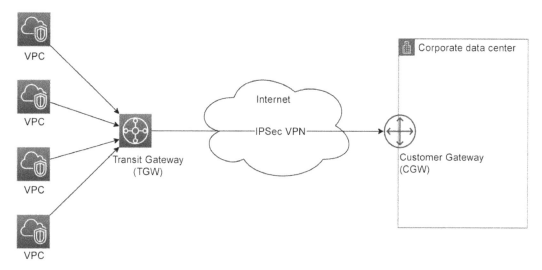

Figure 2.2: VPN connection between TGW and on-premises equipment

Complex organizations usually end up managing multiple VPCs that require inter-VPC communication, connectivity to the internet, and/or connectivity to your on-premises infrastructure. They then often leverage the TGW service to have a clean hub-and-spoke network model (more on this in the section dedicated to TGWs at the end of this chapter).

It is worth noting that AWS Managed VPN also provides redundancy and automatic failover, therefore it is highly recommended to connect your VGW or TGW to two separate CGWs on your end. By doing so, you establish two separate VPN connections, and if one of your on-premises devices fails, all traffic will be automatically redirected to the second VPN connection (see *Figure 2.3*). It allows you to nicely handle failover, as follows:

- In case of an unexpected failure of your on-premises router sitting behind your CGW

- When you need to perform maintenance on your network equipment and must take one of two VPN connections offline for the duration of the maintenance operation

This is illustrated in the following diagram:

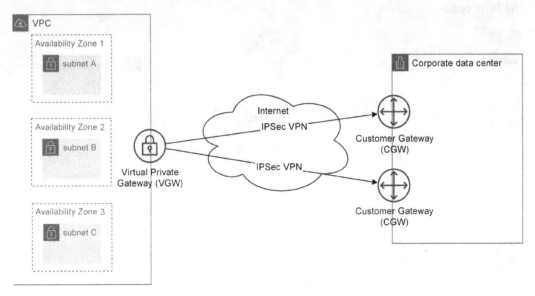

Figure 2.3: VPN connection redundancy for failover

AWS Managed VPN offers both dynamic and static routing options. Dynamic routing leverages **Border Gateway Protocol (BGP)** to pass routing information between the VGW on AWS and your on-premises CGW. It allows you to specify routing priorities, policies, and weights in your BGP advertisements and to influence the network path between your networks and AWS. It is worth noting that when using BGP, both the

IPsec and BGP connections must be terminated on the same CGW device(s). Both the BGP-advertised and static route information tell gateways on each side which tunnels are available to re-route traffic in case of failure. That said, the BGP protocol brings more robustness to the table thanks to the live detection checks it performs, so using BGOP-capable devices will make your life easier when dealing with failover from the primary to the secondary VPN connection upon failure.

AWS Managed VPN is a great approach when you need to connect one on-premises location with your AWS environment, but what about situations where you need to interconnect several remote offices together and with your AWS environment?

AWS VPN CloudHub

AWS VPN CloudHub is a hub-and-spoke VPN solution to securely connect multiple branch offices together and a VPC on AWS. It leverages the AWS Managed VPN service, but instead of creating CGWs for a single on-premises location, you create as many CGWs as you have remote branches/ offices that need a VPN connection and connect all of them to the same VGW on AWS. The result is a simple low-cost hub-and-spoke VPN setup that can be used for communicating securely from one branch/office to another and between your branches/offices and your AWS environment.

The following diagram illustrates this:

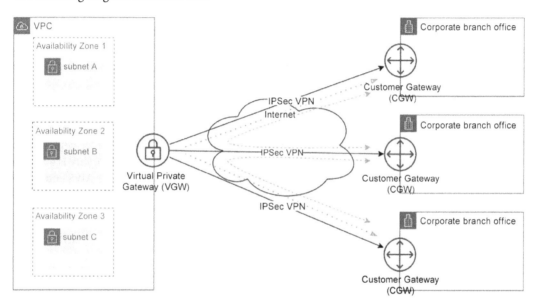

Figure 2.4: Hub-and-spoke VPN

> **Important Note**
> The remote sites must not have overlapping IP ranges.

Redundancy and failover mechanisms follow the same principle as for AWS Managed VPN. For greater reliability, it is recommended to use multiple CGW devices on your on-premises locations.

It is worth noting that the **AWS VPN CloudHub** construct is compatible with AWS DX, which will be covered in the next section. For instance, on the hub-and-spoke model represented in the previous diagram, one of your on-premises environments could connect to AWS using an AWS DX connection while the other two on-premises locations use a VPN connection over the internet.

Now that you've seen which managed services AWS provides to establish a VPN connection, you can consider cases where an organization may prefer or need to bring its own VPN software solution.

Software VPN

An additional alternative consists of connecting your on-premises network equipment to a **software VPN** appliance running inside a VPC on AWS. This is the right option if, for some reason, you want or need to manage both ends of the VPN connection. You can select between several partner solutions or open-source solutions that provide VPN software appliances that can run on Amazon **Elastic Compute Cloud (EC2)** instances.

The major difference between this option and AWS Managed VPN is that in this case, you must manage the software appliances entirely, including updates and patching at **operating system (OS)** and software levels. Another essential point to note is that a software VPN appliance deployed on an Amazon EC2 instance is, per se, a **single point of failure (SPOF)**. Thus, reliability is an extra complexity that you must deal with, whereas it is handled for you by AWS, on the AWS end of the connection, when using the Managed VPN solution.

This concludes the section on VPN connections, but as you will see now, a VPN is not the only way to establish a private connection between your on-premises infrastructure and your AWS environment.

Introducing AWS DX

Using a VPN connection when you get started makes a lot of sense. It can be up and running in no time and will likely cause no big change in your network topology.

However, it is not always the best option. For cases where internet connectivity unreliability becomes a business risk, **AWS DX** offers the right alternative by offering low latency and consistent bandwidth connectivity between your on-premises infrastructure and AWS.

In a nutshell, a DX connection ties one end of the connection to your on-premises router and the other end to **a virtual interface (VIF)** on AWS. There are three different types of VIFs: **public VIFs**, **private VIFs**, and **transit VIFs**. Public VIFs are used to connect to AWS services' public endpoints. Private VIFs are used to connect to your own AWS environments within a VPC. Transit VIFs allow you to end the connection on a TGW.

Various Flavors of AWS DX

You can use AWS DX provided that one of the following applies:

- Your network is co-located with an existing AWS DX location; see `https://packt.link/Awm60` for a current list of these.

- You leverage an AWS DX partner, a member of the AWS Partner Network (APN); see `https://packt.link/6OyGq` for a current list of these.

- You work with an independent service provider to connect to AWS DX.

There exist three types of DX connections. That said, only the first two types listed in the following section are recommended when you require a consistent connection capacity, which is eventually the main reason to set up a DX connection.

Dedicated Connection

This type of connection, available as 1 **gigabit per second (Gb/s)**, 10 Gb/s, or 100 Gb/s ports, consists of a dedicated link assigned to a single customer. Dedicated connections can be combined to further increase your bandwidth by using **link aggregation groups (LAGs)**. Link-speed availability can vary per DX location, so it is best to consult the list of DX connections from the AWS documentation at the link mentioned previously.

Dedicated connections support up to 50 private or public VIFs and 1 transit VIF.

Hosted Connection

This type of connection, available from 50 **megabits per second (Mb/s)** to 10 Gb/s, consists of a connection provided by an AWS DX partner and is made available on a link shared with other customers. AWS makes sure that the sum of all hosted connections' capacities per link does not exceed the network link's actual capacity.

With up to 500 Mb/s capacity, hosted connections support one private or public VIF. Hosted connections of 1 Gb/s or more support one private, public, or transit VIF.

If you require more than one VIF, either obtain multiple hosted connections or use a dedicated connection.

Hosted VIF

Some AWS DX partners provide hosted VIFs, which consist of a VIF made available to you in your AWS environment while the underlying DX connection is managed in a separate account by the provider. However, it is worth noting that AWS does not limit the traffic capacity on hosted VIFs. Therefore, the underlying DX connection capacity can be oversubscribed, which could result in traffic congestion, and therefore it is not a recommended option when you're looking for a consistent capacity to connect your on-premises and AWS environments.

AWS DX Connectivity Overview

The following diagram shows an overview of **end-to-end** (E2E) connectivity when setting up an AWS DX link between your on-premises and AWS environments:

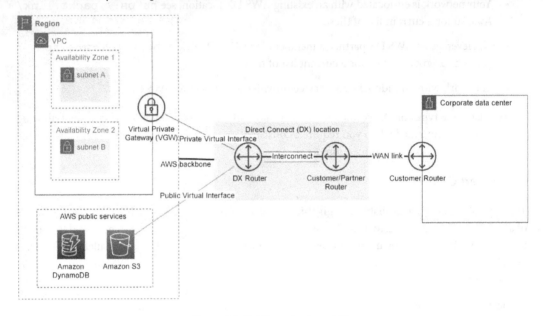

Figure 2.5: Public and private VIFs

In the case of a private VIF, the VIF can be attached either to a VGW in a VPC in the same region as your DX connection or to a **DX gateway (DX GW)**. An AWS DX GW is a globally available resource on AWS that can be accessed from any region. Its role is to help connect multiple VPCs, possibly in multiple AWS regions, through AWS DX.

It is important to note that a single DX dedicated connection can support up to 50 public or private VIFs. When using private VIFs, you have a choice either to connect those VIFs directly to your VPCs or to use a DX GW in between. Because each DX GW can connect on the other end up to 10 VGWs (so, 10 VPCs), using a DX GW allows you not only to connect to 500 VPCs through a single DX connection, but those VPCs can also be in multiple AWS regions.

Additionally—and this will be the focus of a later section in this chapter—you can also leverage an AWS TGW to simplify routing in cases where you have a large number of VPCs (in the 100s or 1,000s). A single TGW can support up to 5,000 (VPC) attachments today.

Large and complex organizations typically have an AWS environment spanning more than one AWS region, whether this is because they operate in multiple geographies or to follow some regulatory recommendations, or for **disaster recovery (DR)** purposes.

The following diagram summarizes the various options available:

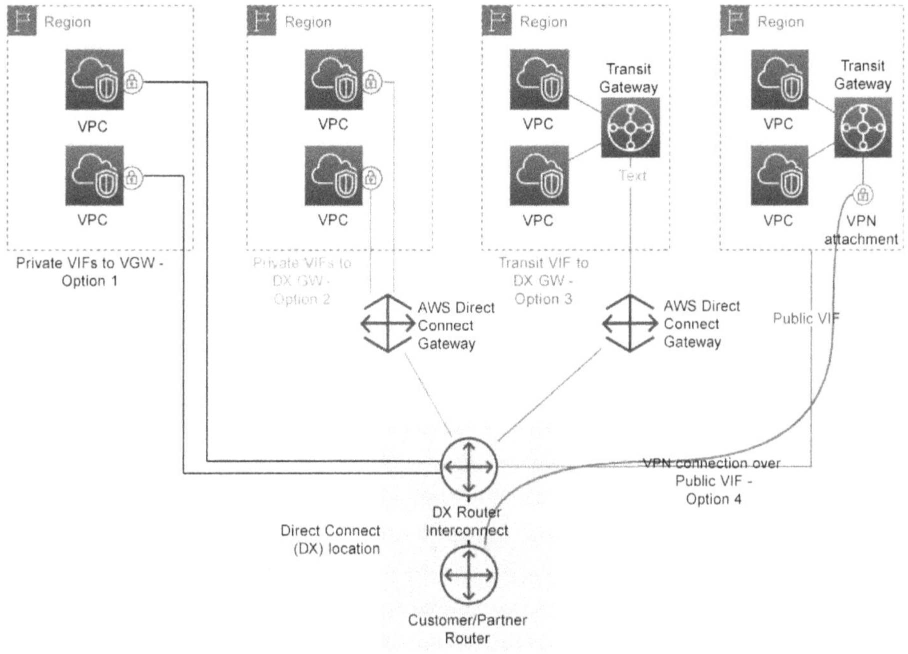

Figure 2.6: DX options summary

Such complex organizations adopt either a private VIF to DX GW (Option 2) or a transit VIF to DX GW (Option 3) or sometimes a combination of the two, essentially because an AWS DX GW and a TGW make their life so much easier. A VPN connection over a public VIF (Option 4) can be used to enforce E2E encryption as an extra security measure over public VIFs when **MACsec (IEEE 802.1AE Media Access Control (MAC) security standard)** encryption over DX is not available at your preferred DX location.

Now, you may be wondering when to use IPsec encryption and when to use MACsec encryption over DX. The first consideration is connection speed. MACsec encryption is available at speeds (10 Gb/s and 100 Gb/s) that cannot be reached with a single VPN connection (maximum 1.25 Gb/s). So, if you require encryption on links of 10 Gb/s or more, then MACsec, if it is available at your DX location, is a much easier solution for encryption. Alternatively, you could think of aggregating multiple VPN IPsec connections to work around the throughput limit, but that increases the operational complexity. The second consideration is technology. IPsec encryption is an E2E connectivity encryption mechanism that works at layer 3 of the **Open Systems Interconnection (OSI)** model (that is, IP). MACsec encryption, on the other hand, is a hop-by-hop encryption mechanism at layer 2 of the OSI model (that is, MAC). In this case, every network hop is responsible for encrypting the data frames until the next hop, and so on so forth. Both encryption mechanisms operate at different layers and are not mutually exclusive, but you can use either of the two or both simultaneously. MACsec encryption brings an additional protection layer to your security arsenal.

Additional Considerations for Resiliency

As a best practice, it is recommended to have at least two separate connections at two different DX locations. In this case, you end up with two DX connections. This will provide resiliency against connectivity failure due to a device failure, a network cable cut, or an entire location failure.

To achieve maximum resiliency, use at least two separate connections terminating on distinct devices in at least two DX locations attached to two different regions. In this case, you end up with at least four DX connections and are protected not just against a single device failure, a network cable cut, or an entire location failure, but also against an entire geography failure.

Either as an alternative to additional DX connections or as an additional resiliency protection measure, you can also create a VPN connection as a backup connectivity option.

Now that you know how to set up hybrid network infrastructure, you are ready to learn how to create a hybrid storage infrastructure between your on-premises locations and your AWS environment.

Cost Factor

On top of the already mentioned reasons to opt for a DX connection, such as network bandwidth consistency and throughput, the cost is obviously an important aspect not to be ignored or discarded too quickly.

For occasional usage and low data volume transmission between your on-premises environment and AWS, in many cases, a VPN connection is good enough, and this is what organizations typically begin with when they start using AWS. After all, all organizations already have broadband internet access nowadays, so setting up a simple IPsec connection is usually straightforward. However, when additional requirements come in, such as improving network consistency and reliability or benefiting from a higher network throughput, organizations start looking into AWS DX.

And beyond technical requirements, the overall cost of the solution should also be estimated. For AWS Managed VPN, you pay for the number of hours the connection is active (which varies per AWS region) and for data (volume) that you transfer from AWS to your on-premises environments, also known as **Data Transfer Out (DTO)**. For AWS DX, you pay a price per port hours a DX connection is up (which varies per AWS region and connection capacity) and for data (volume) that you transfer from AWS to your on-premises environments. Data sent into AWS bears no costs.

DTO costs for data sent over a VPN connection are the same as for data sent over the internet from your AWS environment and vary per AWS region. DTO costs for data sent over a DX connection vary per combination of AWS region and AWS DX location. The closer those two are to each other, the lower the DTO costs. For instance, DTO costs are lower when transferring data from any AWS region in Europe to any DX location in Europe than from any AWS region in Asia to any DX location in Europe (and vice versa, by the way). That said, DTO costs for traffic sent over DX are always lower than DTO costs for traffic sent over the internet (VPN or not), and sometimes even an order of magnitude lower.

Thus, besides mere technical requirements, in situations when large volumes of data (**terabytes (TB)** or beyond) need to be transferred from your AWS environments to your on-premises environment(s), it can become significantly more beneficial financially to leverage a DX connection.

Introducing AWS Storage Gateway

AWS Storage Gateway is a service that provides a series of solutions to expand your storage infrastructure into the AWS cloud for purposes such as data migration, file shares, backup, and archiving. It uses standard protocols to access AWS storage services such as Amazon Simple Storage Service (S3), Amazon S3 Glacier, Amazon Elastic Block Store (EBS) snapshots, and Amazon FSx.

There are three different flavors of Storage Gateway as listed here:

- File Gateway
- Volume Gateway
- Tape Gateway

The following section dives into the details of each.

File Gateway

File Gateway is nowadays further split into two distinct types: **S3 File Gateway** and **FSx File Gateway**.

S3 File Gateway

Initially the only available type of file gateway when AWS Storage Gateway launched, S3 File Gateway allows you to store files on S3 transparently accessible from your on-premises environment through the **Network File System (NFS)** and **Server Message Block (SMB)** protocols. S3 File Gateway does a one-to-one mapping of your files to S3 objects and stores the file metadata (for example, **Portable Operating System Interface (POSIX)** file **access control lists (ACLs)**) in the S3 object metadata. The files are written synchronously to the file gateway local cache before being copied over to S3 asynchronously.

Concretely, S3 File Gateway comes either as a preset hardware appliance or as a software appliance that you deploy in your on-premises environment. The software appliance consists of a **virtual machine (VM)** that can run either on **VMware Elastic Sky X (ESX)**, Microsoft Hyper-V, or a Linux **kernel-based VM (KVM)** hypervisor (but also on Amazon EC2 instances, should you need to).

See the following diagram for an illustration of how S3 File Gateway works:

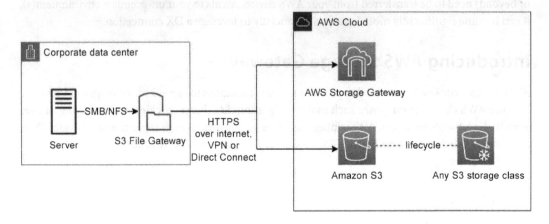

Figure 2.7: Amazon S3 File Gateway

Once deployed and configured, your servers on-premises can use it like any other *file share* through the NFS and SMB protocols. Multiple elements can condition the performance of your gateway, but key factors are **CPUs**, **local disk size**, and **network capacity**.

The CPU resources and network capacity available to the appliance will directly influence the amount of data the gateway can process in parallel. The local disk size assigned to the file gateway will condition the cache size (on the hardware appliance, this is obviously constrained by the amount of physical storage available, so it is best to think it through before ordering the appliance). The cache size is to be determined such that it provides enough capacity to store your most frequently accessed files so that they benefit from low-latency access. On the software appliance, you can always add more cache capacity (additional storage volumes) later if you realize that your cache is undersized.

In terms of security, it remains your responsibility to control and manage access to the S3 bucket(s) sitting behind the gateway and to follow best practices. Therefore, remember to set up the right permissions (**Identity and Access Management (IAM)** role identity-based policies and/or S3 bucket policies) accordingly, to follow a least-privileges approach.

Because the files are ultimately stored as objects on S3, you also have the freedom to use the rich set of capabilities Amazon S3 provides to manage their life cycle, such as life cycle policies, versioning, cross-replication rules, and so on.

Finally, back up your file gateway storage. AWS Backup integrates with AWS Storage Gateway, so you can back up your file gateway storage to AWS. AWS Backup stores the gateway backup on Amazon S3 as EBS snapshots that can later be restored either on-premises or on AWS.

FSx File Gateway

Amazon FSx File Gateway is a recent addition to the AWS Storage Gateway family to provide access to Amazon FSx for **Windows File Server** file shares on AWS from your on-premises environment. The idea is very similar to S3 File Gateway, which is that you can access the data on AWS through either a physical hardware appliance or a software appliance that you deploy on-premises either on VMware ESX, Microsoft Hyper-V, or a Linux KVM hypervisor (but also on Amazon EC2 instances, should you need to).

There are a few major differences from the S3 File Gateway service as outlined below:

- Your files, managed through FSx File Gateway, will be available through the SMB protocol only (you cannot use NFS).

- You need to have previously deployed an Amazon FSx for Windows File Server filesystem in your AWS environment.

- You must have access via VPN or DX from your on-premises environment to that Amazon FSx for Windows File Server filesystem on your AWS environment.

For the above reasons, the use cases for each gateway type are also slightly different.

You would use Amazon S3 File Gateway when you want to access files you have stored on S3 from on-premises or want to make files you store on-premises available on S3 for further processing on AWS. In this case, you can then leverage all the services AWS provides to run all sorts of data analytics, including machine learning capabilities, to analyze the data on S3.

You would rather use Amazon FSx File Gateway when you want to move on-premises network file shares accessed through the SMB protocol to the cloud and keep accessing them seamlessly from your on-premises environment. Think of cases such as user home directories, team file shares, and so on.

See the following diagram for an illustration of how Amazon FSx File Gateway works:

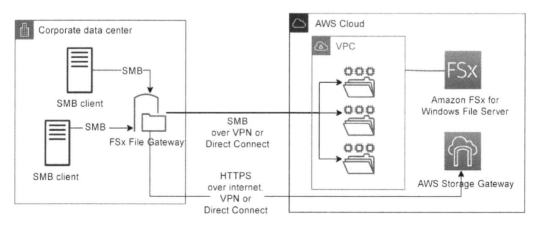

Figure 2.8: FSx File Gateway

Amazon FSx File Gateway is integrated with AWS Backup, so you can also manage and automate backups centrally, like Amazon S3 File Gateway. Additionally, you can activate Microsoft snapshotting technology and Microsoft Windows Shadow Copy on your version of Amazon FSx for the Windows File Server filesystem to allow users to easily view and restore files and folders on your file shares from a snapshot.

Volume Gateway

Volume Gateway allows you to create storage volumes on S3 that offer a block storage interface accessible from your on-premises environment through the standard Internet **Small Computer Systems Interface (iSCSI)** protocol.

Concretely, Volume Gateway comes either as a preset hardware appliance or as a software appliance that you deploy in your on-premises environment. The software appliance consists of a VM that can run either on VMware ESX, Microsoft Hyper-V, or a Linux KVM hypervisor (but also on Amazon EC2 instances, should you need to).

You have the choice between two operations modes for your Volume Gateway Service. Either you cache a portion of the data (**cached volume**) or keep a full copy of the volume (**stored volume**) locally on the gateway.

With cached volumes, as illustrated in *Figure 2.9*, you can reduce the amount of storage you need on-premises by limiting it to store the most frequently accessed data. In this scenario, Volume Gateway stores all your data on storage volumes on Amazon S3 and retains only the most recently accessed data on your local cache storage on-premises for low-latency access. You can additionally take incremental backups, also known as snapshots, of your storage volumes in Amazon S3. These snapshots are also stored in Amazon S3 as Amazon EBS snapshots. If you need to recover your data after an incident, these snapshots can be restored to a storage volume on your gateway.

Alternatively, for cases such as application migration to the cloud or DR in the cloud, you can create a new Amazon EBS volume from one of your EBS snapshots (provided the snapshot is not larger than 16 **tebibytes (TiB)**) and then attach it to an Amazon EC2 instance.

See the following diagram for an illustration of how Volume Gateway works with cached volumes:

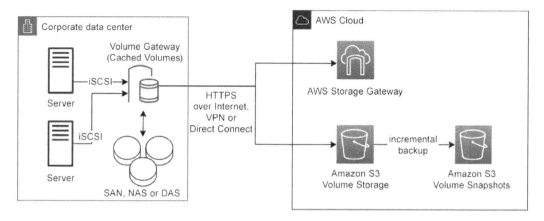

Figure 2.9: Volume Gateway (cached volumes)

With stored volumes, as illustrated in *Figure 2.10*, you retain your data entirely on-premises for low-latency access. In this case, Volume Gateway makes use of your local storage for storing your entire set of data and creates a backup copy of your volumes to Amazon S3 to provide durable offsite backup. The backup copy is performed asynchronously through Amazon EBS snapshots on Amazon S3:

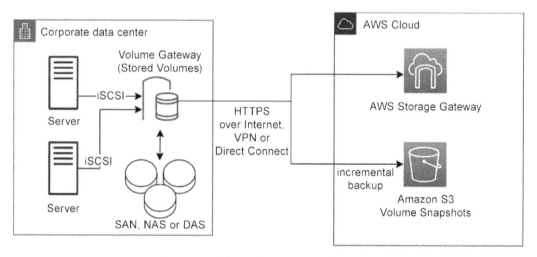

Figure 2.10: Volume Gateway (stored volumes)

Volume Gateway can serve multiple use cases, such as the following:

- Hybrid cloud storage for file services (expandable cloud storage for on-premises file servers)
- Backup and DR (offsite durable storage with DR capability in the cloud)
- Application data migration (application ready to start in the cloud with a copy of the data)

Now, you may be wondering how to choose between cached volumes and stored volumes. Well, they serve slightly different use cases, don't they? On the one hand, Cached Volumes gives you the opportunity to keep your most frequently accessed data on-premises for low latency access, while storing everything else—that is, cold(er) data—on Amazon S3. Thus, they let you keep the storage hardware you need on-premises to a minimum. They are a great solution when only a limited portion of your overall data is frequently accessed and when reducing your on-premises storage footprint and related costs is important to you. Maybe you need to expand your overall storage capacity but don't want to do so on-premises. Occasional longer data access times must also be acceptable in this case (when the requested data is not in the local cache).

On the other hand, Stored Volumes keeps your entire dataset on-premises in local storage for low latency access. They are particularly adapted for cases where longer data access cannot be tolerated and where the focus is not on reducing your on-premises storage infrastructure footprint or costs as much as it is on improving the durability of your data and providing an additional option for DR in the cloud.

Tape Gateway

Tape Gateway offers a **virtual tape library (VTL)** service backed by storage on Amazon S3 and accessible on-premises through the standard iSCSI protocol.

Concretely, Tape Gateway comes either as a preset hardware appliance or as a software appliance that you deploy in your on-premises environment. The software appliance consists of a VM that can run either on VMware ESX, Microsoft Hyper-V, or a Linux KVM hypervisor (but also on Amazon EC2 instances, should you need to).

As illustrated in the following diagram, Tape Gateway provides a VTL infrastructure that scales seamlessly, without the burden of having to operate or maintain the tape infrastructure on-premises. It integrates with the most popular backup solutions on the market, so chances are high that you can keep using your existing backup application. Now, the major difference from your previous physical tape solution or VTL solution is that Tape Library will store your virtual tapes in the cloud on Amazon S3. When your backup application sends data to the tape gateway, the data is first stored locally on the gateway and then copied over to the virtual tapes on Amazon S3 asynchronously:

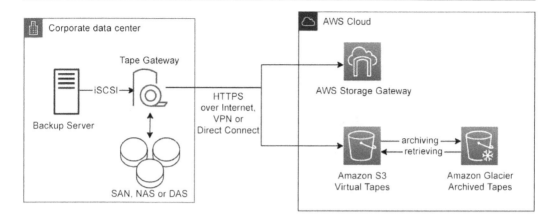

Figure 2.11: Tape gateway

Just as with any VTL solution, Tape Gateway proposes the concepts of a tape drive and media changer. Both the tape drive and media changer are available to your backup application as iSCSI devices.

The tape archive also offers the possibility to archive your tapes. When your backup application instructs Tape Gateway to archive a tape, the tape will be moved to a lower-cost storage tier using Amazon S3 Glacier or Amazon S3 Glacier Deep Archive.

Additional Considerations

To wrap up what was just covered, AWS Storage Gateway offers three different types of gateways to enable a hybrid storage architecture across your on-premises infrastructure and your AWS environment. You leverage each of these three types depending on the use case at stake—File Gateway when setting up a hybrid file server infrastructure, Volume Gateway when expanding your block storage infrastructure to the cloud, and Tape Gateway for replacing your physical tape infrastructure with virtual tapes on AWS.

The following section will take you through a few additional considerations to better plan the actual implementation of such a hybrid storage infrastructure.

Resiliency

The gateway, hardware, or software appliance is by default a SPOF. So, what are your options to deal with any type of failure, for instance, if a component crashes or at least stops responding, whether it is due to the appliance, the hypervisor, the network, and so on?

In the case of a software appliance that you deploy on VMware ESXi, you have an option to enable **high availability (HA)** using VMware HA. AWS Storage Gateway provides a series of application health checks that VMware HA can interpret to automatically recover your storage gateway when the health-check thresholds you specify are breached. That will cater to most failure cases.

This option is most useful when organizations cannot tolerate a long interruption of service or any data loss.

Quotas

As with any other AWS service, AWS Storage Gateway is bound by certain **quotas**. These quotas can be soft or hard limits constraining the service. Different quotas apply depending on the flavor of storage gateway that you implement. Here is an indication of the main quotas for each different type, but remember to check the AWS documentation to have the latest and most up-to-date figures:

- **File Gateway quotas** concern the maximum number of file shares per gateway (10), the maximum size of an individual file in the share (5 TB), the maximum path length (1,024 TiB). Note that one file share maps exactly to one Amazon S3 bucket. Adding more file shares will add more S3 buckets onto your AWS environment, so you also need to make sure you will not be exceeding your Amazon S3 quotas.

- **Volume Gateway quotas** are the maximum size of a volume (32 TiB for cached volumes; 16 TiB for stored volumes), the maximum number of volumes per gateway (32), the maximum size of all volumes per gateway (1,024 TiB for cached volumes; 512 TiB for stored volumes).

- **Tape Gateway quotas** concern the minimum and maximum sizes of a virtual tape (100 gibibytes (GiB) -> 5 TiB), the maximum number of virtual tapes per virtual tape library (1,500), the total size of all tapes in a library (1 **pebibyte (PiB)**).

This concludes the first half of this chapter, which focused on the creation of a hybrid infrastructure across on-premises infrastructure and AWS. In the second half of this chapter, you will investigate how to enhance communication first between your private environment on AWS and AWS services or third-party services offered on AWS, and secondly, within the realm of your AWS environment.

The following sections will describe how you can improve communication between your private environment on AWS and AWS services or third-party services offered on AWS.

Leveraging VPC Endpoints

AWS offers a highly available and scalable technology called **AWS PrivateLink**. AWS PrivateLink enables you to privately connect any of your VPCs either to the supported AWS services or to VPC endpoint services (that is, services powered by AWS PrivateLink that are hosted in other AWS accounts, whether by you or by a third party). For example, many of the services that AWS partners offer on AWS Marketplace support AWS PrivateLink nowadays.

Using AWS PrivateLink, you can then avoid exposing the traffic between your VPC and the target service on AWS to the internet; the E2E communication does not leave the AWS network.

Now, how does this work?

To use AWS PrivateLink, you simply create a VPC endpoint that will serve as an entry point to reach the destination service. This is illustrated in *Figure 2.12*:

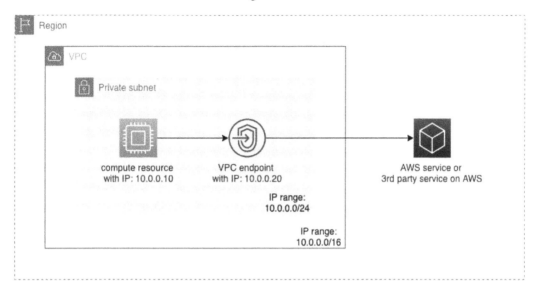

Figure 2.12: VPC endpoint

As illustrated in the preceding diagram, a VPC endpoint does not require a public IP address, an internet gateway, a peering link, a VPN, or a DX connection to be able to reach the destination service using AWS PrivateLink. The traffic always stays within the boundaries of the AWS network.

VPC endpoints are highly available and scalable virtual devices that you create in your AWS environment. There are currently three types of endpoints, as outlined here:

1. Interface endpoints
2. **Gateway Load Balancer (GWLB)** endpoints
3. Gateway endpoints

The following sections discuss each of these in detail.

Interface Endpoints

Interface endpoints, powered by AWS PrivateLink, are entry points for the traffic targeting a supported AWS service or a VPC endpoint service.

Concretely, an interface endpoint consists of an **Elastic Network Interface (ENI)** with a private IP address taken from the address range associated with the subnet in which it is created.

It is recommended to enable the private **Domain Name System (DNS)** (which is the default option) when you create an interface endpoint as this will make it easier to reach out to the supported service. Specifically, it will allow you to make use of the default DNS name of the service and still go through the interface endpoint leveraging private connectivity. Doing so avoids your applications from becoming aware of and having to use the endpoint-specific DNS name; instead, they can keep using the default (public) DNS name of the supported service. The following diagram illustrates this:

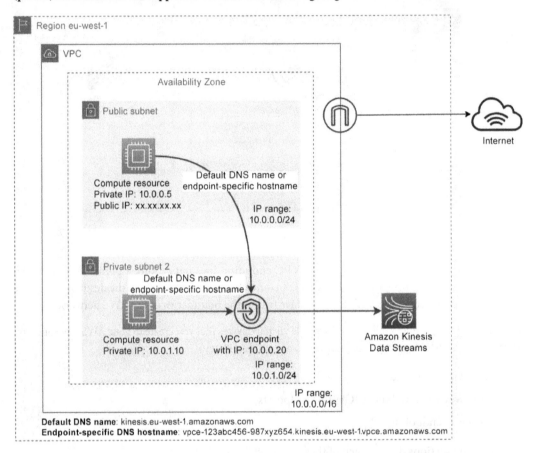

Figure 2.13: VPC interface endpoints and DNS names

You can enforce security best practices with interface endpoints in several ways.

First, you can associate security groups with interface endpoints and control which resources can use your endpoints. Secondly, you can associate IAM resource-based policies—called endpoint policies—with your interface endpoints to control which principals (users or roles) under certain conditions are allowed to use the endpoint.

Furthermore, interface endpoints can also be used in a hybrid cloud scenario where they can be accessed from your on-premises environment. The following current limitations are worth noting:

- An interface endpoint can only be created in one subnet per Availability Zone (AZ).

- Not all AWS services support interface endpoints: the list keeps growing on a regular basis, but it is recommended to check the AWS documentation for the latest update.

An interface endpoint is the principal type of VPC endpoint you will come across but, as previously mentioned, it is not the only one. The following sections present the other two types, starting with the latest and newest sort—GWLB endpoints.

GWLB Endpoints

GWLB endpoints are a new type of endpoint, recently added following the introduction of the GWLB service. GWLB provides inline traffic analysis for when you want to use specific virtual appliances for security inspection on AWS.

GWLB endpoints, powered by AWS PrivateLink, provide private connectivity to your gateway load balancers. A GWLB endpoint effectively consists of an ENI with a private IP address taken from the address range associated with the subnet in which it is created. To make use of this type of endpoint, you need to make sure to add the necessary routes in your subnet and gateway route tables to direct the traffic through the GWLB endpoint.

See an example of this in *Figure 2.14*:

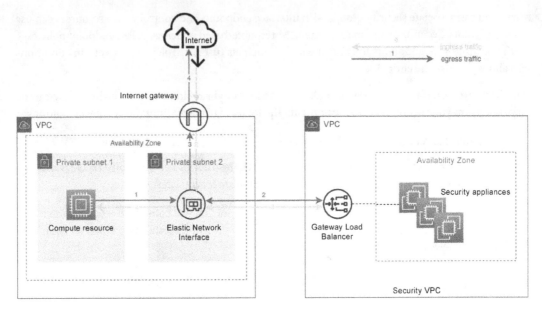

Figure 2.14: GWLB endpoint

The current limitation worth noting is that, at the time of this writing, this type of endpoint does not support endpoint policies and security groups.

Gateway Endpoints

A gateway endpoint is the first type of endpoint that launched on AWS, and it has been supporting connectivity to only two AWS services ever since: Amazon S3 and Amazon DynamoDB.

A gateway endpoint is a routable object that you must add to your VPC or subnet route table to be able to leverage it, like an internet or NAT gateway on AWS. On top of that, you can specify custom access permissions for your gateway endpoint by attaching endpoint policies to it.

See an example of this *Figure 2.15*:

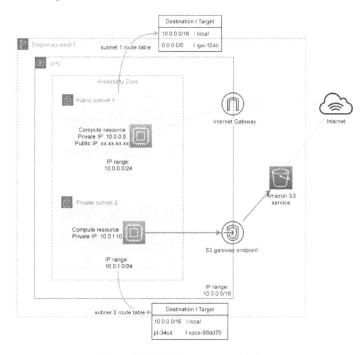

Figure 2.15: VPC gateway endpoint

You can attach several AWS gateway endpoints to any VPC. You will separate gateway endpoints, one for each service (S3 or DynamoDB) that you want to access, and then if you require different access permissions for different groups of resources, you may even have different gateway endpoints for the same service within the same VPC. If you use multiple endpoints for the same service in the one VPC, you will need to set different routes to use each of these endpoints in different route tables (for each service, you can only have a single route in every route table).

The following current limitations are worth noting:

- Cross-region is not supported. Gateway endpoints can only be used to reach out to AWS services in the same region as the VPC where they are set.

- Endpoint connections do not extend beyond the boundaries of a VPC. You cannot leverage the gateway endpoint defined in your VPC to access a service behind

- that endpoint from another VPC or from your on-premises environment, whichever network topology you may have (VPC peering, Transit Gateway, VPN, DX, and so on).

Before moving on to the next section, consider the two key aspects: resiliency and cost.

Additional Considerations

There are a few attention points when using private endpoints. You want to be mindful of resiliency and cost aspects.

AZs

Services offered by third-party providers, whether in your own organization or beyond, may not always be available in each AZ within a given AWS region.

An interface endpoint is mapped to an AZ upon creation. Therefore, it is important, especially for third-party services, to validate in which AZs they are available and to use AZ **identifiers (IDs)** to identify AZs uniquely and consistently across accounts. Remember the difference between the following:

- An AZ name (for example, eu-west-1a) that does not necessarily map to the same AZ in two different AWS accounts

- An AZ ID (for example, euw1-az1) that always refers to the same AZ across all AWS accounts

So first, you must use AZ IDs to make sure that you deploy endpoints in the right AZs where the service is also available. Secondly, it is recommended as a best practice to always deploy endpoints in at least two AZs for HA purposes.

Pricing

Gateway endpoints are provided at no charge, other than the cost generated for using the service and transferring data.

Endpoints powered by AWS PrivateLink—that is, interface endpoints and GWLB endpoints—are priced against two dimensions: the time the endpoint exists (per hour, for each AZ where the endpoint is deployed) and the amount of data that goes through it (per GB).

For enterprises, it becomes cost-efficient to centralize interface endpoints—for example, in a VPC within a central shared services or network services account—and to share them within the rest of the organization. This allows not just better control over connectivity aspects, but by avoiding duplicated interface endpoints (times the number of VPCs in use), you are able to save on costs as well, especially if the number of accounts in your organization grows significantly over time.

You are now ready to investigate yet another service that can help you optimize your organization's network infrastructure, AWS Transit Gateway.

Introducing AWS Transit Gateway

AWS Transit Gateway is a central hub construct to interconnect multiple VPCs on AWS and on-premises networks together.

Its purpose is to do the following:

- Avoid finishing with a spaghetti network topology, which is likely to happen if you start peering all your VPCs one to another.
- Share common network functions across multiple VPCs such as internet and on-premises connectivity (either via VPN or AWS DX), VPC endpoints, and DNS endpoints.
- Keep those essential network functions separate from the rest of your AWS environment and in a central place managed by your network experts.

AWS Transit Gateway Overview

AWS Transit Gateway is a regional network construct, so in the case where you need to operate in more than one AWS region, you would end up with (at least) one TGW in each region. If you need to establish connectivity between VPCs in different regions, you have the option to create a cross-region peering connection between two TGWs.

TGWs are highly available by design, so you do not need to rely on more than one TGW for the resiliency purposes of the network transit hub. That said, when you attach a

VPC to a TGW, you need to specify on which subnet(s) in which AZ(s) you want that attachment to be effective. So, although the TGW is highly available, it is a best practice to specify subnets in more than one AZ when attaching a VPC to make the VPC attachment itself highly available. That said, resources deployed in a subnet within a specific AZ can only reach a TGW if there exists a TGW attachment to a subnet within the same AZ. In other words, even if you specify a route in a subnet's route table to reach the TGW, if there is no TGW attachment to a subnet in the same AZ, then the TGW will not be reachable from that subnet. So, it is key to make sure to tie one subnet in each AZ to a TGW attachment wherever your resources need access to the TGW. It is usually recommended to use a separate subnet for that in each AZ, with a small **Classless Inter-Domain Routing (CIDR)** range (for example, a /28) so that you keep more IP addresses for your own resources. This allows you to have distinct network ACLs for the subnets where you deploy your resources and the subnets associated with the TGW, and you can also use separate route tables for those two types of subnets.

For organizations that intend to use stateful network appliances on their AWS environment, a specific mode called appliance mode can be enabled on the TGW.

The idea is to enable that appliance mode on the VPC attachment corresponding to the VPC where the appliance is deployed. It has then the effect of routing ingress and egress traffic through the same AZ in that VPC (for the sake of **statefulness**), which is not guaranteed otherwise.

Another important consideration for complex organizations that may have an AWS environment spread across multiple AWS regions is that you will not be charged extra for additional TGWs. Indeed, TGW usage is priced along two dimensions: per VPC attachment and per volume of traffic (GB) going through the TGW. So, unless you decide to attach some VPCs to more than one TGW, these costs will stay the same. TGW peering does not affect the costs either since there is no extra cost for peering, and the TGW traffic costs are not accounted for twice but only at one of two peered TGWs (typically at the sending TGW). The only additional costs in the case of cross-region peering between two TGWs would be **inter-region data transfer charges**.

Routing with AWS Transit Gateway

AWS Transit Gateway supports both dynamic and static routing. By default, the network elements (VPCs; VPN or DX connections; peered TGWs) attached to a TGW are associated with its default route table, unless otherwise specified. You naturally have the choice to organize routing as you please by creating additional routing tables and then associating each network element attached to the TGW with the routing table of your liking.

The routes that are defined in those routing tables can be defined statically or dynamically. When you attach a network element to a TGW, you specify whether you want the routes coming from that element to be automatically propagated to the TGW route table associated with that element. If you prefer not to, you must specify routing statically to and from the TGW.

Routes can be propagated automatically both from your on-premises networks connected to the TGW via VPN or DX and from your VPCs attached to the TGW. In the first case, routes are advertised back and forth using BGP between the TGW and your on-premises network equipment on the other end of the VPN or DX connection. In the case of VPCs, the routes are propagated from the VPCs to the TGW but not back to the VPCs from the TGW. You then need to update your VPCs' route table, creating static routes for your VPCs to communicate with the TGW.

One more thing worth mentioning on routing is that Transit Gateway cannot handle VPC attachments when some VPCs contain IP addresses overlapping with each other. Thus, when you want to attach a set of VPCs (or on-premises networks) that may have overlapping IP addresses to a TGW, you need to deal with the overlapping IP addresses first. Going into more details on how exactly to do this goes beyond the scope of this chapter, but make sure to find a solution to that problem before attempting to connect these networks to a TGW. Multiple solutions exist out there, such as **network address translation (NAT)**, leveraging **IP version 6 (IPv6)** instead of **IP version 4 (IPv4)** addresses, or leveraging a third-party solution to do the magic for you (typically through NATing).

Summary

This chapter started with a discussion of the various options available and how best complex organizations can communicate in a hybrid cloud setup between their on-premises network and their AWS environment. You also looked at specific AWS cloud storage solutions that enable hybrid cloud solutions and extend the capabilities of your on-premises infrastructure.

You then reviewed how complex organizations can make the most of the connectivity options offered by AWS to improve their network security, reliability, and performance postures.

All the network constructs that you have reviewed in this chapter constitute the core components that complex organizations will inevitably leverage when laying out their AWS environment network topology.

The next chapter will focus on how you can best organize and structure your resources within your AWS environment, covering (among other things) topics such as AWS Organizations, service control policies (SCPs), and AWS Control Tower.

Further Reading

You can check out the following links for more information about the topics that were covered in this chapter:

- Hybrid Connectivity whitepaper: `https://packt.link/QMK80`

- Building a Scalable and Secure Multi-VPC AWS Infrastructure whitepaper: `https://packt.link/WQ8wS`

3
Designing a Multi-Account AWS Environment for Complex Organizations

Determining a strategy to deploy your resources across multiple **Amazon Web Services** (**AWS**) **accounts** is essential for governance purposes. This can bring benefits not just for billing but also for security and compliance purposes.

This chapter will explain how to organize your resources using multiple AWS accounts for your organization. We will discuss how to approach billing and resource isolation and how to increase security across an entire organization or individual **business units** (**BUs**). You will also examine the various services that AWS provides to assist you with that.

The following main topics will be covered in this chapter:

- Deciding on resource and billing isolation
- Establishing a billing strategy for multiple accounts
- Introducing AWS Organizations
- Setting up **service control policies** (**SCPs**)
- Leveraging AWS Control Tower

The first section of this chapter will discuss the importance of selecting resources and billing isolation strategies.

Deciding on Resource and Billing Isolation

In the complex environment of cloud computing, managing resources efficiently is crucial. As businesses grow, they face the challenge of organizing resources across different projects, teams, and business units. This complexity is further compounded when it comes to billing, where visibility and accountability for cloud expenditure become paramount. The decision on how to isolate resources and manage billing has implications for governance, security, and operational efficiency.

The first decision that an organization needs to make when starting to use AWS is deciding how to organize its AWS resources. Although not crucial in the initial stages if you're just experimenting and dipping your toes in the cloud, this decision becomes paramount for large organizations to avoid potentially painful refactoring later. AWS provides several structures to help you with that, such as AWS Organizations, **organizational units (OUs)**, accounts, **virtual private clouds (VPCs)**, and subnets.

Elements of Structure

Before going any further, review each of the elements that can be used to structure your AWS environment.

Organization

You must have noticed that the term *organization* has been used a number of times already in this chapter. It refers to a business entity that could represent an entire company or a portion of it. AWS also happens to provide an account management service called **AWS Organizations** (more on this later in this chapter, in the *Introducing AWS Organizations* section). In the scope of the AWS Organizations service, an organization represents the structure in which your AWS accounts are grouped.

To avoid any ambiguity, the type of *organization* (the business entity or the AWS resource) will be specified in this book.

OUs

When setting up your organization, AWS Organizations lets you define one or more **OUs** to help you manage your AWS accounts within your organization. An OU refers to a group of multiple AWS accounts and other OUs in a Region.

Account

An **AWS account** (an account, in short) is a virtual environment where you access AWS services and deploy and use AWS resources. The resources you deploy in one account are isolated from the resources deployed in any other account unless you explicitly provide cross-account access (for more on cross-account access, see *Chapter 1, Determining an Authentication and Access Control Strategy for Complex Organizations*). There is no limit to the number of accounts that you can create and use. By default, you can create up to five AWS accounts within your organization created through AWS Organizations.

That default quota can naturally be raised to match your needs. It is not uncommon for complex organizations (intended here in the sense of companies or BUs) to manage hundreds of accounts within a single organization defined in AWS Organizations.

VPC

A Virtual Private Cloud (**VPC**) is a virtual network environment that you define in an AWS account. You can use this to deploy AWS resources, such as **virtual machines** (**VMs**), containers, databases, and more. VPCs provide yet another layer of isolation for your AWS resources since communication from one VPC to another within the same AWS account is disabled unless you explicitly set it up (for more on inter-VPC communication, see *Chapter 2, Designing Networks for Complex Organizations*). As you most certainly remember, a VPC is a regional construct and spans all the **Availability Zones** (**AZs**) of an **AWS Region**.

Subnet

A **subnet** is a network subdivision of a VPC that belongs to a single AZ. It is yet another layer of isolation for your AWS resources.

The next section will discuss how you can best use the above concepts to build proper isolation for a large organization.

Striking the Right Balance for Resource Isolation

Considering all the possibilities, what is the best approach to resource isolation? Well, first, it depends on your requirements. Do you require strong segregation of workloads? What sort of segregation? Is it at a team level or at a department or BU level? Or is it mostly between production and non-production environments? Do the teams in your organization trust each other when it comes to sharing resources? The following paragraph presents an overview of the possible isolation levels, starting from the strongest possible isolation to the weakest one.

To start with, an AWS account provides the highest degree of isolation for resources on AWS. As mentioned earlier, AWS resources in one account are isolated from the rest of the AWS universe and from resources belonging to other accounts in your organization. Further, if something goes wrong in one of your accounts (for instance, a security incident), the affected perimeter—the blast radius—is limited to the affected account.

Therefore, it is strongly recommended, as a best practice, to distribute and separate your resources and workloads across multiple accounts. You may ask, *how should I do that?*

The first natural separation is between *production* and *non-production* resources. As a best practice, you should not put non-production and production resources and workloads in the same account (think again about what would happen in case of an incident). Then, depending on your organization's requirements and security policies, opt for a particular account structure. You may be required to use a certain type of structure based on certain industry regulations for all or part of your BUs. For instance, you may have to strongly isolate regulated workloads (for instance, workloads regulated by the **Health Insurance Portability and Accountability Act (HIPAA)** or **The Payment Card Industry Data Security Standard (PCI-DSS)**) from non-regulated workloads. You may also choose to mirror the structure of your enterprise (for instance, *BU*, *department*, *team*, or *application*) in your account structure.

Figure 3.1 illustrates an account structure as was just discussed:

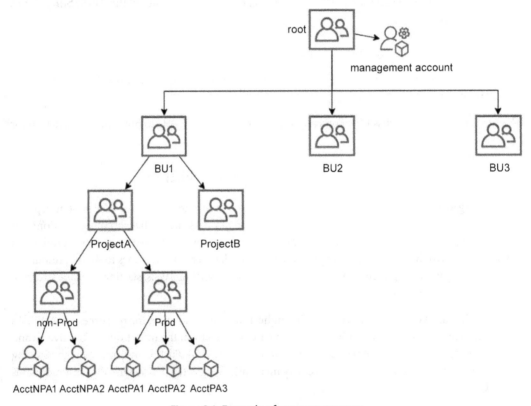

Figure 3.1: Example of account structure

A second method involves achieving strong isolation of resources at a network level, even within the same account, by assigning them to different VPCs—that is, different virtual networks. By default, resources located in different VPCs cannot reach each other; there is no route that connects multiple VPCs by default. If you need inter-VPC connectivity, you must enable it, and the various mechanisms to do so were discussed in *Chapter 2, Designing Networks for Complex Organizations*. You can decide to only allow VPC-to-VPC communication on a case-by-case basis—for instance, when dealing with VPCs hosting resources that belong to the same application or are linked to applications that must talk to each other. Figure 3.2 presents an example of a VPC organization:

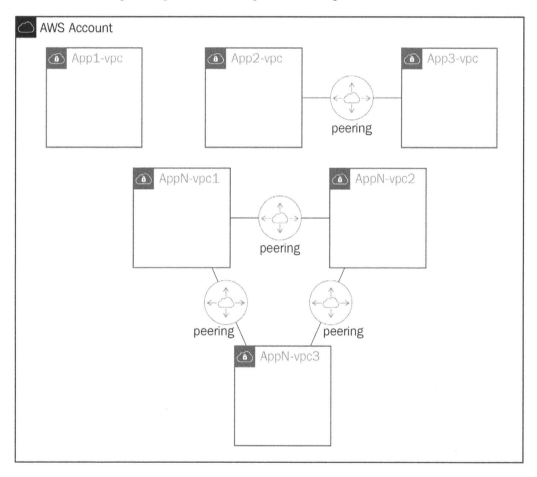

Figure 3.2: Example of VPC organization

A third approach would be to distribute your AWS resources further across separate subnets within a given VPC. However, unlike inter-VPC connectivity, which is disabled by default, inter-subnet connectivity is enabled by default. Two resources in separate subnets can reach each other unless you put countermeasures in place, such as **network access control lists** (**NACLs**) or **security groups**, to prevent this from happening. So, because of this, it is apparent that this approach is suitable for resources that should be protected and segregated from each other but not for those that require a stronger level of isolation than VPCs or accounts can offer. Thus, as a general practice, start by organizing (and isolating) your AWS resources and workloads first at an account level, then at a VPC level, and then, only as a last resort, at a subnet level.

Now that you are familiar with the various approaches to resource isolation in AWS, you can learn about the billing features in AWS, as discussed in the next section.

One Bill or Multiple Bills

By default, when you create a standalone AWS account, you must provide a payment method (for instance, a credit card). For large organizations, it usually doesn't make sense for them to receive as many bills as they have AWS accounts. They usually demand a *consolidated bill* across all the accounts in their AWS environment. Consolidated billing is a feature that was introduced with AWS Organizations. It allows you to designate one account (the management account) as the payer account for all the member accounts in your organization. The bill in this case is centralized by default. You can not only easily track what each account in your organization spends but also combine the usage from all your accounts and share the various discounts you may be entitled to (volume pricing discounts, Reserved Instances discounts, Savings Plans discounts) across all your accounts. This usually results in lower charges for your organization as opposed to the charges for consumption on individual accounts.

There may also be cases where large organizations want to have separate bills—for instance, a multinational corporation composed of several separate companies (or legal entities) may prefer to have their bills consolidated on a company (or legal entity) basis. In such cases, you could do two things. You could set up separate organizations by company (or legal entity) so that you automatically get the centralized bill exactly as needed. Alternatively, you could keep a single organization but split the single centralized bill into separate bills per company (or legal entity), filtering by the list of accounts that belong to each company (or legal entity). AWS indeed creates a separate bill for each account, according to the usage of that account. When choosing the first option (separate organizations), you automatically end up with multiple payer accounts (one per organization) while, in the second case, you still have a single payer account paying the bill for the entire organization but can easily charge back individual companies (or legal entities).

The following section will take you through how to determine billing strategies based on your requirements.

Establishing a Billing Strategy for Multiple Accounts

As was detailed in the previous section, centralized billing brings a lot of benefits, and for most organizations, it is a given. That said, there are a number of best practices to consider when building your account structure and formulating a billing strategy.

First, avoid using an individual's email address as the account email address to ensure continuity of communication from AWS. Instead, use a group alias for a functional mailbox. Individual email addresses can lead to issues related to the availability of the individual as well as business continuity for the enterprise. Suppose the individual leaves the company. The process of regaining root access to the account associated with that individual can be painstaking for the enterprise. Therefore, it is better to use a group alias as an account email address and for all the email notifications configured on your accounts.

Second, tag your AWS resources following standards across all your accounts to standardize how your enterprise categorizes, controls access to, and reports on these resources. Leverage **cost allocation tags**, which can be used with **AWS Cost Explorer** and **Cost and Usage Report** (**CUR**) data. You can define a number of tags (for instance, cost center, department, or project) to be used as cost allocation tags. You can then filter billing data along with those tags for chargeback, reporting, or analysis purposes. Refer to the AWS *Tagging Best Practices* whitepaper for additional guidance (see the *Further Reading* section).

Third, use consolidated billing to make AWS cost management easier.

Consolidated billing is one of the key features provided by AWS Organizations. When you set up an organization in AWS Organizations (see the next section), the **management account** automatically centralizes the bill and pays the charges for all the organization **member accounts**. Consolidated billing is most suitable when you have to handle billing for a large number of accounts (ranging from tens to hundreds—or even thousands—of accounts). It offers a centralized bill for all the accounts belonging to the same organization. So, unless you have particularly good reasons to do this differently, do leverage consolidated billing and, if needed, use cost allocation tags to be able to return the charges back internally to the various teams, departments, or BUs within your organization, as you see fit.

The following section discusses AWS Organizations and how it can help with account management in depth.

Introducing AWS Organizations

As was mentioned earlier, AWS Organizations is an account management service. Its role is to help large and complex organizations handle their AWS environment more efficiently. You can use AWS Organizations to manage security policies across accounts and filter out unwanted access, automate the creation of new accounts through its **application programming interfaces** (**APIs**), organize accounts into OUs, and consolidate billing across multiple accounts.

When you set up an organization with AWS Organizations, the AWS account that you use to set it up becomes the **management account** of that organization. As you invite other accounts to join your organization or directly create new accounts in your organization, these accounts then become **member accounts**.

There are two major modes of working with AWS Organizations: either with all features enabled or with consolidated billing only. Consolidated billing only provides a central consolidated bill of all the member accounts of the organization and does not provide access to advanced functionalities such as centrally managed security policies. To benefit from these advanced features, you must enable all features in your organization.

Managing Policies Across Accounts and Filtering out Unwanted Access

As mentioned in the previous section, AWS Organizations helps you take security and governance to the next level, provided that you enable all features. It does this by means of specific policies, but instead of managing these policies at the individual account level, it allows you to handle them at the organization level.

There are two main types of policies that you can enforce at the organization level—authorization policies and management policies.

Authorization Policies

Authorization policies consist of SCPs, which offer the capability to centrally manage the maximum permissions that can be granted to **Identity and Access Management (IAM)** entities (users or roles) across either your entire organization or one or more specific OUs. The concept of SCP is, in a way, similar to that of IAM permissions boundaries, which we covered in *Chapter 1, Determining an Authentication and Access Control Strategy for Complex Organizations*, whereby the intention is not to grant permissions directly but to limit the perimeter of the permissions that can be granted to IAM entities. The major difference between the two is that SCPs apply at the organization or OU level while permissions boundaries apply at the more granular level of IAM entities.

Note that SCPs are not enabled by default—you must enable them in your organization to be able to use them. We will dive deeper into SCPs in the dedicated section titled *Setting up SCPs* later in this chapter.

Management Policies

The other type of policy handled by AWS Organizations is management policies, which later subdivides into **artificial intelligence (AI)** services opt-out policies, **backup policies**, and **tag policies**.

Management policies are inherited from the root of your organization down to the account level. The effective policy being applied at the account level is the result of the merging of all the policies from the root of your organization down to the account. How these policies are merged depends on inheritance operators. Inheritance operators define how inherited policies merge into the account's effective policy. The visual editor in the AWS Organizations management console only lets you use the `@@assign` operator. The other operators (`@@append` and `@@remove`) are considered advanced features and are available only when you author the policy in **JavaScript Object Notation (JSON)**. The `@@assign` operator simply overwrites the policy settings inherited from the parent levels with those that you specify. The `@@append` operator adds the settings you specify to those inherited from the parent levels. The `@@remove` operator removes the specified settings from the settings inherited from the parent levels.

You can optionally include child control operators to specify which control operators can be specified in child policies. This gives you the opportunity to prohibit certain behavior across your entire organization or across certain OUs (for instance, to avoid specific settings being removed at a lower level within your organization, in child OUs or accounts). Imagine, for instance, that you want to enforce regular backups of all your Amazon **Relational Database Service (RDS)** databases across your entire organization. You would typically define a backup policy at the root level and specify child control operators in a way that ensures child OUs or accounts do not make use of the `@@remove` operator. Please refer to the AWS documentation at `https://packt.link/buM81` for more details and example policies on how to do this.

Whether you're building a backup or tag policy, it is recommended to start with a simple policy, to check the resulting effective policy on the target account(s), and to test its results thoroughly. Then, you can gradually add more complexity as required, always following the same procedure and limiting the scope of the policy change each time to make tests and troubleshooting easier and more efficient.

To check the effective policy applied to your account(s), you can, from the command line, run the command given below. This will yield the managed policy that would be applied on your account's concerned resources:

```
aws organizations describe-effective-policy --policy-type <POLICY-
TYPE>
```

Here, `<POLICY_TYPE>` is either `BACKUP_POLICY`, `TAG_POLICY`, or `AISERVICES_ OPT_ OUT_POLICY`.

When you run it from your organization's management account, you can also add `--target-id <managed-account-id>` as an option, to specify the **identifier** (**ID**) of an account within your organization about which you want to find out the effective managed policy.

AI Services Opt-Out Policies

This type of policy lets you decide whether you allow AI services to collect data when they're being used across your organization. Some AI services provided by AWS, such as **Amazon Lex**, **Amazon Polly**, **Amazon Rekognition**, and more (for a complete list, please consult the AWS documentation) may store and collect some data for service improvement. If you have concerns regarding compliance with regulations applying to the business domain in your organization (for instance, healthcare or finance), you may well avoid any issue by opting out systematically from sharing any data with such AI services.

On the other hand, if you don't have any such regulatory or compliance issue, or if you only process anonymized data with such AI services, you can contribute to and accelerate the service's improvement by opting in. This is what AWS Organizations AI services opt-out policies are about. You can set up such policies at an OU level or across your entire organization. An opt-out policy that is attached to an OU applies to all accounts belonging to that OU and all child OUs across all AWS Regions.

Backup Policies

Backup policies are meant to centrally manage and enforce backup plans across your entire organization, so that you don't have to set this up account per account. This can help you simplify the backup process management overhead and enforce governance hygiene across all your accounts.

AWS Backup allows you to create backup plans specifying how to back up your AWS resources. A backup plan definition includes various settings, such as the backup vault to store the backup, the backup schedule (including the frequency and the time window during which it occurs), the life cycle for expiration or transition to cold storage, and the AWS Region where the resources to be backed up can be found. You can then apply a backup plan to groups of AWS resources identified by tags. Backup policies combine all components of the policy into policy documents written in JSON and can be attached to individual accounts, OUs, or your entire organization.

AWS Backup determines the policy to apply (or the effective backup policy) by applying inheritance rules from the top of your organization down to the account where it is being used.

Ensure that backup policies at each level are complete. You can define incomplete backup policies at parent OU levels, delegating the responsibility to child entities (OUs or accounts) to make those policies complete. This is, however, not an effective idea, especially in large organizations, where you easily risk ending up with an incomplete backup policy at child levels as well. An incomplete effective backup policy in one account would simply cause the backup process to fail to back up the account's resources, leading to disastrous consequences. Thus, it is best to always define complete backup policies at every level so that child entities inherit default settings that they may (or may not) override, depending on your requirements regarding inheritance control operators.

Another best practice is to limit the number of backup plans per policy and have a single backup plan per backup policy. This will make it a lot easier to ensure proper functioning (and troubleshooting) of the backup policy. Should a change in one of your backup policies break things up at some point, its impact would be limited to the single backup plan attached to that policy. All your other backup plans would remain unaffected.

It is also recommended to increase your protection against accidental deletion of your backups by storing backup copies in a separate account, a separate Region, or both. For this, you can use the `copy_actions` instruction in the backup plan rules of your backup policies. Note that you can also use the `$account` implicit variable, representing the account where the backup is taking place, to more easily compile the **Amazon Resource Name** (**ARN**) of your target backup vault, either in the same Region or in another Region. Multiple backup policy examples are proposed in the AWS documentation, so make sure to check them out before rolling up your own.

Tag Policies

Tag policies provide a means to centrally decide which tags are attached to the AWS resources across your organization. A tag policy consists of rules that define for each tag the tag key, including the capitalization preference (for example, `costcenter` or `CostCenter`), tag values that are valid (this is optional), and whether non-compliant tagging operations should be prevented. Tag policies combine all the various components of the tag policy into policy documents written in JSON and can be attached to individual accounts, OUs, or your entire organization.

First, tag policies can be used to standardize tagging across your entire organization. They can then be used to define naming conventions, such as deciding which tag name capitalization to apply (for example, `CostCenter`, `Costcenter`, `costCenter`, or `costcenter`) as well as tag name structure. Tag names can contain alphanumerical characters plus some special characters (+, -, =, ., _, :, /, and @). The `aws:` prefix is reserved by AWS for their own tags (for instance, `aws:createdBy`) so you can't use it for your custom tags. However, you could define your own naming convention based on a format such as `<company>:<team>:<key>` (for instance, `examplecorp:marketing:campaign202101`). Keep in mind though that a tag's total key length (so, in this case, `<company>:<team>:<key>`) is limited to 128 Unicode characters in **Unicode Transformation Format 8 (UTF-8)**. Further, security policies are also limited in size, which varies depending on the type of policy, as we saw in *Chapter 1, Determining an Authentication and Access Control Strategy for Complex Organizations*. So, it's in your own best interest to limit the tag names' length if you plan to use them in security policies.

Second, tag policies can be used either to enforce tagging correctness or simply to report on incorrect tagging. Be aware that not all AWS resources are supported for tag enforcement, so check the AWS documentation to make sure that the behavior you expect is supported (for example, tag enforcement is supported on Amazon **Simple Storage Service (S3)** buckets but not on S3 objects). Alternatively, you can simply report on compliant and/or non-compliant resources across your entire organization regarding compliance with your tag policies.

Automating the Creation of New Accounts through APIs

An organization within AWS Organizations is essentially a group of accounts put together. But the AWS Organizations service provides you with more than just the facility to group accounts—you can also manage those accounts. You can decide how these accounts are structured within your organization (more on this in the *Organizing Accounts into OUs* section), invite accounts to join your organization, or remove accounts from your organization. Inviting and removing accounts gives the organization's administrator the flexibility to shape the composition of the organization as needed. This can come in handy in situations when a re-organization happens within your organization, or in cases of mergers, acquisitions, or divestitures. Finally, AWS Organizations allows you both to create and close an account.

To create member accounts, you need to be logged in to the management account of your organization. Once a new account is created, it is automatically joined to your organization. This means that you don't need to invite that account to join afterward, as it automatically becomes a member account of your organization. It is worth noting that upon creation of a member account, AWS Organizations only copies the necessary information (account name, company name, contact email) from the organization to the newly created account so that it can operate within the organization. It does not populate all the information that a standalone account would need to operate on its own (for instance, there is no payment method provided, since billing and payment are centralized). This consideration becomes important if a member account created by the management account of an organization needs to leave the organization. When you don't need a member account anymore you can simply close it, again from the management account.

All these operations can be performed from the AWS management console but can also be called through the AWS Organizations API or by using the AWS **command-line interface (CLI)** or AWS **software development kits (SDKs)** available by the savvy system administrator who wishes to automate as many of the management tasks as possible.

Account creation can also be automated. Automating account creation means you can automatically provision an account on demand for anyone in your organization and, at the same time, provision that account with the right scope and context from the get-go. For instance, you may want to set up a new member account in a specific OU. From that OU, the new member account will inherit properties such as SCPs associated with that OU and its parent OUs.

Using the AWS CLI, you would perform the following action to create an account:

```
aws organizations create-account --email bob@example.com --account-
name "Account for Bob"
```

The above command would create an account named `Account for Bob` associated with the specified email at the root of your organization. If you want to move this account to a specific OU within your organization, you will typically perform a subsequent action, such as the following:

```
aws organizations move-account --account-id 121212121212
--source-parent-id r-examplerootid123 --destination-parent-id
ou-examplerootid123-exampleouid456
```

Organizing Accounts into OUs

AWS Organizations also offers the possibility to organize your AWS accounts in a logical and hierarchical structure that best reflects your own internal organizational structure. This can be done by creating OUs that follow the structural model of your choice.

Consider the following examples of different structures.

In *Figure 3.3*, the account structure is modeled to match the organization's own hierarchical structure. At the top of the diagram is the root node. This represents the overarching entity, which is the organization itself. From here, the branches extend, shaping the entire structure. Underneath the root node are the business units (BU1, BU2, and BU3). These could represent a department, team, or functional area. Under BU1, we have ProjectA, which is an initiative with defined goals. It could involve product development, process improvement, or strategic planning. The project has different stages Non-Prod which suggests a non-production stage, such as planning or testing and Prod which is the live environment where the final work is carried out.

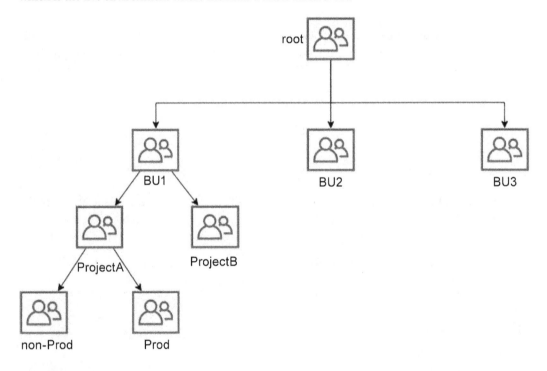

Figure 3.3: BU style

In *Figure 3.4*, the structural backbone is driven by the various environments used to deploy the applications, such as development (dev), testing, user acceptance testing (UAT), or production:

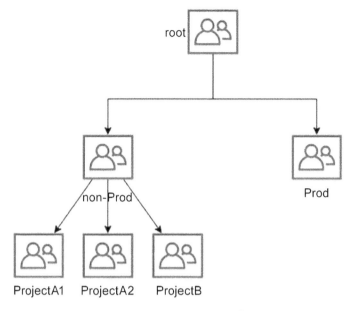

Figure 3.4: Environment style

In *Figure 3.5*, the driving structural element is the project. Every OU below the root maps to a project and contains the accounts bound to that project:

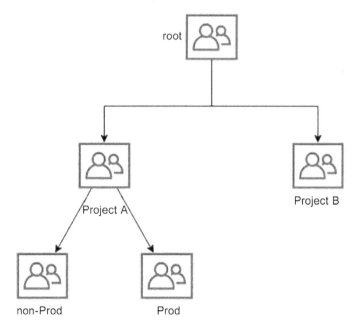

Figure 3.5: Project style

Which account structure would you rather choose? There is no single correct answer to that question. Some organizations are more comfortable with an account structure that reflects their own organization hierarchy (that is, BU style), while some others prefer to adopt a structure driven by environments (environment style), by projects (project style), or a combination of these. Currently, AWS allows you to have up to five levels of hierarchy in an organizational structure, not more. It should be sufficient in most cases but take this into account when planning your account structure as it could be a limiting factor.

Now that you have seen how to organize and isolate your resources on AWS, investigated billing, and discovered what AWS Organizations can do for you, you are ready to dive deeper into SCPs.

Setting up SCPs

As mentioned earlier, the intention behind SCPs is similar to that of IAM permissions boundaries, that is, to limit the perimeter of what is allowed to be done at an account level, an OU level, or an organization level.

SCPs offer central control over that maximum set of permissions that accounts in an OU or across your entire organization can have. However, it is important to understand that SCPs do not grant any permission to IAM entities (users and roles) in your accounts; they can only limit what the entities are allowed to do.

You can attach multiple SCPs (up to five) at any one time to the same organization, OU, or account. SCPs add up from the root structure down to each OU until the account level. Remember how SCPs work: they limit the scope of the possible permissions entities can have at any level. Thus, to determine what account IAM entities (roles or users) are entitled to do, you must travel down the organization tree and look at the intersection of all the permissions defined by the SCPs at each level until you reach the desired account in the organization hierarchy. An SCP at a lower level cannot add permissions that were denied by an SCP at the level above.

Consider the organizational structure, with a root account and BUs, as shown in *Figure 3.6*:

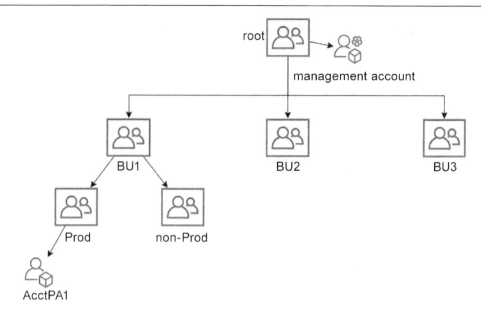

Figure 3.6 – Example organizational structure

Now, consider that the permissions defined in the SCPs attached at each level are assigned, as illustrated in *Figure 3.7*, to the root structure (permissions A, B, and C), to OU BU1 (permissions C and D), and to OU Prod (permissions C, E, F, and G):

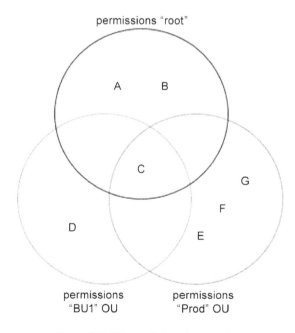

Figure 3.7: OU permissions intersection

Then, the maximum permissions allowed to IAM entities (users or roles) in your account located in the `Prod` OU is the intersection of all the permissions provided by the succession of SCPs from the root down to the OU to which the account belongs—that is, permission C. So, even though your account belongs to the `Prod` OU, which is entitled to permissions C, E, F, and G, permission C is the only one allowed as permissions E, F, and G are not allowed by the two parent levels.

How to determine an account's maximum permissions from SCPs

If an entity belonging to your account need access to a given AWS service, access to that service must be authorized by every SCP at every organization level above the account (from the root down to the OU to which the account belongs).

In practice, SCPs are expressed in two ways, either as **deny lists** or as **allow lists**. The most common way of using SCPs is as deny lists, mostly because they are low maintenance. Whenever AWS adds a new service, if you manage SCPs as allow lists, you will need to update them to be able to use the new service. If you use SCPs as deny lists, this won't be the case unless you want to block the new service. Deny lists are also usually shorter and, as such, fit more easily within the maximum possible size of an SCP (5,120 bytes at the time of this writing, but do check the AWS documentation for the latest value).

Using SCPs as Deny Lists

AWS Organizations, by default, attaches a managed SCP named `FullAWSAccess` to every root and OU structure upon creation. It is up to you to define additional SCPs at each level to limit the permissions as needed, by adding deny statements in the policies of these SCPs. For example, the following SCP blocks access to the Amazon RDS service:

```
{
  "Version": "2012-10-17",
  "Statement": [
    {
      "Sid": "DenyRDSAccess",
      "Effect": "Deny",
      "Action": "rds:*",
      "Resource": "*"
    }
  ]
}
```

You can leverage SCPs to prevent all sorts of actions, such as the following:

- To ensure that no one in your organization deploys anything in AWS Regions outside of specific pre-approved Regions

- To prevent users or roles in your organization from making damaging changes (especially to resources that may be managed centrally by a specific team, such as your security policies or your network topology)

- To prevent member accounts from leaving the organization

- To prevent users from disabling AWS Config or changing its rules

For more examples, please see the link to the AWS documentation in the *Further Reading* section.

Using SCPs as Allow Lists

If you want to use SCPs as allow lists, you must replace the AWS-managed `FullAWSAccess` SCP mentioned previously with one or more SCPs that explicitly allow only those services and actions you want to authorize. Note that when you remove the `FullAWSAccess` SCP, all actions are implicitly denied. The code in the following example grants explicit permissions to perform all actions on Amazon **Elastic Compute Cloud (EC2)** and Amazon S3 services, but implicitly denies access to any other service:

```
{
    "Version": "2012-10-17",
    "Statement": [
        {
            "Effect": "Allow",
            "Action": [
                "ec2:*",
                "s3:*"
            ],
            "Resource": "*"
        }
    ]
}
```

As we have seen, SCPs are extremely powerful as their action scope is potentially broad (single or multiple OUs, or even an entire organization). They can prove especially useful in protecting your key resources from being altered but also efficiently limiting the scope of what users and roles can do across multiple accounts.

Account Management at Scale with AWS Organizations

Beyond creating, structuring, managing accounts, and enforcing global security, backup, and tagging policies, AWS Organizations also allows you to scale the management of multiple other AWS services throughout an entire organization.

Nowadays, many different AWS services and features integrate with AWS Organizations to allow the central management of the resources they handle. We have already seen a few of them, such as backup and tagging. Another one, Control Tower, will be the subject of the next section. You will soon learn about several others in the coming chapters of this book, such as **AWS Config**, **AWS Security Hub**, **Amazon GuardDuty**, **AWS Service Catalog**, **AWS Single Sign-On** (**AWS SSO**), and **AWS CloudTrail**, to name a few. This list is by no means exhaustive, so for a complete list, please check the AWS Organizations documentation, as AWS Organizations keeps evolving and the list of integrated AWS services keeps growing.

Before we close this section on AWS Organizations, make sure to familiarize yourself with the quotas affecting it by looking at the AWS documentation at `https://packt.link/5hQUP`. Some of the quotas cannot be altered (you can have only one root in an organization and its depth is limited to five levels), while others can be raised by submitting a ticket to AWS Support (for instance, the number of accounts in an organization is ten by default; this is, luckily, only a soft limit).

The following section will introduce a service that is meant to enforce governance best practices at scale—**Control Tower**.

Leveraging Control Tower

Control Tower is an AWS service that addresses all the aspects covered earlier in this chapter in a prescriptive way. It is an opinionated service that allows you to automate the setup of your baseline environment—in other words, your *landing zone*. Control Tower does this by following a set of best practices coming from the collective experience of AWS. This experience was built over the years by working with thousands of customers who needed to set up a secure AWS environment to govern their AWS workloads more easily with central rules for security, operations, and compliance.

On top of these best practices, Control Tower relies on multiple other AWS services such as, but not limited to, **AWS Organizations**, **AWS Config**, **AWS Service Catalog**, **AWS SSO**, and **AWS CloudTrail**.

You can either set up Control Tower in a brand-new organization (as defined in AWS Organizations) when starting afresh or use it in an existing organization that you already have in place. In the latter case, there are some considerations to ensure that your existing organization will not be impacted before you actually set up the Control Tower. The bottom line is that it's always easier to start with a new organization, but there may be situations where you'd prefer not to start from scratch.

Note that existing accounts are not automatically enrolled in Control Tower when you set it up in an existing organization. These existing accounts need to be explicitly enrolled to be governed by Control Tower, which will, upon enrollment, baseline each account according to best practices and apply any security guardrails that you have enabled. There are a number of prerequisites to follow for enrollment. When onboarding existing accounts, it is best to always refer to the Control Tower documentation to check those prerequisites. Depending on the number of accounts that you need to enroll, you may want to automate the enrollment process as much as possible (repeating the various steps manually is also error-prone). For more details on how to do this, refer to the *Enroll Existing AWS Accounts into AWS Control Tower* blog in the *Further reading* section at the end of this chapter.

What does Control Tower Deliver Exactly?

Control Tower delivers a number of things:

- First and foremost, it will create what we call your **landing zone**—that is, your multi-account AWS environment set up according to AWS best practices. It follows the recommendations from the **AWS Well-Architected Framework** (see `https://packt.link/yhehg` for more details). This environment will contain some essential elements to start with and provide room for extension so that you can later add additional OUs and accounts where you can deploy your resources.

- Second, Control Tower will provide and enforce **guardrails** across your entire organization (with the exception of your management account). These guardrails are rules of two types—*preventive* or *detective*. Each rule type is further declined into **Mandatory**, **Strongly recommended**, and **Elective** kinds.

- Thirdly, Control Tower will create an **account factory** that you can configure and leverage to automatically provision new accounts with pre-approved configurations. The idea is to facilitate account provisioning to do the heavy lifting for you.

- Finally, Control Tower comes with a dashboard that lets your AWS system administrators oversee the landing zone operations and control their status.

How does Control Tower Operate?

Upon setup, Control Tower deploys a certain number of resources in your organization. It leverages **CloudFormation** templates through **stacks** and **stacksets** to deploy and manage these resources. The following steps will further explain the process:

- First, Control Tower will create a root for your new organization or reuse your existing organization root, depending on your specific case.

- It will then create two OUs, security (always) and sandbox (optional) under that root structure. Within the Security OU, Control Tower will create two accounts—a log archive account and an audit account. The log archive account is meant to serve as a central repository for the CloudTrail and CloudWatch logs delivered from the various accounts inside your landing zone, to track API activities and resource configurations. The audit account is a restricted access account dedicated to your security and audit teams (think compliance here) to review the accounts in your landing zone.

- Control Tower will then set up AWS SSO with a user directory, preconfigured groups, and SSO access. These preconfigured groups can then be used to organize the users who, depending on their roles, will need to perform specific tasks in your landing zone accounts. If you already have a third-party **identity provider** (**IdP**) such as Okta in place, you can also integrate that IdP with AWS SSO and synchronize your users between the two, using **System for Cross-domain Identity Management** (**SCIM**), for instance. Integration with **Microsoft Active Directory** (**MS AD**) is slightly different in the sense that, in that case, Control Tower will not manage the SSO directory.

- Eventually, Control Tower will apply a number of preventive guardrails to enforce best practices. Preventive guardrails are meant to block unauthorized actions that could lead to policy violations from happening.

- Control Tower will also apply some mandatory detective guardrails to detect configuration violations. Note that preventive guardrails apply in all AWS regions, while detective guardrails only apply in those AWS regions where Control Tower is supported. Note also that mandatory guardrails are enabled by default; while strongly recommended, elective guardrails are not enabled by default, so it's for you to decide whether to activate them.

Preventive guardrails are implemented using SCPs, while detective guardrails are implemented using **AWS Config** rules and **AWS Lambda** functions.

The following is an example of a mandatory preventive guardrail implemented by Control Tower using SCPs, which prevents any change to CloudTrail settings by Control Tower itself:

```
{
  "Version": "2012-10-17",
  "Statement": [
    {
      "Sid": "GRCLOUDTRAILENABLED",
      "Effect": "Deny",
      "Action": [
        "cloudtrail:DeleteTrail",
        "cloudtrail:PutEventSelectors",
        "cloudtrail:StopLogging",
        "cloudtrail:UpdateTrail"
      ],
```

```
      "Resource": ["arn:aws:cloudtrail:*:*:trail/awscontroltower-*"],
      "Condition": {
        "ArnNotLike": {
        "aws:PrincipalARN":"arn:aws:iam::*:role/
          AWSControlTowerExecution"
        }
      }
    }
  ]
}
```

The example below illustrates detective guardrails. It configures a mandatory guardrail implemented by Control Tower on the Security OU that detects whether public read access is enabled on the S3 buckets within the accounts belonging to that OU:

```
AWSTemplateFormatVersion: 2010-09-09
Description: Configure AWS Config rules to check that your S3
buckets do not allow public access
Parameters:
  ConfigRuleName:
    Type: 'String'
    Description: 'Name for the Config rule'
Resources:
  CheckForS3PublicRead:
    Type: AWS::Config::ConfigRule
    Properties:
      ConfigRuleName: !Sub ${ConfigRuleName}
      Description: Checks that your S3 buckets do not allow public
read access. If an S3 bucket policy or bucket ACL allows public read
access, the bucket is noncompliant.
      Source:
        Owner: AWS
        SourceIdentifier: S3_BUCKET_PUBLIC_READ_PROHIBITED
      Scope:
        ComplianceResourceTypes:
          - AWS::S3::Bucket
```

In the preceding example, the intention is to report any S3 bucket with public read access enabled in the log archive and audit accounts.

Summary

You explored a number of key topics in this chapter—how to best organize and isolate your AWS resources, which billing strategy to implement, how to leverage AWS Organizations and OUs to structure your AWS environment, how to enforce security best practices and protect your AWS environment using SCPs, and finally, how to automate governance at scale with Control Tower.

The topics covered in the first three chapters are part of the foundational skills that you must know inside out to be a successful AWS solutions architect. We will build upon these foundations as we progress on to more complex and advanced topics in the rest of the book.

In *Chapter 4, Ensuring Cost Optimization*, you will explore the various mechanisms and services that you can leverage to keep your AWS bills under control.

Further Reading

For more information on the topics covered in this chapter, consult the following documentation:

- AWS Multiple Account Billing Strategy: `https://packt.link/A1Klo`
- Tagging Best Practices whitepaper: `https://packt.link/z5sL2`
- Establishing your best-practice AWS environment with AWS Organizations: `https://packt.link/CwhEF`
- Several examples of SCPs: `https://packt.link/ejmG1`
- Enroll Existing AWS Accounts into AWS Control Tower: `https://packt.link/1VSRp`.

4

Ensuring Cost Optimization

In the previous section, we explored various AWS pricing models and how each of these models varies from service to service. In addition to this, you also learned about the importance of evaluating your costs and explored a few AWS services that help customers understand, evaluate, and forecast their spending on the cloud.

In this section and chapter, you will explore various options and opportunities that a solutions architect can undertake to ensure that costs are evaluated and optimized throughout the lifecycle of a particular workload. You will learn about certain design principles using the **Well-Architected Framework** (**WAF**), along with taking a deeper dive into how you can ensure governance, reporting, and monitoring of your costs using a combination of design strategies as well as AWS services and tools.

The following main topics will be covered in this chapter:

- Cost optimization principles
- Establishing governance with tagging
- Monitoring with alerts, notifications, and reports

Cost Optimization Principles

Cost optimization, for any and all workloads, is a continuous process of refinement and improvement. This section will discuss some key design principles that all solutions architects as well as cloud FinOps (which stands for financial operations) teams should keep in mind when optimizing costs for their workloads:

- **Consumption-based pricing model**: Unlike traditional hosting solutions that require large upfront payments, the cloud simplifies the pricing of resources based solely on their consumption. This method of pricing helps reduce the unwanted sprawl of resources and helps teams design more optimal solutions that can scale in and out based on specific requirements. A typical example of this would be dev and test workloads that do not need to run 24x7. Such workloads can easily be spun up in the cloud, utilized, and then terminated once the tasks are over.

- **Setting up FinOps CoEs**: FinOps is all buzz nowadays in most organizations, and it should be as well, considering the fact that it aims to bridge the gap between IT operational and financial teams within an enterprise. Setting up FinOps CoEs (meaning financial operations centers of excellence) enforces collaboration and a sense of shared responsibility between teams when it comes to IT resources and costs to procure, run, and manage them. The primary goal of such CoEs is to promote cost optimization best practices, define controls, and implement cost awareness across the organization, as well as enforce guardrails for cost monitoring and management.

- **Attribute expenses with owners**: A FinOps CoE can define, as well as put in place, controls to monitor cost expenditure across the cloud by attributing each resource to a particular team, project, or user. This not only ensures ownership of the resources but also provides a better understanding and awareness of where costs are actually spent, and how they can be optimized in the future.

- **Continuous measurement and monitoring**: As mentioned at the beginning of this chapter, cost optimization is not a one-stop process but rather a continuous operation that involves constant measurement and monitoring of resources and usage patterns. Analyzing these metrics proactively on a regular basis ensures that individual workloads are constantly optimized by either redesigning applications to leverage more cloud-native services or modifying the underlying resources of an application, such as CPU, memory, storage, and so on, based on actual utilization patterns.

- **Adopting cloud native**: A key design principle of cost optimization is to look for opportunities to leverage as many cloud-native services as possible. Why is this so important? Well, in most traditional scenarios, the majority of the costs required to run an application sprout from operational teams that are required to manage and maintain the underlying hardware, ensure its uptime, monitor it, and patch it, all manually. Cloud storage takes over the responsibility of such operational tasks from your IT teams, so they can focus on what is important from a business perspective, such as creating and rolling out new application functionalities and so on, without having to manage any of the underlying, undifferentiated tasks of maintaining infrastructure, such as server maintenance and patching or storage management. This can only be made possible if applications and workloads are constantly monitored and optimized to leverage as much of the cloud's native services as possible.

Having explored the key cost optimization principles, you are now ready to examine some of the mechanisms using which you can implement cost governance across your organization, starting with tagging.

Establishing Governance with Tagging

Organizing resources in a meaningful way helps IT teams understand cost spending and overall usage patterns of even the most complex workloads, and this organization of resources can be achieved by leveraging a simple concept called tags. This section covers tagging in depth, as well as discussing some essential best practices and strategies to keep in mind when it comes to tagging resources in your own AWS cloud.

So, what are tags? Tags are key-value pairs of metadata that help identify resources in your AWS account. Each tag's key is a unique identifier and each key can have only one value associated with it. You can create tags and assign them to almost all AWS resources that you create throughout your AWS accounts, including IAM users, roles, EC2 instances, RDS databases, S3 buckets, and so on and so forth. The most important tagging feature that AWS provides specifically for cost management is AWS cost allocation tags.

A cost allocation tag is an AWS feature that helps monitor your resource usage and costs granularly. Once the tags are activated in the AWS Billing console, you can use them to categorize, organize, and track resources as per their usage and expenditure. There are two types of cost allocation tags:

- **AWS-generated**: As the name suggests, these tags are created and propagated by AWS for supported resources for cost-tracking purposes. Resources are tagged with the `createdBy` key automatically, and its corresponding value is one of the following attributes: `account-ID`, `access-key`, `user-name`, or `role`.

  The following are some examples of this:

  ```
  key = aws:createdBy
  value = 1234567890:dummyUser
  ```

  Or

  ```
  key = aws:createdBy
  value = AKIAUITOFQDN5EXAMPLE:dummyIamRole
  ```

- **User-defined**: These tags are created, applied, and managed by individual users, or in most cases, by a centralized FinOps team. These tags can be used to categorize or organize resources as per an organization's needs, such as the following:

  ```
  key = EnvironmentName
  value = Production
  ```

  Or

  ```
  key = Department
  value = DEV01
  ```

Before you learn further about cost allocation tags, quickly review the few simple steps required to enable these tags using the AWS Management Console.

Activating Cost Allocation Tags

In order to activate the AWS-generated cost allocation tags, you need to do the following:

1. Sign in to the AWS Management Console and launch the Billing console by visiting the following link: `https://packt.link/5jZdj`.

2. From the navigation pane on the left, select the `Cost allocation tags` option.

3. From the `AWS-generated cost allocation tags` tab, filter and select the `aws:createdBy` tag, as shown in the following screenshot.

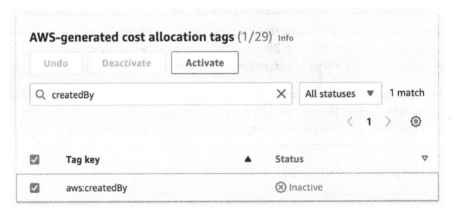

Figure 4.1: AWS-generated cost allocation tag enablement

4. Select the `Activate` option to complete the process. Note that it will take approximately 24 hours for the tag to activate.

The same process can be followed to enable and activate a user-defined cost allocation tag as well:

1. Simply log in to the AWS Billing console as before and select the `Cost allocation tags` option from the navigation pane on the left.

2. Next, select the appropriate tag from the `User-defined cost allocation tags` tab and, once again, click on the `Activate` option to enable the tag in the cost allocation/billing reports.

> **Note**
> Once the tags are activated, they are propagated only to newly created AWS resources that were spawned after the tags were enabled.

Creating Cost Allocation Tags

Once the tags are activated and enabled, you can visualize them using AWS Cost and Usage Reports, or various other tools such as Cost Explorer, AWS Budgets, and so on.

AWS also offers different options when it comes to creating tags for your resources, such as the AWS Management Console, AWS APIs, and, quite recently, AWS Tag Editor. In this section, you will explore AWS Tag Editor at a high level and understand some common best practices and considerations to keep in mind when it comes to creating cost allocation tags.

> **Note**
> You, as a user/administrator, cannot edit AWS-generated cost allocation tags, their keys, or their values.

AWS Tag Editor is perhaps the simplest way of tagging and managing resources using the AWS Management Console:

1. To get started with AWS Tag Editor, simply log in to the AWS Management Console and select the `Search` option at the top of the page.

2. Next, type in the `tag editor` keywords and select the `Resource Groups & Tag Editor` option to launch the service.

3. This will open up the `AWS Resource Groups` dashboard. From here, select the `Tag Editor` option from the navigation pane on the left, as demonstrated in the following figure:

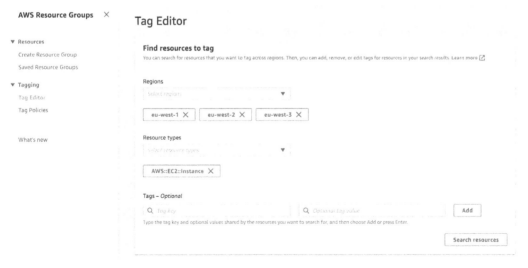

Figure 4.2: AWS Tag Editor dashboard

4. Now select the Regions and the specific AWS resource types that you wish to find and tag based on your requirements using the appropriate drop-down options provided. Once completed, select the Search resources option to view the final list of resources.

5. In order to apply tags to the newly discovered resource, select it from the Resource search results pane, and select the Manage tags of selected resource option to continue.

6. On the Manage tags page, add subsequent tags to your resource using the Add tag option. Once done, simply select Review and apply tag changes for the changes to come into effect, as shown in *Figure 4.3*:

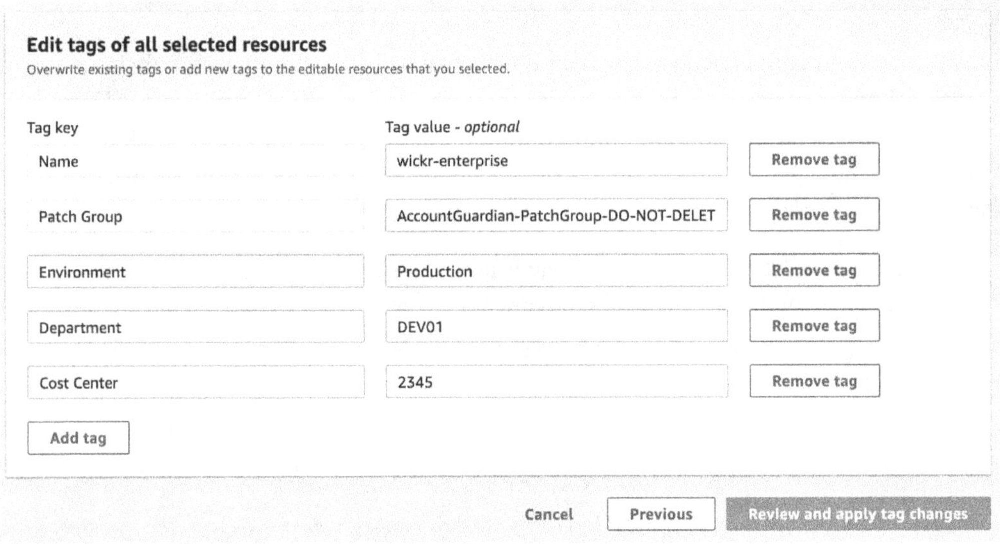

Edit tags of all selected resources

Overwrite existing tags or add new tags to the editable resources that you selected.

Tag key	Tag value - *optional*	
Name	wickr-enterprise	Remove tag
Patch Group	AccountGuardian-PatchGroup-DO-NOT-DELET	Remove tag
Environment	Production	Remove tag
Department	DEV01	Remove tag
Cost Center	2345	Remove tag
Add tag		

Cancel   Previous   Review and apply tag changes

Figure 4.3: Applying tags to resources

Now that was simple, wasn't it? You can follow the same steps to tag other AWS resources, such as databases, ENIs, Cloud9 environments, CloudTrail trails, CloudWatch alarms, and much more! For a full list of supported AWS services along with their resource types, please take a look at the documentation provided here: https://packt.link/wkiSF.

Tagging Strategies and Considerations

The following are some key tagging strategies and considerations to keep in mind when considering cost-optimization exercises:

- Governance is always a key factor when it comes to enforcing tagging strategies across enterprise organizations, and the best way to do this is by creating a cross-functional team that is solely responsible for defining, maintaining, and enforcing tagging requirements for all workloads on the cloud. These requirements can essentially be documented as a standard to be used across the entire organization.

- Once the team is created, start by defining consistent tagging values to be used across all workloads on the cloud. These values can be based on factors such as the following:

  - **Owner of the resource**: Identifying who is responsible for the resource

  - **Deployment stack**: Describing an environment such as development or staging

  - **Cost center**: Identifying, tracking, and charging back the department that is utilizing the resources

  - **Project/application**: Describing and grouping the resources required to run a particular project or application

  - **Compliance**: An optional but important value that can help identify workloads based on security compliance requirements such as the **Health Insurance Portability and Accountability Act (HIPAA)** for healthcare data, and **Payment Card Industry Data Security Standard (PCI-DSS)** for handling credit card information

- Enforce a mix of both AWS-generated as well as user-defined tags using AWS Organizations **Service Control Policies (SCPs)**.

- Keep the tags consistent and up to date with the help of automation such as AWS CloudFormation templates or AWS Systems Manager Automation.

- Tag all resources wherever and whenever possible.

- If a resource is not tagged, enforce remediation.

- Propagate tags across related resources whenever possible, for example, tagging an EC2 instance and propagating those tags to the underlying EBS volume, and so on.

- Create monthly reports of untagged resources and follow up with the resource owners for remediation.

- Analyze costs and usage on a monthly basis with the help of tools such as AWS Cost Explorer.

With this, you now have a good understanding of tags as well as tagging strategies and considerations. In the next section of this chapter, you will go through a few simple mechanisms using which you can set up proactive alerts, notifications, and reports for monitoring and optimizing your AWS costs.

> **Note**
>
> Tagging the cost center helps track and allocate costs within an organization. It associates cloud resources with specific departments or projects. For instance, if the marketing team uses certain VMs, tagging those resources with the marketing department's cost center ensures accurate billing.

Monitoring with Alerts, Notifications, and Reports

Now you have selected the necessary tags and enforced a tagging strategy for your workloads on AWS. In this section, you will now look at how to use these tagging mechanisms to generate simple billing alerts, notifications, and reports using AWS tools and services.

First, you will learn how to create a fairly simple billing alarm using Amazon CloudWatch. But before that, there are a few prerequisite steps that you need to complete, starting with enabling billing alerts.

Enabling Billing Alerts

Before you go ahead and create billing alarms using Amazon CloudWatch, first enable billing alerts using the AWS Billing console by following these steps:

1. Sign in to the AWS Management Console and launch the AWS Billing dashboard using the following link: `https://packt.link/BeKTf`.

2. Next, select the `Billing Preferences` option from the navigation pane on the left.

3. Click on the checkbox next to the `Receive billing alerts` option and click on `Save preferences` to complete the process, as shown in the following screenshot:

☑ **Receive Billing Alerts**

Turn on this feature to monitor your AWS usage charges and recurring fees automatically, making it easier to track and manage your spending on AWS. You can set up billing alerts to receive email notifications when your charges reach a specified threshold. Once enabled, this preference cannot be disabled. <u>Manage Billing Alerts</u> or <u>try the new budgets feature!</u>

▶ **Detailed Billing Reports [Legacy]**

> **Save preferences**

Figure 4.4: Enabling billing alerts using the AWS Billing dashboard

Creating a Billing Alarm

Once the billing alerts are enabled, the next step in the process is to create a simple billing alarm using Amazon CloudWatch. Follow these steps to do this:

1. Launch the Amazon CloudWatch console using the following link: `https://packt.link/43h8J`. Ensure that your Region is set to `US East (N. Virginia)` before you carry on with the next steps.

> **Note**
> All billing data is consolidated and stored in the US East (N. Virginia) Region.

2. Next, from the navigation pane, select the `Alarms` option and then click on `All Alarms`.

3. Click on the `Create alarm` option.

4. On the `Specify metric and conditions` page, click on `Select metric` and then use the `Metrics` search field to search for billing-related metrics as shown:

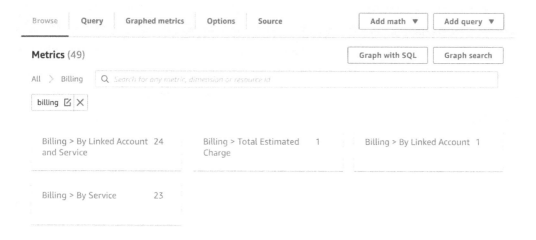

Figure 4.5: Total estimate billing alarm configuration

5. Select the `Billing > Total Estimated Charge` metric to continue.

6. Finally, select the currency as `USD` and click on `Select metric` to complete the process.

7. Once again, on the `Specify metric and conditions` page, provide a suitable `Threshold` value against which your billing alarm monitors and triggers. This example sets the alarm to trigger if the threshold value of $500 is crossed for a period of 6 hours; you can configure the values and conditions here as per your requirements.

8. Once the thresholds are set, select Next to configure an SNS topic to send out notifications in case the alarm is in a triggered state. You can choose to create a new SNS topic here as well.

9. Once completed, click on Next to provide a suitable name and description for your alarm, and there you have it! Your very own billing alarm has been created using just a few simple steps!

Setting Up Notifications

The last piece of the puzzle is enabling and configuring notifications for your billing alerts and alarms. This is important for a number of reasons, such as cost management, security, and creating automated workflows. This can also be achieved with a few simple clicks from the AWS Management Console:

1. To get started, log in to the AWS Management Console and launch Amazon **Simple Notification Service (SNS)** using the following URL: https://packt.link/6Q1U7.

2. Once done, select the Topics option from the navigation pane.

3. On the Topics page, click on Create topic to get started.

4. On the Create topic page, select either a FIFO (First in First out) topic, which guarantees message ordering, or a Standard topic, which performs message ordering on a best-effort basis. It's best to select the Standard topic, which will provide you with multiple endpoint options to subscribe to, such as Amazon Kinesis Data Firehose, AWS Lambda, email, HTTP, SMS, and so on. With the FIFO (First in First out) topic, you can only select Amazon SQS as a subscription.

5. Once selected, provide a suitable name for your topic, followed by an optional display name.

    That's pretty much it!

6. With all the settings completed, click on Create topic to complete the process.

7. Now, from your newly created Topic page itself, select the Create subscription option provided under the Subscriptions tab.

8. On the Create subscription page, start by selecting the newly created topic from the Topic ARN drop-down list.

9. Next, select the appropriate protocol for your notification delivery. For this example, select Email as the default protocol and provide the FinOps team's email address in Endpoint, as shown in *Figure 4.6*:

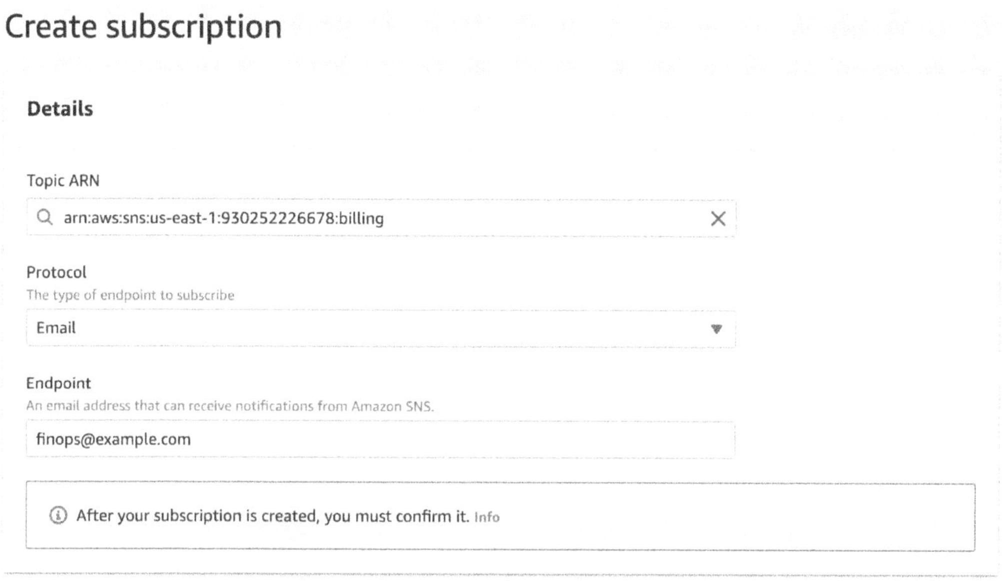

Create subscription

Details

Topic ARN

🔍 arn:aws:sns:us-east-1:930252226678:billing   ✕

Protocol
The type of endpoint to subscribe

Email   ▼

Endpoint
An email address that can receive notifications from Amazon SNS.

finops@example.com

ⓘ After your subscription is created, you must confirm it. Info

Figure 4.6: Billing notification configuration using Amazon SNS

10. Once completed, select the `Create subscription` option to complete the process.

> **Note**
>
> You will need to confirm the subscription using the provided endpoint (email ID in this case) once the subscription is created.

You could also create additional subscriptions under the same topic for other AWS billing- and pricing-related news and information, such as notifications of when the price of a service changes and so on.

To do so, simply follow the same steps as before while creating a subscription, with the only exception being the custom topic ARN for any AWS service-related pricing changes, which is `arn:aws:sns:us-east-1:1234567890:price-list-api`.

Viewing Reports

The last and perhaps most important part of any cost optimization exercise is to continuously measure and monitor the overall consumption of resources across your AWS environments, and this can be achieved with the help of granular reports generated using AWS Cost Explorer.

AWS Cost Explorer provides you with out-of-the-box reports with insights into the most common usage and spending patterns, such as `Daily costs`, `Monthly costs by service`, `Reserved Instance report`, and `Monthly EC2 running hours costs`. Each of these default reports is non-editable; however, you can use them as boilerplates to create your very own custom reports too:

1. To view the default reports, sign in to the AWS Management Console and go to the following URL: `https://packt.link/4Avly`.

2. Next, select the `Reports` option from the navigation pane to view a list of pre-created, default reports.

3. Select the `Monthly costs by service` report to get started.

4. On the `Cost Explorer` page, use the various filters to filter by the usage of AWS resources across your accounts, Regions, and in some cases, even Availability Zones. You can also group the graphs as per your requirements by selecting the `More` option.

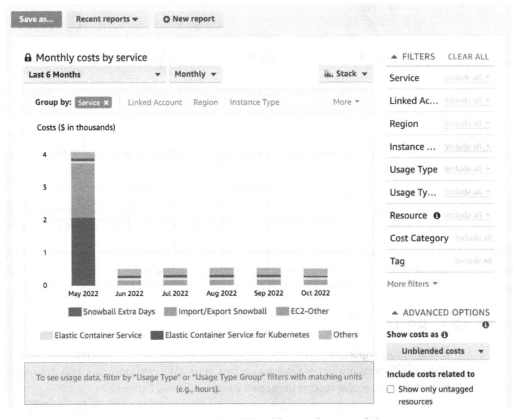

Figure 4.7: Cost Explorer "Monthly costs by service" view

5. Once done, save the report using either the `Save as...` option provided or, alternatively, download the data in the form of a CSV. You can use other AWS services such as Amazon Athena or Amazon QuickSight to create further fine-grained reports using these CSVs and share them across your organization as well.

Summary

This chapter started off by discussing a few key cost-optimization principles that are essential for any and all organizations to follow in order to understand, manage, and forecast costs on the cloud. You then reviewed some tagging best practices and strategies, as well as diving into how you can enable and configure cost allocation tags using AWS Tag Editor. Finally, you learned how to further optimize costs by setting up simple alarms, alerts, notifications, and reports using various AWS services.

In *Chapter 5*, *Determining Security Requirements and Controls*, you will explore how to control access to resources across multiple AWS accounts in your organization, including how to apply security and compliance controls and go through the measures and the AWS services.

Further Reading

- Cost optimization – Well-Architected Framework: `https://packt.link/KxO91`

- AWS cost categories: `https://packt.link/4yx6X`

5

Determining Security Requirements and Controls

Designing secure workloads is essential to protect your data and systems and to be able to respond to security threats in a timely and successful manner. When you design a new solution on **Amazon Web Services** (**AWS**), security is the first topic you want to focus on. Whether the application that you design is a public application or for internal use only, it is paramount to establish a trustworthy, secure foundation on AWS for your application to run in a safe environment.

In this chapter, you will learn how to control access to resources across multiple AWS accounts in your organization when building a new solution. You will look at applying security and compliance controls and go through the measures and the AWS services to leverage for the following topics:

- Managing identity and access
- Protecting your infrastructure
- Protecting your data
- Detecting incidents
- Responding to incidents

Managing Identity and Access

The first security aspect to consider for your application is the credentials and the associated permissions that are necessary both for the application components to do their jobs and for the end users to be able to interact with the application.

IAM Users and Roles

The complete set of features available on AWS for **identity and access management (IAM)** was already discussed in *Chapter 1, Determining an Authentication and Access Control Strategy for Complex Organizations*, but here a quick recap of the major concepts for this chapter's discussion. If, at any moment, you feel you need more information to fully understand this part, please refer to *Chapter 1, Determining an Authentication and Access Control Strategy for Complex Organizations*.

When designing your application on AWS, you will leverage one or more AWS services to fulfill its mission. Accessing those services requires permissions, thus you must provide the necessary credentials for your application to work. IAM provides two types of entities— *users* and *roles*—for this purpose. IAM users are allowed access to AWS resources either through the AWS Management Console (by providing a username and password) and/or programmatically (using an *access key* and a *secret access key*) from the **command- line interface (CLI)** or one of the AWS **software development kits (SDKs)**. IAM users are given permissions either by being directly assigned IAM policies or by being assigned to an IAM user group. IAM roles are like IAM users in the sense that they also provide access to AWS resources and define what the end user or application assuming that role can do on AWS. There are two major differences, though. The first one is that a role is not associated with a single individual or application, but it can be assumed by multiple entities (users or roles). The second major difference is that IAM roles provide temporary credentials as opposed to IAM users' long-lived credentials.

As much as possible, it is good practice to use temporary credentials, which means preferring IAM roles to IAM users. Temporary credentials take a thorn out of your side since you don't need to store those credentials for more than the duration of the session for which they are valid. On the other hand, if you need to use long-lived credentials, you also need to secure their storage, which requires particular care, can be tricky, and inevitably increases your security risk exposure.

What can temporary credentials be used for, then? Well, they can be used for pretty much everything on AWS. IAM roles can be used for end users, applications, and AWS services themselves. There are a few exceptions where you need to rely on IAM users and long-lived credentials.

One such notable exception is, for instance, if you need to create Amazon **Simple Email Service (SES) Simple Mail Transfer Protocol (SMTP)** credentials for sending emails via SMTP—this can only be achieved by an IAM user (at least, at the time of writing this).

IAM roles are the standard mechanism through which AWS services call each other's **application programming interfaces (APIs)**. Whenever one service—for example, Amazon **Elastic Compute Cloud (EC2)**—needs to call another service's APIs—for example, Amazon **Simple Storage Service (S3)**—then the caller service must assume a service role that gives them enough permissions to invoke the called service's APIs they want to use. It is also recommended to use IAM roles to invoke any AWS service APIs when calling them from your own application.

Last but not least, if your application needs to authenticate end users before granting them access, in many cases you will be interested in leveraging identity federation, relying on a third-party **identity provider** (**IdP**). This is true whether we are talking of a public-facing application or an internal application. In the former case, you will rely on an external IdP such as **Apple**, **Facebook**, **Google**, or **Amazon**. In the latter case, you will want to rely on a pre-existing corporate IdP, such as **Microsoft Active Directory** (**MS AD**), not only to avoid impacting your admin's identity management process but also to avoid duplicating identities. There can, of course, be exceptions if you wish to start from a blank slate, free of ties with any pre-existing corporate environment.

The next section dives a little bit more into service roles and end user identity federation.

AWS Service Roles

AWS service roles allow AWS services to access resources in other AWS services on your behalf. A service must assume a service role to perform actions, on your behalf, on other AWS services. In some cases, AWS services provide a predefined service role out of the box—these are called **service-linked roles**. The list of services supporting service-linked roles can be found at `https://packt.link/80RzZ`

> Note:
> The list of supporting service-linked roles keeps evolving over time so, when in doubt, please make sure to consult the AWS documentation.

The above documentation page will tell you whether a given AWS service supports using service-linked roles or whether it supports other forms of temporary credentials. When a given AWS service does not offer a service-linked role but supports IAM temporary credentials, it is then up to you to set up an IAM role for that service—aka a service role—when needed. You then have to refer to the specific service documentation to understand which permissions you need to associate with the service role. The advantage when a service provides a service-linked role is that it does create the service role automatically with the right set of permissions on your behalf.

Creating a service-linked role using the **AWS CLI** is extremely easy. You simply run a command such as the following:

```
aws iam create-service-linked-role --aws-service-name SERVICE- NAME.
amazonaws.com --description "My service-linked role to support Service
XYZ"
```

So, for instance, if you were to create a service-linked role for Amazon **Relational Database Service (RDS)**, you would run something like this:

```
aws iam create-service-linked-role --aws-service-name rds.amazonaws.
com --description "My service-linked role to support RDS"
```

This would result in the creation of a role called `AWSServiceRoleForRDS` automatically associated with the right permissions required by Amazon RDS to work.

> **Note**
>
> It is unlikely that you will be required to create the above service-linked role manually for Amazon RDS as the service does create it for you upon database creation if it does not exist in your account. However, if you do need to create one—for instance, if you have deleted the service-linked role by accident—then you can use the preceding command to re-create it manually.

Now, for AWS services that do not support service-linked roles, you have to provide service roles yourself. Unfortunately, in such cases, you have to do the heavy lifting, which service-linked roles do in the background for you. You have to first create a trust policy, using a command such as the following:

```
aws iam create-role --role-name MyServiceRole --assume-role- policy-
document file://MyServiceRole-Trust-Policy.json
```

The trust policy instructs IAM that a given service, as defined in the trust policy **JavaScript Object Notation (JSON)** document, is trusted to assume that specific role you just created. For instance, if you have created a role to be assumed by Amazon EC2, the trust policy would look like this:

```
{
  "Version": "2012-10-17",
  "Statement": {
    "Effect": "Allow",
    "Principal": {"Service": "ec2.amazonaws.com"},
    "Action": "sts:AssumeRole"
  }
}
```

Then, you need to attach a permissions policy to the role so that the trusted service has the right privileges for performing its task, using a command like this:

```
aws iam put-role-policy --role-name MyServiceRole --policy-name
MyServicePolicy --policy-document file://MyService-Policy.json
```

Suppose you want to grant an Amazon EC2 instance permissions to access Amazon S3 to list the contents of a specific bucket. In that case, your policy document would look something like this:

```
{
  "Version": "2012-10-17",
  "Statement": {
    "Effect": "Allow",
    "Action": "s3:ListBucket",
    "Resource": "arn:aws:s3:::my_bucket"
  }
}
```

Optionally, you can also add permissions boundaries or tags to define the maximum permissions the role can ever have.

Note that if you intend to use the role with Amazon EC2 instances or an AWS service that will use Amazon EC2, the role must be packaged in an instance profile that can be later attached to an Amazon EC2 instance. So, in our case, two additional commands must be run for that. The first command is to create an instance profile and the second one is to add the role to the instance profile. Both commands are presented below:

```
aws iam create-instance-profile --instance-profile-name MyEC2Profile
aws iam add-role-to-instance-profile --instance-profile-name
MyEC2Profile --role-name MyServiceRole
```

Using Federation for Access Control and Authentication

What we are going to look at more specifically now is how to manage end user access for a new solution that you design for AWS, whether it is for public access or internal use only.

User federation was introduced in *Chapter 1, Determining an Authentication and Access Control Strategy for Complex Organizations*. In *Chapter 1*, user federation was discussed from the perspective of an organization willing to manage access to their AWS environment leveraging either their corporate IdP, such as MS AD or a third-party IdP. You learned how such an organization could achieve **single sign-on (SSO)** using either **AWS Single Sign-On (AWS SSO)** or IAM depending on their specific use case.

For this scenario, assume that your solution has a web or a mobile frontend of some sort and requires authentication in place. For this case, there is one service of choice on AWS: **Amazon Cognito**. Briefly, Amazon Cognito provides authentication, authorization, and user management for web and mobile applications. It provides end users with sign-in functionality either through a third-party IdP such as Facebook, Amazon, Google, or Apple or directly with a username and password. Amazon Cognito has two main components: *user pools* and *identity pools*.

User pools are actual user directories. You typically leverage user pools when you need to provide sign-up and sign-in functionalities to a mobile or web application, such as when you want to manage end user data or need a custom authentication flow for your application.

Identity pools allow you to federate identities with IdPs and provide end users access to AWS services. You typically use identity pools to give end users direct access to AWS resources, such as Amazon S3 objects, or to generate temporary AWS credentials for unauthenticated users.

Both components can be used independently of each other or together. For instance, identity pools could be used to federate identities leveraging user pools as an IdP and provide access to AWS resources to end users registered and authenticated with user pools.

Figure 5.1 presents an overview of Amazon Cognito in the following diagram:

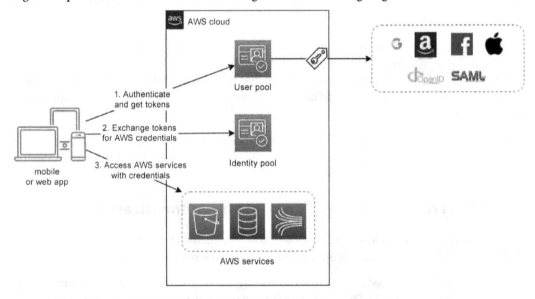

Figure 5.1: Amazon Cognito concepts overview

Now that you are aware of the key concepts of managing identity and access in AWS, you can proceed to examine the many ways in which you can protect your infrastructure.

Protecting your Infrastructure

Before you dive into infrastructure protection, first recall a key principle of AWS—the shared responsibility model. Security, along with compliance, is considered a shared responsibility between AWS and the customer. Essentially, AWS is responsible for the security *of* the cloud and you, the customer, are responsible for the security *in* the cloud.

The AWS shared responsibility model is represented Figure 5.2:

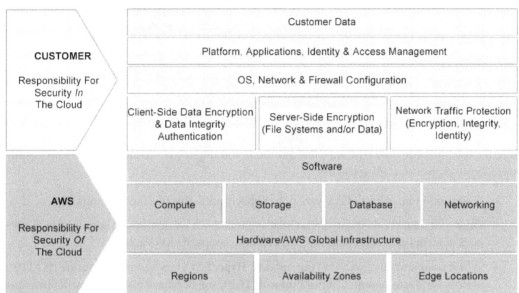

Figure 5.2: AWS shared responsibility model

AWS secures the infrastructure for supporting the services they provide—that is, both the facilities (data centers) and the hardware equipment running in these facilities, whether it is compute, storage, or networking. The customer is in charge of securing their AWS resources on top of what AWS already does. This naturally varies from service to service.

If you are using an **Infrastructure-as-a-Service** (**IaaS**) layer service such as Amazon EC2, you must ensure that the **virtual machines** (**VMs**) you create on top of it are secure, that is, make sure the **operating system** (**OS**) is patched to the latest level of security. The same goes for any software package or library installed on it and for protecting access to these VMs both from a networking perspective and from an identity and access control perspective. If you use an AWS service where the infrastructure is abstracted away and managed by AWS—for instance, **Amazon S3**—you obviously don't have to patch the OS or the software since AWS manages such tasks, but you still have to ensure the protection of the resources deployed with that service. For instance, in the case of Amazon S3, you must ensure that access to the objects you store in your S3 buckets matches your security guidelines and compliance requirements.

Essentially, you are fully responsible for the security of the resources you deploy on AWS. This may seem daunting at first to some but really, it is not—remember that you have sole and full control of your own resources.

Keep in mind that whichever resource you plan to deploy on AWS, you must make sure it operates in a safe environment. You need to take care of the management and operational aspects that support the life cycle of that resource, which can mean quite different things depending on the type of the resource and the AWS services used.

Therefore, for any new solution that you design on AWS, the first thing that you need to do is to have a clear understanding of the security and compliance requirements that apply. Why is this important? These requirements primarily dictate the security measures that you will put in place. You can, for instance, set up the most elaborate protection system against **distributed denial-of-service (DDoS)** attacks from the outside world, but it is going to be of little use if you are designing a solution for internal use only. The following are a few questions you might want to answer before taking any measures to protect your infrastructure:

- Is your solution, or part of it, meant to be publicly accessible (via the internet)?

- Does your solution integrate with external applications or services hosted outside of your AWS environment (such as running on-premises or in a separate AWS account)?

- Which communication protocols (for instance, **HyperText Transfer Protocol Secure (HTTPS)**, WebSocket, **Transmission Control Protocol (TCP)**) will be offered to external parties (whether inside or outside your organization) for accessing your solution?

- Is access to your solution limited to a set of well-known, pre-approved parties or will it be open to anyone?

- Does your solution have to abide by a security framework of any kind (such as the United States **National Institute of Standards and Technology (NIST)**)?

This is by no means an exhaustive list of questions but only serves to illustrate some of the key aspects that will guide you to having a thorough understanding of your solution's security and compliance requirements. Once you have a better view of your solution's security scope, you can see how these requirements can be met on AWS.

The second aspect that you want to consider is the AWS corpus of best practices concerning security. For this, the starting point and essential guidebook for any solutions architect is the *Well-Architected Framework* security pillar. It contains a wealth of recommendations to help you set up a secure foundation on AWS. There are various additional sources out there, including blogs and whitepapers, so please check the *Further reading* section at the end of this chapter for the links to resources.

Your environment security is, overall, as strong as your weakest system. But where do you start?

Nowadays, the **zero-trust** security model has is becoming increasingly popular. In the traditional network security model, corporate perimeters were protected and every system inside the corporate perimeter was considered to be trusted, while every system outside that perimeter was considered untrusted. The zero-trust model takes the approach that no system should be trusted based solely on its network location.

The zero-trust model is not new, but its recent popularity is mostly due to the fact that it is particularly adapted to a cloud or hybrid IT environment. In such an environment, you naturally find IT systems that do not follow the traditional concept of a corporate perimeter as well as modern architectural models—such as microservices architecture or event-driven architecture—that involve numerous interactions between IT systems belonging to separate network perimeters.

The zero-trust model thus advocates a dynamic approach to security, whereby individual interactions between systems are authorized based on the context of the session in which they happen, leveraging fine-grained authentication and authorization mechanisms.

Therefore, your primary objective in a zero-trust architecture is to prevent any unauthorized access from happening. The following subsection dives deeper into what you can do to achieve this objective on AWS.

Protecting the Network

You may now be wondering why protecting the network is important even though it was just mentioned that zero-trust concepts recommend not to trust systems based on their location. Now while zero trust advocates not to solely use the location of a system to decide whether it can be trusted or not, it certainly does go against leveraging network security best practices to protect your environments.

Thus, protecting your virtual networks (**virtual private clouds** (**VPCs**) or **subnets**) on AWS is still good practice and a must-do. It is recommended to apply multiple levels of control for both inbound and outbound traffic. This typically includes using subnets, route tables, NFW, **web application firewalls** (**WAFs**), security groups (stateful firewalls), and network **access control lists** (**ACLs**) (stateless firewalls).

First, you want to layer your virtual networks in such a manner that resources are placed according to their connectivity requirements. A VPC can be split into multiple subnets. Remember that a VPC is a regional construct, while subnets are attached to a single **Availability Zone** (**AZ**). Each subnet can have an associated route table that specifies how traffic flows within the subnet. Resources that do not require internet access should be placed in a network without access to or from the internet. So, resources such as a backend system or a database should be placed in a private subnet, such as a subnet without access to or from the internet. As a corollary, resources that need access to or from the internet—for instance, a load balancer—should be placed in a virtual network that can provide connectivity to the internet, in this case, a public subnet.

Further, VPCs and subnets on AWS provide natural means of isolating your AWS resources depending on their business function, your organizational structure, and other factors. Complex organizations will end up with hundreds or even thousands of VPCs and interconnect them by means of an **AWS transit gateway** (**TGW**). If they have resources deployed across multiple AWS Regions and require connectivity between VPCs across regions, they will then typically have one TGW per region acting as a regional hub, each peered with one or more other TGWs sitting in other regions. If you are unsure about *hub-and-spoke* network architectures on AWS, please refer to *Chapter 2, Designing Networks for Complex Organizations,* for more details.

Similarly, these organizations may require connectivity from AWS back to their on-premises environment and vice versa. Again, the TGW can act as a central hub where you would terminate your **VPN** or **Direct Connect** connections.

You can see a diagram outlining this concept in *Figure 5.3*:

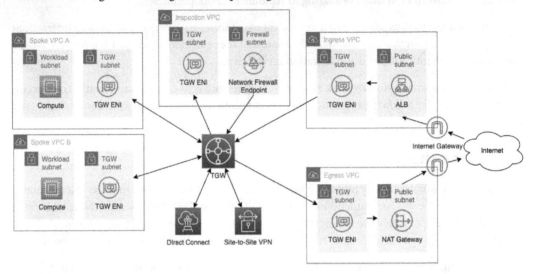

Figure 5.3: Illustration of a centralized setup with TGW and inspection via Network Firewall (NFW)

As per the preceding diagram, you may end up with the TGW acting as a central hub both for East–West communication (inter-VPC) and for North–South connectivity (to/from the internet and to/from on-premises). In a zero-trust security approach, you want to manage and control the traffic flowing through all these connections. First, you may want to place some network firewalls (through **AWS NFW** or using your own self-managed appliances) on some or all these paths, to control and/or inspect traffic. Second, you want to control routing to decide how traffic flows. With a central hub (TGW) in place, you have routing tables at the level of the hub to control how traffic is routed in and out of the hub, into the VPCs, and toward the internet and/or your on-premises environment.

You also have routing tables in your VPCs and subnets where you can also decide to put additional control depending on your overall network topology. For instance, consider a situation where the central hub (TGW) handles East–West communication, sending inter-VPC traffic for inspection through a set of network firewalls. In this case, you may also want to put specific routing in place at the VPC level for internet-connected VPCs in order to force all ingress and/or egress traffic from and/or to the internet through the same or another set of network firewalls for inspection.

When dealing with layer 7 communication (such as HTTP/S; WebSocket) from the **Open Systems Interconnection** (**OSI**) model, you may also want to leverage a specific type of firewall to protect your web application against malicious attacks such as **cross-site scripting** (**XSS**), **Structured Query Language** (**SQL**) injections, and many more. For that, you can use **AWS WAF** or your own self-managed WAF if you prefer to. When using AWS WAF, you benefit from its out-of-the-box integration with other AWS services, such as **Amazon CloudFront**, **Application Load Balancer** (**ALB**), **Amazon API Gateway**, and **AWS AppSync**, which are all services commonly used for designing web applications on AWS.

However, ideally, you also want to block any malicious attack upfront before it even enters your VPCs to protect your internet-facing applications. For that, AWS provides **AWS Shield**, which comes in two flavors: *Standard* and *Advanced*. AWS Shield Standard is available at no extra cost to all AWS customers. It protects your AWS environment against the most common types of DDoS attacks at **International Organization for Standardization** (**ISO**) layers 3 and 4, such as **synchronize** (**SYN**)/ **User Datagram Protocol** (**UDP**) **floods** and **reflection attacks**. For enhanced protection, AWS Shield Advanced brings extra features to counter larger and more sophisticated attacks as described below:

- It protects your selected resources running on Amazon EC2, **Elastic Load Balancing** (**ELB**), Amazon CloudFront, and Route 53 by providing application traffic monitoring as well as advanced and automatic mitigation.

- It provides notification and reporting on layer 3/4/7 attacks for better visibility.

- It gives you access, if you are subscribed to Business- or Enterprise-level support, to the AWS **Shield Response Team** (**SRT**), whose security experts can review your architecture and distill best practices, assist with custom mitigation during attacks, and run some post-attack analysis for you.

- Finally, it also brings some potentially significant economic benefits, providing AWS WAF at no charge for your AWS resources protected by AWS Shield Advanced and offering credits to compensate for the extra costs caused when scaling your resources due to a DDoS attack.

Additionally, if you are looking to apply a variety of protections across multiple AWS accounts belonging to your **AWS Organizations** organization, **AWS Firewall Manager** can help simplify administration and maintenance tasks. This is particularly welcome for large organizations where managing these resources on an individual account basis could become a daunting task. AWS Firewall Manager helps you pilot protections with AWS WAF, AWS Shield Advanced, AWS NWF, Amazon Route 53 Resolver DNS Firewall, and Amazon VPC security groups. The idea is that you can apply a protection baseline to your entire organization or whenever you deploy a new application in your AWS environment, to make sure it is compliant with your security rules.

Now that you are aware of the critical aspects of network protection, you are ready to dive into the protection of compute resources that you deploy in your AWS environment.

Protecting the Compute

What should you do to ensure the protection of your application's Amazon EC2 instances, containers, **AWS Lambda** functions, databases, and so on?

Well, to start with, you want to design an AWS environment that has proper resource isolation. There are multiple means of achieving this isolation, as we have seen in *Chapter 3, Designing a Multi-Account AWS Environment for Complex Organizations*. This resource isolation provides the first level of defense by segregating resources that belong to different business domains or functions, teams, or applications, and placing them in separate networks (VPCs or subnets), or even separate accounts. Then, beyond the network protection that was discussed earlier, it is recommended that you provide a more granular level of protection on each resource by assigning it to a security group (a stateful firewall) in charge of restricting access to it from relevant sources only (for instance, **Internet Protocol (IP)** ranges or security groups) and through specific protocols and ports.

Performing Vulnerability Assessments

Regularly scan and patch your application's code and its dependencies against any known vulnerability. Adopt best practices for software life cycle management such as CI and/or CD and integrate vulnerability scanning as part of your CI/CD pipelines.

Further, don't forget to do the same for the infrastructure supporting your application. As part of the shared responsibility model, you are responsible for securing—and thus patching—your Amazon EC2 instances, the **Amazon Machine Images (AMIs)** used to launch those instances, your containers, the container images used to run those containers, and so on.

Some tools provided by AWS can assist you with patching automation, such as **AWS Systems Manager Patch Manager**, especially if you need to manage a large fleet of EC2 instances. Alternatively, you can also use EC2 Image Builder for maintaining your AMIs, including patching and security policy enforcement. You can also leverage **Elastic Container Registry (ECR) Image Scanning** to check your container images for known vulnerabilities.

Reducing the Attack Surface

Harden your application's components as much as possible, limiting the software packages installed on your OSes and the libraries used in your application's code. Leverage the industry best practices and recommendations for hardening OSes and software, such as the resources from the **Center for Internet Security (CIS)**. Several organizations create their own hardened AMIs and container images in line with their own security best practices and standards. They may validate them using the CIS benchmarks or other industry standards. Check your application's code for **common vulnerabilities and exposures (CVEs)**, using tools such as **Amazon CodeGuru** or third-party code analyzers.

Leveraging Managed Services

Leverage the shared responsibility to your advantage by letting AWS do most of the heavy lifting for you. Using managed services—for instance, Amazon RDS or Amazon **Elastic Container Service (ECS)**—you can delegate as many of the administrative and security maintenance tasks to AWS as possible, reducing your maintenance tasks. The more managed AWS services you use, the less maintenance for you to do, and the more the time you can spend on improving your solution design following AWS best practices in terms of security.

Automating the Protection

Finally, automate all security maintenance tasks as much as possible. Repetitive manual tasks are error-prone and should be avoided at all costs. It may take you a bit more time to put the automation in place at first, but it will pay off from the very beginning as your AWS footprint grows. Automate the deployment of compute resources, taking an **infrastructure-as-code (IaC)** approach and using **AWS CloudFormation** or equivalent technology. Then, also automate configuration management tasks, using either built-in solutions such as **AWS Systems Manager** or third-party solutions if you are used to these (such as Chef, Ansible, and so on), or any combination of them.

Now that we have covered network and compute protection, we can proceed to discuss data protection aspects.

Protecting your Data

To start with, here are a few questions you should answer before you take any measures to protect your data:

- Is the classification of the data used by the solution clearly established?

- Does the solution process any **personally identifiable information (PII)**?

- Does part or all of the solution have to comply with a specific regulation (such as the **Health Insurance Portability and Accountability Act (HIPAA)**, the **Payment Card Industry Data Security Standard (PCI-DSS)**, or the **General Data Protection Regulation (GDPR)**)? If so, which one(s)?

This is by no means an exhaustive list of questions but is only given to illustrate a few examples of some of the questions that should guide you to clearly understand your data security and compliance requirements. Do not rush into solutions, and start looking at what your options are until you have the answers.

Data Classification

This is the first step in protecting your data. You need to figure out the type and classification of data your solution needs to handle. Data classification consists of categorizing the data to be processed according to its criticality and sensitivity. It is one of the key factors that will help you determine the degree of protection to apply as well as the data life cycle management.

You want to collect as much information as possible regarding the data at stake, such as the following:

- Who owns the data?
- Who has access to it, and which are the permissions for each authorized entity (end user; application)?
- Does it contain any PII data or confidential/sensitive information? Or is the data meant to be publicly available?
- What are the regulatory compliance requirements?
- Does any of the data require some transformation (such as anonymization)?
- What is the expected data retention period? Do you need to delete the data, or part of it, beyond that period?

Then, establish and maintain a data classification system documenting these properties for each type of data your solution must process. If needed, you can rely on tooling to automatically extract some of these properties from the data itself. For instance, **Amazon Macie** can analyze your data and alert you if it detects any PII data.

To ease access control management later, it is also recommended that you tag your data accordingly. You want to use any tag that can help you grant or block access to the data using **attribute-based access control** (**ABAC**)—for instance, a team's **identifier** (**ID**) or a project's ID. Please refer to *Chapter 1, Determining an Authentication and Access Control Strategy for Complex Organizations,* for more details on ABAC.

Now that you have a clearer picture of the data to be processed by your solution, it is time to look at how to protect that data.

Protecting Data at Rest

The first task is to protect the data at rest, that is, where it is stored. AWS best practices recommend that you encrypt the data—no exception. Your data must be encrypted, whether you decide to use object storage, file storage, block storage, databases, or anything else. Many AWS services (storage, compute, and others) integrate with AWS **Key Management Service** (**KMS**), which they use for cryptographic operations (such as generating data keys, data encryption, and decryption). A notable exception is Amazon S3, which not only uses KMS for encryption but, for historical reasons, also has its own built-in server-side encryption mechanism (**SSE-S3**).

Data Encryption

The following are the main reasons why you should always use encryption at rest:

- The risk, when failing to encrypt data, of sensitive information leaking or falling into the wrong hands bears a much higher cost than the cost of encrypting all data, sensitive or not.

- AWS storage services support data encryption at rest out of the box.

- There is no noticeable penalty (in terms of compute) for encrypting your data. Computer chips have long undergone significant improvements over the years and can now deal with cryptographic operations very efficiently.

The following subsections discuss the various options for encrypting the data, starting with AWS KMS and AWS CloudHSM. AWS KMS integrates seamlessly with many AWS services, so it is your go-to service to take a standardized approach on data encryption in your AWS environment.

AWS KMS and AWS CloudHSM

AWS KMS relies on a fleet of **hardware security modules (HSMs)** to store cryptographic material—that is, the keys being used to perform cryptographic operations such as encrypting and decrypting your data. Those HSMs consist of specific tamper-proof security hardware validated under the US NIST **Federal Information Processing Standard (FIPS)** *140-2* program, in charge of assessing cryptographic modules. Such HSMs are managed by AWS on your behalf and validated at *FIPS 140-2* level 3.

For some organizations, however, it may not be enough to meet their own corporate security requirements and they may require that their cryptographic material be secured within HSMs that they manage themselves. For such requirements, AWS provides a service called **AWS CloudHSM** where you fully control your own HSMs. Such HSMs are validated at *FIPS 140-2* level 3.

You can manage and interact with your HSMs, deployed with AWS CloudHSM, through a set of command-line tools and libraries. The AWS CloudHSM Client SDK provides standard mechanisms such as, for instance, a **Public Key Cryptography Standard (PKCS)** #11 API to perform cryptographic operations following the **Organization for the Advancement of Structured Information Standards (OASIS)** standard or a **Java Cryptography Extension (JCE)** provider to offload cryptographic operations to AWS CloudHSM from Java applications.

However, you can also leverage AWS CloudHSM through AWS KMS. With AWS KMS, you can indeed opt to use a *custom key store* for which the keys are stored on an AWS CloudHSM cluster that you control. It allows you to keep leveraging the AWS services integrated with AWS KMS while making sure that the HSMs protecting your cryptographic material remain fully under your control if that is required by your organization's corporate security policies.

When to Use AWS CloudHSM

You may now be thinking: CloudHSM sounds like the cherry-pick for cryptography, so why would I want to use anything else?

First, you need to consider whether you have an actual use case for AWS CloudHSM. For instance, does your corporate security require that you store your keys on HSMs fully under your control? Or are you looking to offload **Secure Sockets Layer** (**SSL**)/TLS processing from some of your web servers to an HSM? If the answer to these questions is no, then you should not use AWS CloudHSM, simply because you don't have a need for it.

The second consideration to make concerns costs. When using AWS CloudHSM, you have to set up a cluster composed of two or more HSMs for resiliency purposes. This cluster bears a non-indifferent run cost associated with it (slightly over 2,000 US dollars (USD) per month, at the time of writing this).

The third consideration is that AWS KMS, as with any other AWS service, has some associated quotas, among which are the number of concurrent cryptographic operations that can be performed within a specific AWS Region at any one time in a given AWS account. When you use keys managed by KMS, the default value for the number of operations is in the thousands or even tens of thousands of operations per second, depending on the AWS Region where you operate. When you use a custom key store with KMS in combination with AWS CloudHSM, the default value for that quota is somehow lower (1,800 operations per second at the time of writing). On top of that, differently from other KMS quotas, that specific quota cannot be increased in the case of a custom key store. Thus, if you use a custom key store and need to process a very large number of concurrent cryptographic operations per second, you might be throttled by KMS at some point. So, another account in the same Region or the same account in another AWS Region is entitled to the same quota managed separately.

The vast majority of KMS users out there who don't use custom key stores remain of course unaffected by this non-adjustable quota. That said, if you reach the specific quota for your KMS setup, your requests for cryptographic operations will be throttled by KMS and you would need to handle that situation by finding a means of reducing your request rates.

Regardless of the approach you select, what really matters in the end is that you encrypt your data. There is no excuse not to do so since AWS offers you all the tools to do so.

Trust but Control

Now, how can you make sure that all your data is encrypted at rest in your AWS environment? Some services such as Amazon S3 or Amazon EC2 allow you to enforce data encryption by default. For the services that don't, and also to make sure that part of your data has not been left unencrypted, you should verify that encryption at rest is used systematically on storage services such as Amazon S3, Amazon RDS, or Amazon **Elastic Block Store** (**EBS**) volumes.

For that purpose, it is recommended to leverage **AWS Config** managed rules to check that your data is indeed encrypted at rest across the various AWS storage services, but we will talk a bit more about AWS Config in the *Detecting incidents* section of this chapter.

Beyond encryption, you also want to take some additional measures to protect your data, as discussed in the subsections below.

Limiting Data Access and Visibility

First, keep people away whenever it is feasible. End users should consume the data as much as possible through an interface of some sort, such as a custom **user interface** (**UI**), a custom API, or another AWS service UI or API. Allowing access to the data directly where it is stored should be the exception, not the rule.

Second, make sure that only authorized people access your data. Based on the data classification you established earlier, on a need-to-view or need-to-edit basis, you should establish fine-grained access control to your data following the least-privilege principle. It means, on the one hand, managing, authenticating, and authorizing the end users and, on the other hand, filtering access to data, using, for instance, row-level or column-level access filtering.

Third you need to take particular care of any sensitive, confidential, and PII data. For such data, you may want to put additional measures on top of encryption. In many cases, even the end users who are authorized to access the data may not need to view the actual data and access to fictitious data may be good enough for the tasks they need to perform. Think of PII data. For such sensitive information, you can leverage some extra protection such as data tokenization, which consists of masking the actual data that is then replaced by some similar but fictitious data. The tokenization process can take place upfront (tokenized data is preprocessed and stored along with the rest of the data) or on the fly (for instance, leveraging **Amazon S3 Object Lambda** when the data is stored on S3). The tokenization mechanism to be used depends mostly on your use case and whether you need frequent access to the same data (in which case, on-the-fly tokenization may not be the most efficient approach) or you need the tokenization process to be reversible so you can retrieve the original data from the tokenized data (in which case, a simple hash function won't work).

Finally, you also want to regularly audit your data access logs, from **AWS CloudTrail** or other logs, such as S3 access logs. We will cover this in more detail in the *Detecting Incidents* section of this chapter.

Protecting Data in Transit

As you can now imagine, protecting your data at rest is not enough—you also need to ensure its protection when it is in transit. When data needs to be provided to end users or exchanged between applications or services in the context of your solution, you are responsible for protecting its integrity and confidentiality.

First, use secure protocols, such as TLS, that enforce **end-to-end** (**E2E**) encryption whenever your data's integrity and confidentiality are at stake while it is being transported from one system to another. AWS services provide HTTPS endpoints using TLS to encrypt all communications.

Second, manage the life cycle of your TLS certificates and limit access to the bare minimum. Prefer a managed service such as **AWS Certificate Manager** (**ACM**) to automate and delegate most maintenance tasks. ACM can be used to maintain both your public and private certificates and it integrates natively with AWS services such as AWS ELB, Amazon CloudFront, or AWS API Gateway, handling automatic certificate renewal for the resources they protect.

Third, enforce encryption in transit by blocking unsafe protocols, such as HTTP, using, for instance, security groups to protect your resources. In this case, you can also force a redirect of HTTP to HTTPS with Amazon CloudFront or an ALB sitting in front of your application.

Last but not least, you also want to regularly audit your data access logs from AWS CloudTrail or other logs such as VPC flow logs. We will cover this in more detail in the *Detecting incidents* section of this chapter.

We have now covered different approaches and various AWS services that you can use to protect your infrastructure and your data. No matter how secure you think your solution is, you should nevertheless prepare for the worst-case scenario and be able to answer this question: *How will the solution behave in case of a security incident?* The next section discusses this.

Detecting Incidents

Even after you have put all the necessary measures in place to protect your infrastructure and your data, you are only halfway through ensuring security. Despite all the protections implemented, some incidents can still occur. It can be any type of incident—a security breach, a data leak, a system misconfiguration, a configuration change, or unexpected behavior. If you don't do anything to check for such incidents, they will go undetected most of the time, causing potentially acute damage to your business.

The following subsection discusses the various approaches to incident detection.

Picking the Right Tool for the Right Task

First, activate AWS CloudTrail on all your accounts. AWS CloudTrail logs keep a record of all activity (such as who made what request, at what time, and from which IP address) that took place within your account, whether the related actions come from the AWS Management Console, the AWS CLI, or by using AWS SDKs. As we have already discussed in *Chapter 3, Designing a Multi-Account AWS Environment for Complex Organizations*, it is recommended, especially in complex organizations, to centralize the CloudTrail logs from all your accounts in a specific account for the usage of your security and audit teams.

Second, leverage **AWS Config** to continuously monitor and record any configuration change in your AWS resources. AWS Config also allows you to review the configuration history of your resources. AWS Config integrates with CloudTrail to correlate configuration changes to events that took place within your account. To monitor and detect configuration changes, you create AWS Config rules that get triggered whenever compliance with the specified rules is breached. AWS Config rules can be deployed standalone or as part of a conformance pack. Multiple conformance packs are available to group-related rules together, whether it is by service affinity (such as *Operational Best Practices for Amazon S3*) or by compliance affinity (such as *Operational Best Practices for HIPAA Security*). In a multi-account organization, AWS Config allows you to centralize both the rules management and the collection of findings to provide complete visibility and control of the compliance status against your own rules across your entire organization.

Figure 5.4 shows some examples of incident detection and centralized CloudTrail logs:

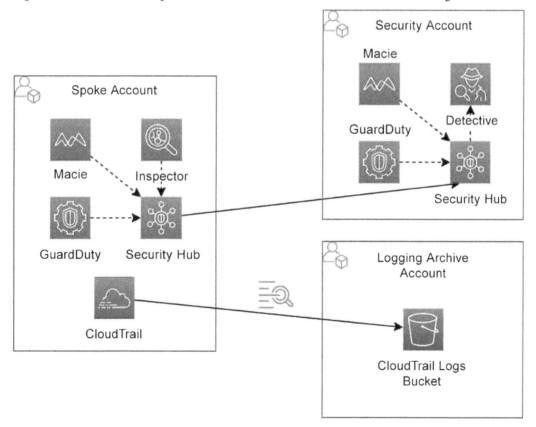

Figure 5.4: Incident detection and centralized CloudTrail logs

Third, turn on automatic threat detection with **Amazon GuardDuty** to identify any malicious activity across your organization. **GuardDuty** leverages **machine learning** (ML) techniques to analyze all events coming through your CloudTrail logs, **Amazon VPC Flow Logs**, and **Domain Name System (DNS)** logs. It automatically identifies and prioritizes threats based on your account's behavior—for instance, compromised credentials or unusual data access.

Finally, leverage **AWS Security Hub** as the central place to aggregate, organize, and prioritize your security alerts or findings. It collects information from multiple sources such as Amazon GuardDuty, Amazon Macie, **Amazon Inspector**, AWS Firewall Manager, and so on, including third-party solutions. AWS Security Hub can conduct automated security checks to verify compliance with industry standards and best practices, such as the *CIS AWS Foundations Benchmark*. Its role is to centralize and prioritize security findings from your AWS environment, across all your AWS accounts. It also integrates with third-party solutions, either to ingest findings they may produce or to act based on its own findings, for instance, alerting a specific group of people to act swiftly in presence of a critical risk or feeding your IT service management system with findings to trigger a specific workflow.

Centralizing and Analyzing Logs

It is essential to consolidate logging from your solution in a central location for further analysis. In that central location, you want to make sure you collect all the logs, whether from AWS services, custom solution components, or third-party services. The objective of aggregating the logs is twofold: analysis and retention. Analysis can consist of forensics or troubleshooting activity. Retention is meant to ensure compliance with your organization's requirements or to satisfy any legal or regulatory obligation to keep activity logs for a minimum period.

As was already mentioned in *Chapter 3, Designing a Multi-Account AWS Environment for Complex Organizations,* you are recommended, especially if you are part of a complex organization, to centralize your various logs—for instance, from CloudTrail and CloudWatch, from all accounts into a central Log Archive account. Think of that central log archive storage as a **write once, read many** (WORM) kind of storage where you keep logging information as immutable data. Typically, you would land all logs onto Amazon S3 and leverage features such as **S3 Object Lock** to guarantee immutability.

However, because you need to make the logs available to the security operations team, internal or external auditors, and possibly a few other actors within the organization, it is important to make sure that they do not contain any sensitive information (for instance, people's names; credit card or social security numbers). All sensitive information should be redacted before landing on the central log archive location, while you can keep a local copy of the logs in the source account, for a shorter period, with the unredacted information if needed.

You can see some examples of centralized logging and log indexing solutions in Figure 5.5:

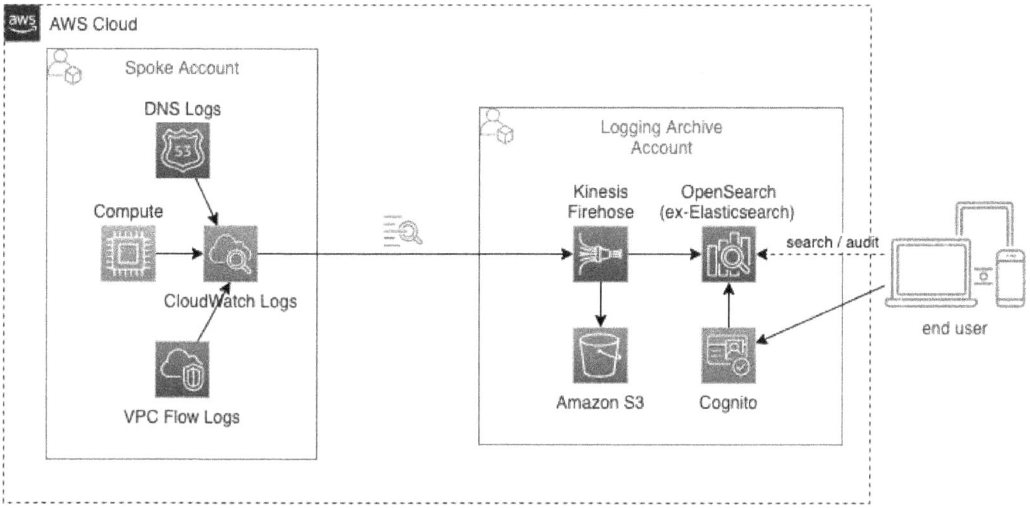

Figure 5.5: Centralized logging and log indexing for search and audit

From the central log archive, the security operations team can put in place their preferred workflow for analysis and monitoring. They will want to feed logs data into their existing **Security Information and Event Management** (**SIEM**) system or index the information to leverage their preferred search tool (for instance, **Amazon OpenSearch**, which was previously known as **Amazon Elasticsearch**). The amount of logging data collected can rapidly become overwhelming, so it is essential to have the right capabilities in place to automate filtering and prioritization of security incidents. Some AWS services such as AWS Security Hub can help you with that, but depending on your organization's context, you're likely to integrate these services with your own tools, for instance, to use a single pane of glass for security incident detection and analysis or to correlate incident data collected from AWS services with some other information coming from third-party tooling.

Now that you have gone through some best practices to detect and analyze incidents, you can review how best you can respond to them.

Responding to Incidents

Detecting incidents is one thing, but being able to respond to them in a timely manner is even more important. Assume that you have put in place the necessary mechanisms to detect and prioritize incidents. What is next?

Next, you want the ability to remediate these incidents. There can be, however, several types of incidents: first, in terms of severity—from minor to major or critical incidents—second, in terms of complexity—from easy-to-fix with a single problem to address to more complex ones caused by multiple intertwined issues. From a priority perspective, you want to address the most critical issues first. How can you tackle all incidents in a timely manner? The solution is a combination of automation—for straightforward issues—and prescriptive guidance.

First, automation is a must-have. It might not look necessary from the perspective of a single solution managed by a single team, but if you consider a security operations team working for a complex organization with tens or hundreds of projects and teams running on AWS in production, automation becomes critical. Incident remediation simply cannot scale if all actions need to be taken manually.

Some incidents will have a single cause and be easy to fix. Think of incidents that are triggered by a violation of your organization's security principles codified as AWS Config rules. Imagine, for instance, that you prohibit the use of S3 buckets with public access, whether read or write. That is straightforward to fix and shouldn't require any manual intervention. Once the AWS Config rule violation has been captured, it could trigger a remediation action that, for instance, first notifies the bucket owner and then ,if no corrective action is taken by the owner within a specific timeframe, it automatically blocks all public access to the S3 bucket at hand. This is just an example and should not be followed to the letter as the best remediation course in such a case. The remediation action that you should take depends on your organization's culture of risk and compliance management, as well as on the type of incident and on the specific context in which it occurs. If you transpose, for instance, the same example in the context of a financial or healthcare institution where the S3 bucket happens to contain sensitive information (such as PII data), a different remediation course of actions will apply, such as immediately blocking all public access and then notifying the bucket owner. The bottom line is you need to adapt the remediation to your own context.

In any case, how can you achieve an effective response strategy on AWS?

First, consider a situation where an incident is caused due to a configuration policy violation. Suppose you have some AWS Config rules activated on your accounts. When one of these rules is violated, AWS Config can trigger a remediation action using **AWS Systems Manager Automation** runbooks. This remediation action can take place manually or automatically as per your requirements. These runbooks specify the actions to be performed on non-compliant AWS resources. AWS Config comes with a set of predefined managed automation runbooks with remediation actions. You can also create and associate your own custom automation runbooks with AWS Config rules. For the example that is being used here, there is a predefined runbook called `AWS-DisableS3BucketPublicReadWrite` that could directly be leveraged to fix the issue. This automation runbook expects two parameters: the S3 bucket name and the IAM role that should be used by the runbook to execute the remediation action. The runbook will use these parameters—in this case, provided by the AWS Config rule—to execute a call to the Amazon S3 `PutPublicAccessBlock` API.

Now, consider a different situation where an incident is caused by a security policy violation. Suppose you have either **Amazon GuardDuty** or **AWS Security Hub** set up within your organization. When they report a finding, you can trigger an event based on that finding, using **Amazon CloudWatch Events** or **Amazon EventBridge**. Amazon EventBridge extends Amazon CloudWatch Events, building upon the same API and bringing integration with third-party solutions, either as events sources or destinations, events replay, and a schema registry. You can then decide what to do with the event, for instance, whether to push it to an event bus in a different account in a different Region, send it to a Lambda function, call an API destination to send it to a third-party solution, or any other options. The routing decision depends entirely on how you intend to handle the remediation.

In some cases, it will not be easy to trigger remediation immediately. Such cases will usually consist of complex incidents, requiring the intervention of the security operations team to investigate the issue further and decide on the best remediation course. The team should have all the necessary instruments at their disposal to act in a timely manner and take the most efficient course of action. This means that you need to put in place an incident management plan for your solution that will consider the most likely scenarios. The incident management plan should document the communication and escalation paths that should be followed in such cases. It should also document the best responses to issue in each scenario, providing clear step-by-step documentation for the operations team to follow, including code snippets and command lines whenever needed.

Summary

This chapter has covered quite a lot of ground in terms of designing secure solutions on AWS.

You learned how to leverage IAM and identity federation in your solution to provide granular access control. You then looked at the best practices to protect your infrastructure resources—using tools such as AWS WAF, AWS Shield, and AWS Firewall Manager—and your data using encryption at rest with AWS KMS and enforcing encryption in transit. The chapter then concluded with a discussion on incident detection and response to prepare for worst-case scenarios, leveraging tools such as AWS CloudTrail, AWS Config, Amazon GuardDuty, AWS Security Hub, and Amazon EventBridge.

In *Chapter 6*, Meeting Reliability Requirements, we will dive into the best practices for designing and implementing reliable solutions on AWS.

Further Reading

- Best Practices for Security, Identity, & Compliance: `https://packt.link/pyBa0`

- The Well-Architected Framework security pillar: `https://packt.link/N2wpN`

- The AWS Security Reference Architecture (SRA): `https://packt.link/7lcji`

- Zero Trust architectures: An AWS perspective: `https://packt.link/KHiRj`

- AWS Best Practices for DDoS Resiliency: `https://packt.link/BZRnk`

6
Meeting Reliability Requirements

This chapter will focus on determining a solution design and implementation strategy to meet reliability requirements. You will explore several architecture patterns and architectural best practices for designing and implementing reliable workloads on AWS.

Designing and implementing solutions with resilient architecture is essential to recover easily and successfully in case of failure. You will look at the following topics:

- Reliability design principles
- Foundational requirements
- Designing for failure
- Change management
- Failure management

Reliability Design Principles

Reliability refers to the ability of a system to function repeatedly and consistently as expected. As you can imagine from that definition, it can mean totally different things depending on the system at hand. Ensuring the reliability of a nightly batch application running on weekdays will be something very different from ensuring the reliability of an application serving requests 24/7.

The reliability pillar of the AWS Well-Architected Framework comprises five design principles to keep in mind when designing a workload for reliability in the cloud.

Principle 1 – Automatically Recover from Failure

"Everything will eventually fail over time," said Werner Vogels, the CTO of Amazon. You can't expect to have humans constantly watching the vital signals, also known as **key performance indicators** (**KPIs**), of each workload you deploy in the cloud and taking action whenever something goes wrong. Although you may need to rely on human assistance in some very specific cases, it is neither scalable nor sustainable. Here, automation is key.

The idea is to monitor the KPIs of your workloads and trigger any necessary processing when one or more thresholds are breached. You may wonder which KPIs you should monitor then. Well, it depends. What is important, however, is to make sure the monitored KPIs reflect the business value of the workload and not technical operational aspects; thus, depending on what is important for your business, you will be watching different things. For instance, if speed is essential, you might be monitoring the number of tasks or requests processed over time, whereas if quality is paramount, it might be more important to monitor the number of requests returning errors or timing out.

Once you have defined the relevant KPIs for your case, it's a matter of taking action when thresholds are breached: sending notifications, tracking failures, and triggering recovery processes to work around or repair the failure(s). The more sophisticated the automation, the more prepared you are to even anticipate failures before they occur.

Principle 2 – Test Recovery Procedures

Having automated recovery procedures is good, but making sure they work is better. If you're new to the cloud, you may be used (in your on-premises environment) to testing your workloads and making sure they work in "normal" conditions, but you may be less used to testing recovery procedures to handle failures.

In the cloud, validating your recovery procedures is as easy as starting a test environment, deploying your workload on it, and carrying out enough tests to simulate various failure types. This will allow you to test, make fixes, or apply changes to your workloads, and to feel more confident that they can handle a real failure when it occurs, thus reducing your risks.

Principle 3 – Scale Horizontally to Increase Aggregate Workload Availability

Large monolithic systems that only scale vertically, that is, by adding more resources such as CPU and RAM, can lead to a complete system failure when unexpected events occur. It is thus highly recommended to design systems that can scale horizontally, that is, by replicating some or all of its components. By replacing a large resource with multiple smaller ones, you inherently reduce the impact of a single failure on the overall system. Doing so at every layer of your system (infrastructure elements, communication system, frontend, and backend components) will allow you to get rid of single points of failure and, as a result, improve the overall reliability of your workload.

Principle 4 – Stop Guessing Capacity

Resource exhaustion is one of the most natural causes of failure for any workload. It occurs when the system's capacity is outrun by the demand set on a workload. Excessive demand could result from a genuine peak in the demand or from malicious usage, such as denial-of-service attacks. An on-premises environment requires you to guess your capacity needs to provision the necessary infrastructure upfront. The cloud gives you elasticity for you to scale as much as you need to meet the demand, provided that your design supports it.

So, no more capacity guessing. Instead, you first need to ensure that your application can scale horizontally; that's most important. Second, you need to watch service quotas to make sure they don't keep your workload from scaling out. Some service quotas are set on your accounts to protect you from over-provisioning resources. However, in some cases, you may need more resources than your current quotas allow. Most of these quotas can be adjusted, provided that you submit a request to increase them on time. So, it is important to monitor them regularly to anticipate any potential limit being reached.

Principle 5 – Manage Change in Automation

All changes made to your infrastructure should be automated. This is a matter of being able to make consistent deployments that you can track and repeat at will. Deploying changes manually is error-prone and should be avoided when managing distributed systems at scale. In the cloud, you want the ability to re-create an exact copy of any specific environment at any time, be it for test or business continuity purposes.

Foundational Requirements

First things first, it's essential to consider your foundations, that is, your AWS environment, which must be able to accommodate the workload requirements. Two elements in particular must be tackled as they could impact the reliability of any workload: resource constraints and network topology.

Resource Constraints

Resource constraints can be further split into two types: service quotas and environmental constraints.

As mentioned previously, service quotas are default predefined values on each AWS account – on the one hand, to protect you from over-provisioning AWS resources, and on the other hand, to protect the AWS cloud from abuse. Different quotas apply to each service and could represent very different items and quantities. Some of them are adjustable and represent soft limits, while others cannot be changed and represent hard limits. To illustrate this, the VPC service has a number of quotas for various features. For instance, your VPC entitles you, by default, to up to five IPv4 **Classless Inter-Domain Routing** (**CIDR**) blocks and a single IPv6 CIDR block. The former limit is adjustable, but the latter isn't. Thus, when designing your solution, you must take those service quotas into account: be very mindful of hard and soft limits, and put in place a mechanism to monitor your usage of the AWS services to detect whenever you're getting close to any relevant quota limit. For soft limits, you can request any quota at stake to be raised by submitting a request via the **Service Quotas console** or API at any time.

Now that you have understood the importance and ways of monitoring and managing service quotas, you can review the second type of constraint that was mentioned earlier, that is, **environmental constraints**. These refer to the constraints imposed by the physical resources supporting the AWS infrastructure. For instance, it could be the amount of storage available on a physical disk used for Amazon EC2 instances, or the network bandwidth available between your AWS environment and your on-premises environment. Those environmental constraints may impact your solution, so it is key to bear them in mind. Imagine, for instance, that you are building an application on AWS that relies on data that is stored in an operational data store located in your on-premises environment. The bandwidth and latency of the network connection between your on-premises and AWS environments will naturally constrain the possible use cases.

Network Topology

It's also paramount to plan your network topology when you architect for reliability. Several aspects of networking were discussed in *Chapter 2, Designing Networks for Complex Organizations*. You can review that chapter if essential networking concepts on AWS, such as VPCs, VPN, **Direct Connect** (**DX**), and **Transit Gateway** (**TGW**), are not entirely clear to you. Now, you can proceed to consider networking from the aspect of resiliency.

When you're laying out the foundations in your AWS environment, you must prepare for the foreseeable future and be ready for the unknown. It could be quite painful to have to revise your entire network topology on AWS after a couple of projects because of decisions made without careful planning and forward thinking.

As a solutions architect, an essential part of your job consists of making decisions supported by a rationale, and not light-heartedly choosing to turn right or left. As much as possible, you want to use two-way doors, that is, to make reversible decisions. One-way doors, or irreversible decisions (or possibly reversible but at a very high cost), should be avoided or at least limited, and in any case delayed until you can't further delay making that choice. Picking an EC2 instance type to deploy an application is an example of a two-way-door type of decision. You should make that decision without further ado, as soon as you have sufficient information to make an educated choice. You should check, possibly before going into production, whether the selected instance type is the optimal choice. If not, changing the instance type is in most cases straightforward and painless, even more so if you put in place the proper mechanisms, such as infrastructure as code, automated build and deployment, and rolling updates. Choosing between a hub-and-spoke and a mesh network topology is the type of one-way-door decision you want to make sure to do right the first time. Although you could object that there is always a way to migrate from one such network topology to the other one and that it is not entirely one-way, it is going to be such a painful exercise that you will likely regret it.

Among the many different things that you, as a solutions architect, will have to question is the right network topology for your organization. You need to lay out a futureproof topology that can accommodate not just your first workload but also future growth, and you must make sure that you can cope with failures.

The following sub-sections present some general recommendations based on best practices, that will help you avoid having to change a topology soon after having implemented it.

Using Highly Available Network Connectivity for Your Public Endpoints

Starting from the assumption that your internet-facing endpoints must be highly available, you have to make sure that whichever component you lay on their path is also highly available, whether it is a *DNS* service, a **content delivery network** (**CDN**), a *load-balancing* capability, or a *gateway* of some sort, for example.

Depending on the AWS services or third-party components that you decide to use, make sure that their availability **service-level agreements** (**SLAs**) and deployment models match the requirements of your network connectivity. The following presents a brief overview of some of those networking building blocks:

- **Amazon Route 53**, an AWS DNS service, is both scalable and highly available out of the box. It provides domain name resolution, registration, and health checks. It is built as a globally distributed service that provides consistent and reliable DNS service, independently of local or regional network conditions. Route 53 uses Anycast routing technology to ensure requests are answered from the optimal location depending on network conditions. Route 53 will then provide the route to your AWS resources, such as EC2 instances, **Elastic Load Balancing** (**ELB**) load balancers, or Amazon S3 buckets. It can also be used to route requests to resources outside of AWS.

- **Amazon CloudFront** is an AWS CDN service. It distributes your content across multiple edge locations across the AWS global network. It can significantly reduce the network latency by delivering content closer to the end users, improve the availability of your content thanks to the distributed nature of the service, and also limit access to your origin servers thanks to edge and regional caches. It is highly available due to its distributed nature.

- The ELB service provides various types of load balancers: **Classic Load Balancer** (**CLB**), **Application Load Balancer** (**ALB**), **Network Load Balancer** (**NLB**), and **Gateway Load Balancer** (**GWLB**). The ELB service allows you to load balance IP-based traffic across multiple **Availability Zones** (**AZs**) within any given Region. The first three types (CLB, ALB, and NLB) offer an SLA of 99.99% availability. The availability of an SLA for the fourth type of load balancer (GWLB) actually depends on how you deploy the service since it relies on your implementation of third-party appliances.

- **AWS Global Accelerator** is another service that builds on top of the Amazon global network. If you need to provide a service to end users globally, Global Accelerator offers a way to deliver that service through a set of static IP addresses and from the optimal endpoint based on your user's location. Those IP addresses remain the same globally, but Global Accelerator will find out the optimal regional resource that can deliver the service. Those regional resources could be ALBs or NLBs, EC2 instances, or elastic IP addresses. It is also well suited for cross-region failover scenarios (single-region failover scenarios are usually better served by ELB load balancers). It is worth noting that, unlike CloudFront, Global Accelerator can also process non-HTTP requests.

The following section presents the private networking aspects at hand when building for resiliency.

Provisioning Redundant Connectivity between Your AWS and On-Premises Environments

When connecting your on-premises environment to your AWS environment, it is highly recommended to make those connections redundant to sustain the failure of any single one of them.

Large organizations are likely to use some form of private connectivity for security purposes; you will then end up choosing between VPN connections over the internet and private connections. You've already looked at these two types of connectivity in *Chapter 2, Designing Networks for Complex Organizations*; please check that chapter out if you're not sure what the various connectivity options are.

The two major private connectivity options are **AWS Managed VPN** and **AWS DX**.

When using AWS Managed VPN, it is highly recommended to connect your **Virtual Gateway** (**VGW**) or AWS TGW to two separate **Customer Gateways** (**CGWs**) on your end. By doing so, you establish two separate VPN connections, and if one of your on-premises devices fails, all traffic will be automatically redirected to the second VPN. A single VPN connection allows for 1.25 Gbps of bandwidth; however, when using TGW, you can leverage **Equal Cost Multi-Path** (**ECMP**) to route traffic across multiple tunnels in parallel, up to 50 Gbps, which is useful when you need to scale beyond the 1.25 Gbps of a single VPN connection.

When using AWS DX, it is recommended to have at least two separate connections at two different DX locations. It will provide resiliency against connectivity failure due to a device failure, a network cable cut, or an entire location failure. You can increase the level of reliability of your DX setup further by adding more connections to additional DX locations. For more details on DX connectivity reliability, you can use the DX resiliency toolkit to help you find the best setup: `https://packt.link/x88RO`.

It is also possible to combine both technologies in scenarios where your SLAs allow it. If you would like to benefit from the consistency offered by DX most of the time but also, in case of DX failure at your DX location, it is acceptable to use a VPN connection over the internet, then setting up a failover from DX to VPN is a cost-efficient option compared to multiple DX connections to multiple DX locations. In such a scenario, if you have a TGW in your network topology, you could set up ECMP to boost the fallback VPN connection, maintaining a decent bandwidth in case of failover, at least for those who are used to working with 10G or 100G DX connections.

Now, the main reason to have this actual connectivity in the first place between your on-premises and AWS environments is to be able to leverage AWS services or your AWS resources from your on-premises environment, or vice versa. One essential mechanism here, which was already covered in *Chapter 2, Designing Networks for Complex Organizations*, is VPC endpoints. VPC endpoints, more specifically interface endpoints, powered by **AWS PrivateLink**, can be used to provide redundant entry points for any traffic targeting a supported AWS service or a VPC endpoint service. Materialized by **Elastic Network Interfaces** (**ENIs**) deployed in your subnets within your VPCs, the VPC endpoints are resilient in nature and the connectivity they provide can be made more highly available by deploying them in subnets in multiple Azs.

For more details on the most common hybrid network resiliency approaches, please consult the *Hybrid Connectivity* whitepaper (you can find the reference in the *Further Reading* section at the end of the chapter).

Ensuring IPv4 Subnet Allocation Accounts for Expansion and Availability

You need to plan for enough IPv4 addresses to be available, not only today but also for future use as well. That may sound obvious but the number of IP addresses needed can be easily overlooked. In the count, you must include all the resources that will be deployed in your VPCs, such as elastic load balancers, Amazon EC2 instances, **Amazon RDS** databases, **Amazon Redshift** clusters, and **Amazon EMR** clusters. You also have to be aware that the first four IP addresses and the last one of every subnet CIDR block are also reserved by AWS for internal use.

When defining your private address space plan, accommodate address space for multiple VPCs in a region, and within each VPC for multiple subnets spanning several Azs. Make sure to leave room for expansion in the future, by keeping unused CIDR block space.

As a general rule of thumb, it is better to size VPC and subnet CIDR blocks by excess rather than the other way around. That's especially true if you plan to accommodate highly variable workloads that will need enough room to scale up. So, it is recommended to deploy large VPC CIDR blocks. Note that the largest IPv4 CIDR block you can create on AWS is a /16 block, providing 65536 IP addresses.

Using Hub-and-Spoke Topologies Instead of a Many-to-Many Mesh

Hub-and-spoke topology is highly recommended if you plan to either connect multiple VPCs to each other or multiple VPCs to an on-premises environment. The main reason is that when you connect multiple address spaces to each other, the number of connections grows with the power of the number of spaces to interconnect, which can rapidly become unmanageable when you expect tens, hundreds, or even thousands of different address spaces (including VPCs and on-premises environments). That's precisely AWS TGW's role, to provide a managed hub-and-spoke solution. It is highly available within a region across multiple AZs, so no need to deploy multiple TGWs, unless your organization requires a clear traffic split between some business entities or if you exhaust the resources a single TGW is able to sustain.

Enforcing Non-Overlapping Private IPv4 Address Ranges Where Private Networks Are Interconnected

It is best to plan for non-overlapping address spaces to avoid running into trouble when interconnecting multiple address spaces (VPCs and on-premises environments). Even if you carefully plan your networking address spaces, there's no guarantee that you won't run into a situation with overlapping IPv4 addresses tomorrow, for instance, as a result of a merger/acquisition. If the issue ever happens, you'll still be able to interconnect the overlapping address spaces but it'll make things somehow more complicated. In such a case, you'll have to rely on NATing, or switching to IPv6, if that's an option, or using VPC endpoints based on AWS PrivateLink or any other workaround to sort it out.

Thus, it is best to avoid running into the issue if you can and to plan accordingly. If that can help, you can find IP address management solutions on the AWS Marketplace to help plan and manage your IP address space.

Designing for Failure

When designing a workload's architecture for a distributed environment such as the AWS cloud, you must design first to try and prevent failures, and second to handle failures. As previously mentioned in the design principles, your workload design must be able to cope with variations in the workload's demand, detect failures, and automatically heal itself.

Before going any further, it is important to note that the AWS Well-Architected Framework, on which this book relies broadly since it is the backbone of AWS best practices, strongly recommends that you avoid building a monolithic architecture and prefer service-oriented or microservices architectures instead. Debating on monolithic versus any other style of architecture is beyond the scope of this book. However, it is noteworthy that monolithic architectures are often wrongly associated with evil *big-ball-of-mud* systems. It is unfair, considering that monolithic systems may have a very neat and structured design that has nothing to do with a randomly structured spaghetti code jungle, also known as a big ball of mud. On the other hand, a poorly built microservices architecture can very well turn into a distributed big ball of mud, which is considerably worse than a well-structured monolith; thus, make no mistake here; distributing the components of your system doesn't make it more organized, it just makes it distributed. Adopting a service-oriented or microservices approach doesn't keep you from bringing structure to your workload. In other words, it doesn't keep the architect – that's you – from properly designing the system to meet its requirements.

Designing Your Workload Service Architecture

As mentioned in the preceding section, the Well-Architected Framework advises building service-oriented or microservices architectures for your workloads on AWS, but that remains a generic recommendation. Which approach should you take? Well, it depends.

Start by assessing your workload requirements and the environment in which it will operate:

- Is there a strong culture in your organization that advocates one or the other architectural style? Your workload will likely not exist standalone in its AWS environment, or at least not forever, assuming your organization will deploy additional workloads on AWS.

- Does your organization plan to build reusable services, such as business functions, that can be later composed into more complex services?

- What type of governance does your organization want to have for the cloud: centralized or decentralized?

- Does your workload have strong consistency requirements? Is it ready to deal with eventual consistency?

- How mature are your operations teams? How experienced are they with operating distributed systems in production? How comfortable are they with dealing with complex distributed systems?

Answering these questions will help you make an informed decision on which architectural style is best for you. Every architectural decision is eventually a trade-off between the pros and cons of each option. Selecting an architectural style is no exception. The rationale driving your choice must consider the highest benefits over time while reducing risks and costs. And, in some cases, it may even mean starting with a monolithic architecture, as counter-intuitive as this may sound.

Whatever your choice is, keep in mind the point that was made previously about one-way versus two-way door decisions. Your workload architecture will evolve over time and it is paramount to keep your options open for as long as possible. So, even if you decide to implement a monolithic architecture to start with, make sure to take a modular approach so that you can evolve toward a different architectural style, such as SOA or microservices, in the future.

I cannot stress enough the importance of looking at best practices, from AWS or other trustworthy sources out there. In particular, you will find a link to a whitepaper about microservices architecture on AWS in the *Further Reading* section at the end of the chapter.

Designing Interactions in a Distributed System to Prevent Failures

One of the aspects that influences a distributed system most is communication between the different elements of the system. There is already quite a number of things that can go wrong in a simple client-server scenario; that number becomes simply mind-boggling in a large microservices architecture. If you wish to have a better grasp of why this is the case in detail, read the *Challenges with Distributed Systems* paper from the Amazon Builders' Library, which is referenced in the *Further Reading* section of this chapter. In a nutshell, in a distributed system, you have to assume that every single line of code in your system that induces network communication may not work as expected. If you have a simple client-server system, it may still be managed without too much trouble, but if you have a system split into tens, hundreds, or more components distributed all over the place, a small glitch in one of these lines of code could cause a complete failure of the entire system. So, the more distributed your system is, the more complex failure handling becomes, which also explains why monolithic architectures are still around (ever wondered why some companies still have mainframes?).

So, how can you design your workload to keep it from being strongly impacted by network issues causing increased latency or even data loss?

Here are some best practices that you should follow.

Identifying Which Kind of Distributed System Is Required

One of the hardest things to handle with distributed systems is time constraints. If you've read the paper from the Amazon Builders' Library that was pointed out earlier, you know that distributed systems with hard real-time constraints have extremely strict reliability requirements.

The following questions also become important to consider. Does your workload have hard real-time constraints? Does it require rapid, synchronous responses within seconds? Or does it have soft real-time constraints for which responses are allowed within a broader time window, for instance, a few minutes or more? Or, on the other end of the spectrum, does it mostly work as an offline system that can handle responses through batch processing.

In any case, having a clear understanding of which type of distributed system you have to build is paramount.

Implementing Loosely Coupled Dependencies

In a distributed system, what you want to avoid at all costs is that a component's failure has a ripple effect and causes a disruption of other components until the entire system fails. So, rule #1 for the solutions architect when designing distributed systems is to loosely couple components that depend on each other.

Loose coupling brings more resiliency since a failure in a given component does not immediately impact other components that depend on it. It also brings more agility, as a positive side effect. When isolating components through loose coupling, components' life cycles become independent of each other. You can start modifying the code of these components, to bring improvements or fix issues. Then, if you respect the contract that has been defined through the interface of a component, changes in their implementation are less likely to affect others. You may wonder why this is not being heavily asserted here. Well, changing a component's implementation may, for instance, make it perform faster or slower, and even though you didn't change its interface, it could still impact other components depending on it. For the same reasons, loose coupling also extends the flexibility and scalability of the system. In a system, not all components perform equally well or need identical resources for processing. But, when components are loosely coupled, you have the opportunity to scale them independently of each other, based on their individual needs.

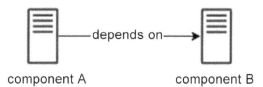

Figure 6.1: Tight coupling

Now, you could think of implementing loose coupling in either synchronous or asynchronous mode. It may be a surprise to some, but loose coupling does not necessarily imply asynchronous communication, so don't confuse the two.

Loose coupling two components means that you add an intermediate processing step, acting as a decoupling mechanism between the two. But the end-to-end communication could still take place synchronously. As illustrated in the following diagram, think of a load balancer, such as an ELB load balancer, between a client and a group of servers, for instance. A client making a request is loosely coupled to each server behind the load balancer, yet the communication happens synchronously.

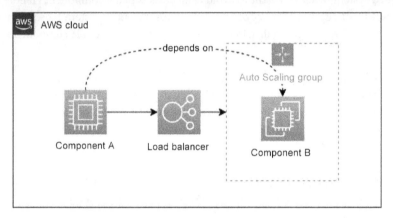

Figure 6.2: Loose coupling (synchronous communication)

That said, to further improve resiliency, it is advised, whenever possible, to use asynchronous communication between loosely coupled components. That's the best approach whenever an immediate response, other than the acknowledgment of the request, is not required. Think, for instance, of a queueing mechanism, such as **Amazon SQS**, buffering the requests between two components, as illustrated in the following diagram. Similarly, you could use some other mechanism, such as a notification system (for instance, **Amazon SNS**), a workflow engine (such as **AWS Step Functions**), or a streaming system (such as **Amazon Kinesis**), to handle asynchronous communication between loosely coupled components.

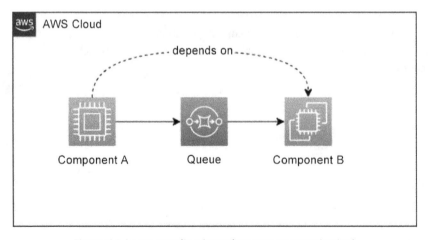

Figure 6.3: Loose coupling (asynchronous communication)

In a similar fashion, you could also think of using an event-driven architecture, where event producers and event consumers are loosely coupled, and an event engine (such as **Amazon EventBridge**) plays the role of an intermediate system.

Making All Responses Idempotent

A system is said to be idempotent if a given request made multiple times leads to the same result as that same request made exactly once. Idempotency facilitates failure handling since, upon a request failure, you can retry the request without taking much care whether it is retried once or multiple times. You can implement such a mechanism relatively easily using an idempotency token, a unique identifier that remains the same even if the request is repeated multiple times. The component receiving the request with that idempotency token should then ensure to provide the same response to repeated identical requests.

That's a powerful mechanism to avoid duplicating processing and data, for instance, in the case of transactions.

Doing Constant Work

This anti-fragility best practice consists of limiting the variance of work done across time, independent of the status of the system. It is based on the premise that variance causes disruption: a system has more chances to fail when it is subject to rapid and important variations in load. To have a better understanding of this idea, read the *Reliability and Constant Work* paper from the Amazon Builders' Library, which is referenced in the *Further Reading* section at the end of this chapter. It explains very well how Amazon Route 53 or **AWS Hyperplane** (the network function virtualization platform underpinning several AWS services) leverages this technique internally to make sure that their critical components will not fail due to changes in load, since they always keep doing the same constant amount of work, day in day out, no matter what.

For a more concrete example, similar to what Route 53 does internally, imagine that a critical component in your workload is in charge of performing health checks on a fleet of servers. The idea of constant work is to design your component and its supporting infrastructure in such a manner that they always do the same job and the same amount of work. For instance, suppose after performing some tests you measure that your health check component can do health checks on 100 servers at a time. Now, if your overall system is made of a fleet of 1,000 servers, you would simply deploy 10 instances of your health check component, each instance performing health checks on 100 servers. But what if your fleet is composed of a number of servers that is not a multiple of 100? If, for instance, you only have a fleet of 250 servers, you would need to deploy three instances of your health check component, each of the three capable of performing health checks on 100 servers. But wait, you only have 250 servers, not 300; how do you split the health checks? At first, it would seem logical to split the 250 servers more or less equally across the three health check component instances.

However, the trick, according to the principle of doing constant work, is to assign 100 servers to the first two health check component instances and the remaining 50 servers to the third health check component instance. To make sure that all your health check component instances keep doing constant work, add dummy data for the 50 missing servers and keep doing health checks on 100 servers, whether they are real or not. The key here is that your health check component keeps performing the same job again and again so that it is not affected by a sudden variation in load.

You just learned how to design interactions between components to prevent failures. However, note that everything will eventually fail over time. So, despite the precautions taken, some failures can occur at any time and you need to make sure that your workload can handle them.

Designing Interactions in a Distributed System to Mitigate or Withstand Failures

First, every component in your workload must behave in a way that does not negatively impact other components. Second, every component in your workload must be able to withstand the failure of one or more other components. Now, how can you achieve this?

The following best practices will help you limit the impact of failures on your workload.

Implementing Graceful Degradation to Transform Applicable Hard Dependencies into Soft Dependencies

This is a principle that has been successfully implemented by many companies running infrastructure on AWS. A concrete example of this is Netflix. Netflix has spoken publicly on numerous occasions about their approach to resiliency, so it is commonly known that, for instance, their **user interface** (**UI**) is designed to sustain the failure of the components in charge of providing information that it relies upon. As you log in to their platform and land on your home page, you see a list of various features (for instance, **Continue watching**, **Recommended for you**, and **Watch again**) each displaying a set of items (such as films, series, and documentaries). When the UI cannot reach one of its dependencies for any reason (component failure, network issue, or anything else), it misses a piece of information, such as all the items of a given category, for instance. In such a case, it would be very inefficient to display an error page to the end user simply because one of the many features provided is temporarily unavailable. It would also cause a very poor user experience. So, instead, the UI component is designed to keep working but will do so in degraded mode. In the example mentioned here, the UI would simply hide the feature related to the unavailable service and maybe serve cached data instead, if that makes sense for the impacted features. For end users, it is a much better experience (compared to landing on an error page) since they can still access the overall service even though some of its features are temporarily unavailable. In some cases, they may not even notice that the UI is operating in degraded mode.

The general idea is that when a component's dependencies become unhealthy, the component itself can keep working, although in degraded mode. What degraded mode means depends entirely on the use case at hand. It will likely mean different things for a video-on-demand UI and a mobile banking application. Yet the same graceful degradation approach can still be applied in both cases.

A popular design pattern that makes use of this principle is the circuit breaker. Consider two components, a client and a server decoupled through a circuit breaker component. The circuit breaker component has two states: open and closed. When the server is healthy and available, the circuit breaker is closed and the traffic flows normally between the client and the server through the circuit breaker. If the server becomes unhealthy, the circuit breaker opens, and the flow is interrupted to spare the client a call to the server that would inevitably result in a failure. Instead, the circuit breaker sends back a response on behalf of the server (depending on the use case, it could serve content from a cache or predefined data, for instance). The circuit breaker will also typically implement a retry strategy to come back to a closed state whenever the server is available again.

Throttling Requests

Throttling is a useful mechanism to respond to an unexpected burst in demand that exceeds a component's capacity. Some of the requests are still served but those over a specific threshold are rejected with a return message that indicates they have been throttled. Why is it important to mention the reason for the rejection? Because you expect the client to take it into account and drop the request, back off, and maybe try again after a while but at a slower rate.

After designing your workload, you should perform stress tests to determine the request rate that each component can handle. You can then use those metrics to define when to throttle incoming requests. You can leverage **Amazon API Gateway** for that.

That said, if you can address requests asynchronously, Amazon SQS and Amazon Kinesis can also be used to buffer requests, so you can handle requests at your own pace, avoiding the need for throttling.

Controlling and Limiting Retry Calls

What should you do when an error occurs? How should your components behave when receiving errors to requests they've made to their dependencies?

The temptation is high, especially when your application is time-constrained, to retry the failed request immediately and to keep retrying several times before dropping the request. There could be multiple reasons for that failure, and retrying immediately and repeatedly several times will inevitably add more stress to the network and the server if the requests ever reach it. This behavior is then amplified by the number of clients retrying at the exact same time.

A better approach is to use exponential backoff to retry after longer intervals; and, to make sure that retries from all clients do not occur at the same time, it is recommended to introduce some jitter. This will effectively randomize the retry intervals. Also, don't forget to limit the number of retries: if a request doesn't go through after a series of retries, it may be a sign of a deeper issue, such as incomplete or corrupt data.

Exponential backoff and retry is a technique implemented, for instance, by the AWS SDKs. That said, whenever you use third-party libraries or SDKs, always verify that this works as expected.

Failing Fast and Limiting Queues

This best practice applies to the server side or receiving end of the request.

For any component receiving a request, it is essential to determine rapidly whether it can handle the request or not. If for some reason it is unable to process the request, for instance, due to a lack of resources available, it should fail fast. The idea is to rapidly free resources to take some stress off the component and allow it to recover. As was mentioned already, one technique that can be used to relieve some of the pressure of the incoming requests is to buffer them, in a queue, for instance. That said, it is recommended not to let the queue grow too much since it will potentially result in increased wait time on the client side and, in the worst-case scenario, processing requests that the client has eventually dropped (**stale requests**). It could even lead to a state where the server tries to catch up with queued, but now stale, requests while fresh requests keep piling up in the queue, where they eventually become stale. That could keep the server from actually fully recovering. So, internal queue depths should be limited and kept under control.

Setting Client Timeouts

This best practice now applies to the client side, or sending end, of the request.

Set timeouts accordingly depending on your use case when you depend on other components since they can become unhealthy and stop responding, as explained in the preceding section. Also, avoid relying on default values since they may not fit your use case.

A good practice is to define timeouts for all external calls, whether they are made to a local or a remote system. The difficulty lies in finding the right value for the timeout. It shouldn't be too high to remain useful, but it shouldn't be too low to avoid seeing an excessive number of timeouts. In the latter case, the risk is generating unnecessary retries that will add stress to the network and/or the server, which would reduce or slow down the flow of responses, causing more timeouts, for instance.

Making Services Stateless Where Possible

When users or services interact with a workload, the series of interactions they perform is referred to as a **session**. A session contains user (or service) data that gets persisted between requests. An application is stateless if it does not require knowledge of prior interactions and does not need to store session information.

Stateless services bring major benefits in terms of scalability and reliability since incoming requests can be handled equally by any instance of the service. Plus, they become good candidates to be deployed on serverless compute platforms, such as AWS Lambda or AWS Fargate.

You may wonder how to handle state information in a stateless service. Well, instead of keeping it as memory or on a disk, you instead offload it to another component of your service, such as a cache system (for instance, **Amazon ElastiCache**) or a database (for instance, **Amazon DynamoDB**).

You can find more details on how to tackle several techniques discussed in the preceding section in the *Timeouts, retries, and backoff with jitter* paper from the Amazon Builders' Library, referenced in the *Further Reading* section at the end of this chapter.

The next section will talk about why and how you should manage changes to your workload.

Change Management

Managing changes to your workload is important since any of those changes can potentially affect its resiliency. So, to ensure that it can handle changes without impact, you must anticipate, but also monitor and control, those changes. To clarify further, all kinds of changes are being considered here, whether they are made to your application code or they affect the environment where your workload operates (such as, for instance, a surge in demand or a change of OS).

Monitoring Workload Resources

Monitoring is critical to ensure that you keep an eye on your workload's behavior at all times. It doesn't mean that you need to have someone watch over a screen 24/7, thankfully.

Logs and metrics are powerful instruments to provide insight into the health of your workload. Firstly, you should make sure to monitor logs and metrics emitted by your workload. Secondly, you should send notifications when thresholds are crossed or significant events hit your workload. Monitoring enables you to identify when your workload's SLAs are breached, when some KPIs or other thresholds are crossed, or when failures occur.

First things first, you must be able to instrument your workload to measure and extract metrics and then to measure and evaluate thresholds, KPIs, or SLAs against your objectives. You should also record essential operational metrics, such as latency, request rates, error rates, and success rates. Absolute numbers or average values won't be sufficient to make an informed decision on the best course of action. Depending on the event occurring, you will likely need to look at various metrics, such as ratios, averages, and percentiles.

How do you go about monitoring on AWS then?

Monitoring on AWS consists of essentially four phases:

1. **Generation** – Monitoring all components of your workload
2. **Aggregation** – Defining and calculating metrics
3. **Real-time processing and alarming** – Sending notifications and automating responses
4. **Storage and analytics** – Keeping logs for further analysis

The following sections present each of these steps in more detail.

Generation – Monitoring All Components of Your Workload

This may sound obvious, but it is essential to monitor all the components of your workload without exception, using either **Amazon CloudWatch** or third-party solutions if you prefer. From the frontend to the backend and the storage or database layer, you should make sure to collect the key metrics for your workload. That includes extracting them from the logs when necessary. Then, define all the thresholds that you want to monitor, typically those for which you want to trigger an alarm when they're crossed.

AWS provides monitoring information and logs in abundance.

Many services, such as, for instance, Amazon EC2, Amazon **Elastic Container Service (ECS)**, and Amazon **Relational Database Service (RDS)**, publish metrics for CPU or RAM consumption, network I/O, and disk I/O. Many other AWS services publish service-specific metrics to CloudWatch. For instance, Amazon API Gateway or AIML services such as **Amazon Rekognition** publish metrics for successful and unsuccessful requests. For a complete list of all the AWS services that publish metrics to CloudWatch, and of the metrics published by each of them, please check out the AWS CloudWatch documentation at `https://packt.link/sBreQ`.

Next to metrics publication, Amazon CloudWatch logs collect logs streamed from AWS services and your own applications. You can also leverage additional logs, such as the following:

- **VPC Flow Logs** to analyze network traffic in and out of your VPCs
- **AWS CloudTrail** to find out about any activity on your accounts that involves AWS service API calls, including actions taken through the AWS Management Console, AWS SDKs, and command-line tools

AWS provides a number of additional services that can come in handy as well:

- **Amazon EventBridge** is a real-time event delivery system where you can listen to events describing changes in AWS services. You can also use EventBridge to publish and listen to your own custom events, for your workload, and for third-party solutions.

- **AWS Personal Health Dashboard** is a service that provides your very own personal health dashboard of the AWS services being used by your workload(s). If a service event potentially impacts your resources in one of the AWS regions or AZs, you will find an event description and a link to your impacted resources.

- **AWS Config** is a configuration management service offering an AWS resource inventory, configuration history, and configuration changes (and notifications of these changes). You can track changes including those that put your workload reliability at risk. A set of config rules following the best practices from the AWS Well-Architected Framework reliability pillar is available out of the box as a Config conformance pack to make your life easier.

What you've learned so far in terms of monitoring allows you to monitor your workload from the inside by processing metrics collected from its various components. Now, if your workload offers external endpoints, you also want to monitor them from the outside. First, you want to verify that your external endpoint(s) can be reached. Second, this also gives you another chance to detect faulty behavior, if for some reason your monitoring failed to report it or, more likely, you failed to capture it. You can conduct this type of active monitoring with synthetic transactions, also referred to as **canaries**. The name takes its origin from the birds that were carried by the miners down the coal mines to detect lethal gas leaks early. The idea is essentially the same, fortunately without harming any actual bird. However, don't overload your endpoints with canary tests. Their purpose is merely to do a health check, not to put your workload under stress. **Amazon CloudWatch Synthetics** enables you to create such canaries to monitor your endpoints. It then deploys your canaries (scripts written in Node.js or Python) to Lambda functions in your account.

Aggregation – Defining and Calculating Metrics

As already mentioned, several AWS services provide service-specific metrics in CloudWatch out of the box. For others, for instance, VPC Flow Logs, you'll have to define metrics yourself by extracting data directly from the logs in CloudWatch. You do that by creating a metric filter that will look for the pattern you specify in the log data in CloudWatch Logs.

Maybe you're also interested in defining your own custom metrics to compute your KPIs and SLAs. Custom CloudWatch metrics can be defined and used to process metrics of any dimension. You can also use math expressions to combine data from multiple CloudWatch metrics.

Real-Time Processing and Alarming – Sending Notifications

Once your workload metrics are defined, you can then set alarms to be triggered when a metric threshold is crossed. When this happens you can notify any team who should be aware of the event. For that, you can use **Amazon SNS**, which can publish a notification message to multiple destinations, such as HTTP endpoints, AWS Lambda, Amazon SQS, Amazon Kinesis Firehose, AWS Chatbot (for delivery to a Slack channel for instance), PagerDuty, email, SMS, or mobile push notification.

Real-Time Processing and Alarming – Automating Responses

Once your event notification is in place, you can use automation on the other end to process the events received. One type of automation that comes immediately to mind is remediation to actually fix any issue behind an event; that's one possibility but not the only one. If the issue cannot be easily remediated, you could, for instance, automatically open a ticket with your organization's ticketing system so that the issue is routed and prioritized accordingly.

Storage and Analytics – Keeping Logs for Further Analysis

So far, you've collected logs and set up metrics, defined thresholds and created alarms, and set up notifications to the various teams and systems. So, you should be good to go, right?

Well, not quite yet. An often-overlooked part of the monitoring process is data management. What are your data retention needs for monitoring data, including logs?

You can use CloudWatch Logs for data retention, and leverage Amazon CloudWatch Logs Insights to run some queries on your log data. However, you might be interested in transferring your logs to your organization's existing log management system, such as Splunk or Logstash, for instance. Alternatively, you may also be interested in leveraging AWS analytics services, such as Amazon Athena or Amazon EMR, for instance, to run analytics on your log data using SQL queries or Spark jobs. To do that, you need first to instruct CloudWatch to transfer your logs to Amazon S3. Once on S3, you have complete freedom to use any of the AWS analytics services or your own third-party analytics tools. Another benefit is that you can then leverage the object life cycle management capability of S3 to optimize the associated storage costs, progressively transitioning your logs through the various S3 storage tiers available down to Amazon Glacier for long-term archival and retention.

Monitoring End-to-End Tracing of Requests through Your System

This was slightly touched upon under canary testing. It's good practice to validate that end-to-end requests perform as expected.

Leverage AWS X-Ray, or third-party equivalent tools, to help you understand how your workload and its underlying components are performing. Tracing can also prove particularly useful for debugging, as this can be a pretty arduous task on distributed systems.

Designing Your Workload to Adapt to Changes in Demand

Unless your workload expects a constant demand forever, which would be quite exceptional and unlikely in most cases, you need to make sure it can scale. And, to reap the most benefits, you want that scaling to be as automated as possible to closely follow the demand. That means scaling up when a surge in demand occurs, but also scaling down with the demand.

Using Automation When Obtaining or Scaling Resources

AWS provides a number of mechanisms to scale resources. When using serverless AWS services, such as, for instance, Amazon S3, AWS Lambda, or Amazon DynamoDB (on-demand throughput), your resources scale automatically with the demand on your behalf. You only need to make sure to configure the services properly for scaling, for instance, Lambda concurrency, and that you don't overrun your service quotas.

On the other hand, **AWS Auto Scaling** lets you automatically scale a number of resources and services, among which are Amazon EC2, Amazon ECS, Amazon DynamoDB (provisioned throughput), and **Amazon Aurora**. It provides two types of scaling: *dynamic scaling* and *predictive scaling*. Dynamic scaling lets you add or remove resources based on the actual utilization as measured with commonly used metrics. For instance, consider an application deployed on EC2 instances behind a load balancer. After running some performance tests, you will realise that, under stress, the application saturates the CPU first. You may then group the EC2 instances in an **Auto Scaling Group** (**ASG**) and define an autoscaling plan where you specify that whenever the average CPU over your ASG goes above 75%, you scale out, and whenever the average CPU over the fleet goes below 30%, then you scale back in. With predictive scaling, AWS Auto Scaling lets you anticipate the needs and scales resources to the expected foreseen capacity. To do that, it analyzes the historical behavior of your workload (for 14 days) against a specific metric and makes a prediction for the coming 2 days. This is useful to make sure the performance of your workload remains constant (if the prediction is correct). And, if needed, you can combine predictive and dynamic scaling for a more effective scaling strategy (for instance, to handle an unexpected surge in demand not anticipated by the predictions). Note that, at the time of writing this, predictive scaling is only available with EC2 ASGs.

Before scaling every resource at every layer of your design, don't forget to optimize your design for your use case(s). For instance, if your workload would benefit from a CDN (such as Amazon CloudFront), leverage it to offload your origin servers, then you only need to set the autoscaling mechanism to handle the residual load reaching your origin servers. Similarly, when using predictive scaling, you can first observe the validity of predictions over time, and then draw out your autoscaling strategy based on those observations.

Obtaining Resources upon Detection of Impairment

When you detect that the availability of your workload is impaired, you should scale the necessary resources to make it available again. For that, it's important that you can detect the health issue and be notified in the first place, for instance, using a canary test. Refer to the discussion in the *Monitoring Workload Resources* section for further details. Then, you should try to handle such conditions automatically and scale accordingly.

Obtaining Resources to Satisfy the Demand

Ideally, you want to leverage the dynamic scaling mechanisms described earlier, to scale your AWS resources proactively before your workload even becomes impacted. Based on the most relevant metric for your use case (for instance, average CPU utilization), you decide how to scale your resources automatically to satisfy the demand, whenever a specific threshold is breached. You could also leverage predictive scaling or a combination of predictive and dynamic scaling.

Load Testing Your Workload

Testing is crucial. Perform some load tests on your workload, and do it early enough. You'd hate to find out the day before a new major feature launch that your workload is not ready to sustain the expected load. You can easily spin up a new test environment on AWS manually, but preferably using infrastructure as code and automated deployment. After all, that's also an opportunity to test that those shiny **continuous integration (CI)**/**continuous deployment (CD)** pipelines of yours are properly working. Then, stress test your workload using synthetic load and identify any breaking points.

On top of that, once your workload is running in production, you can run occasional stress tests, during off-peak periods naturally, to validate that it keeps behaving as expected.

Implementing Change

Implementing changes requires enough discipline to ensure that the outcome remains fully under control. You want to make sure that the introduced changes, whether in the application code or the runtime environment, are not going to threaten the reliability of your workload.

Use Runbooks for Standard Activities such as Deployment

This is about using standard operating procedures to achieve repeatable, predictable outcomes. Runbooks document a series of steps, whether automated or manual, to be performed. They will include things such as how to patch or upgrade your workload environment.

As much as possible, automate the entire process. The less human interaction, the fewer errors. And don't forget to document a rollback process in case something goes wrong along the way. That also should be automated as much as possible for the same reason.

AWS Systems Manager is here to assist you with the creation and management of your own runbooks. In particular, **AWS Systems Manager Automation** lets you automate common maintenance or deployment tasks for AWS services such as EC2, RDS, and S3. For instance, it offers several runbooks managed by AWS, with predefined steps that can be used to perform common tasks such as restarting or resizing EC2 instances, or creating an **Amazon Machine Image** (**AMI**). You can naturally create your own custom runbooks, leveraging any of the available predefined steps and augmenting them with your own scripts.

Integrate Functional Testing as Part of Your Deployment

This is a standard best practice. Functional tests should be part of your CI/CD pipeline(s). Failing any of those tests should stop the pipeline(s) from deploying any further, and trigger a rollback as needed.

Integrate Resiliency Testing as Part of Your Deployment

This is a more advanced best practice, but resiliency tests should also be part of your CI/CD pipeline(s). Just as for functional testing, failing any of the resiliency tests should stop the pipeline(s) from deploying any further, and trigger a rollback as needed.

Deploy Using Immutable Infrastructure

The idea behind immutable infrastructure is that you don't update any component of the infrastructure in place. No OS patching or upgrade, and no software patching. If you need to upgrade or patch any component, for whatever reason, a new version of that component will be deployed and the existing one terminated. How it happens is down to you and may depend upon several factors, including the magnitude of the change, or whether you are allowed some downtime or not. You can adopt a blue-green or a canary deployment strategy to completely or progressively migrate to the new release. Once you're satisfied with the new release's behavior, the old environment is decommissioned.

There are many benefits to immutability. By replacing components with every release, the infrastructure is reset to a known state, avoiding configuration drifts. Deployments are also simplified because you don't need to support updates or upgrades. Deployments are atomic: either they complete successfully, or nothing changes. You build once and deploy (the same code base) to multiple environments, which prevents inconsistent environments. All of this builds more trust in the deployment process and simplifies testing.

Deploy Changes with Automation

You've seen it already in the previous sections and chapters, and you will keep seeing this recommendation again and again throughout this book: automate everything (or as much as possible). Deploying a release in a production environment should not be left to humans, or at least as little as possible should be done manually. Automated processes will never miss one step due to distraction, or execute steps in the wrong order. Humans make these mistakes all the time. So, get rid of the humans in the deployment process as much as you can.

To help you with that, you can leverage the AWS Code suite (**AWS CodePipeline**, **AWS CodeBuild**, and **AWS CodeDeploy**) or your preferred development toolchain.

Failure Management

Should you remember only one thing from this chapter, it is this: "*Everything will eventually fail over time,*" (Werner Vogels, CTO of Amazon.com). Failures are a given and it is better to not be under the illusion that they can be prevented forever, however good your design may be.

Backing Up Data

This is another thing that seems obvious but that is often overlooked; backing up data is paramount, and making sure you can recover with your backup data is even more important. This section will only briefly discuss backups as in *Chapter 7, Ensuring Business Continuity,* will dive deeper into defining a backup strategy.

So, you want to back up your data, your workload configuration, and everything you need to meet the specific business requirements of your workload. Two requirements, in particular, will define your backup strategy: **recovery time objective** (**RTO**) and **recovery point objective** (**RPO**). In some cases, you may not even need a backup. Can you reproduce all your important data from other sources to satisfy your RTO and RPO? This is the first thing to validate because you don't want to go and spend a humongous effort and loads of money on the greatest backup solution ever if you actually don't need one.

If backups are essential for your workloads, you want to define them, set them up, automatically run them, and forget about them. The latter is a figure of speech. You still want to verify that your backup process works fine.

The major backup solutions out there, which you can find on AWS Marketplace, including AWS' own solutions, allow you to run your backups on a schedule.

Take AWS Backup as an example. It provides a fully managed, policy-based backup solution. It lets you centralize, enforce, and automate the backup of data for multiple AWS services across your entire AWS organization.

Last but not least, you want to perform periodic recovery of your data to verify backup integrity and processes. This is one of the most critical aspects of backup because backup data that cannot be recovered is useless. So, it's essential to ensure that your data is recoverable by performing a recovery test.

Using Fault Isolation to Protect Your Data

Fault isolation limits the impact a failure can have on a workload to a limited set of components. This is a similar effect to the blast radius limitation that you want to achieve in terms of security. The idea is always the same: the components located outside of the boundary remain unaffected by the failure. Therefore, it is good practice to create multiple fault-isolated boundaries to reduce the impact of a failure on your workload.

Deploying the Workload to Multiple Locations

Consider a brief recap of how the AWS infrastructure is structured.

At the top, you find the AWS Regions. Regions are geographical locations where data centers are clustered (for instance, Dublin, Ireland, or Sydney, Australia). Each Region is composed of multiple, at least three, AZs. Each AZ consists of one or more data centers and has redundant power and connectivity within a region. AZs are located several kilometers apart, but less than 100 kilometers. They are interconnected via high-throughput, low-latency networking, over redundant fiber links.

Next, you have Local Zones, which are similar to AZs. They can indeed be used as a zonal placement for zonal AWS resources, such as subnets or EC2 instances. However, they are not directly located in the associated AWS Region, but near large industry or IT centers where no AWS Region is present (for instance, Los Angeles, CA, USA). That said, they are still capable of ensuring high-bandwidth, secure connectivity between resources in the Local Zone and resources running in the AWS Region. Local Zones are useful to manage workloads closer to your users for super low-latency requirements.

Last but not least, you find the Amazon Global Edge Network, which consists of edge locations in multiple cities around the world. With over 300 edge locations across the globe, the purpose of this network is to provide access to AWS resources and the AWS network closest to the end user location. Amazon CloudFront, for instance, uses this network to deliver content to end users with lower latency. Several other AWS services and features, such as, for instance, AWS Global Accelerator or **Amazon S3 Transfer Acceleration**, leverage the edge network.

For more details, please consult the AWS infrastructure web page at `https://packt.link/hROPk`.

A very strong suit of the AWS infrastructure, as it is built with redundancy at every layer, is avoiding a single point of failure. Naturally, you also want to avoid having single points of failure in your own workload. So, the first recommendation is to distribute your workload resources at least across multiple AZs.

So, given the properties of the AWS infrastructure, by distributing your resources across multiple AZs, they automatically benefit from strong protection against power outages or disasters such as fires, lightning strikes, floods, or earthquakes.

Some AWS services, for instance, Amazon EC2, are strictly zonal, and when using such a service, your resources share the fate of the AZ they are in. However, resources of the same service running in a different AZ within the same region will not be affected by a failure impacting only the first AZ. On the other hand, some AWS services are regional, such as Amazon DynamoDB, and use multiple AZs in an active/active configuration out of the box. It lets you achieve your availability design goals without having to define the multi-AZ configuration yourself. It is absolutely key to know whether a given service is regional or zonal since this can strongly influence the design of your workload to ensure it meets its reliability requirements.

Note that some services offer APIs that allow you to specify the regional or zonal scope of the request. When you can reduce the scope of a request to a single AZ, the request is processed only in the specified AZ, not only reducing the exposure to disrupt resources in other AZs but also avoiding being disrupted by an event in another AZ. The following AWS CLI example illustrates how to extract some information about Amazon EC2 instances from the `eu-west-1a` AZ only:

```
aws ec2 describe-instances --filters Name=availability-zone,Values=eu-
west-1a
```

Now, you may wonder whether it may be necessary to go a step further and distribute your resources in multiple regions to increase the reliability of your workload. Well, it depends. The following is a quote from the AWS Well-Architected Framework reliability pillar:

> *"Availability goals for most workloads can be satisfied using a Multi-AZ strategy*
> *within a single AWS Region. Consider multi-Region architectures only when*
> *workloads have extreme availability requirements, or other business goals, that*
> *require a multi-region architecture."*

Taking a multi-Region approach seems natural for a disaster recovery strategy since you may want to protect your workload against a large-scale event if that's necessary to meet your recovery objectives. Such a large-scale event could consist, for instance, of an AWS service becoming unavailable across all AZs of a Region, or even worse, more than one AWS service becoming unavailable within a given Region.

In such a case, your RTOs and RPOs, together with the budget at your disposal to implement this disaster recovery protection, will largely influence your solution design. AWS provides multiple capabilities to operate services across Regions. For example, AWS provides continuous, asynchronous data replication of data stored on Amazon S3 using its Cross-Region Replication feature. Amazon RDS Read Replicas and Amazon DynamoDB global tables also support multi-Region setups. With continuous replication in place, your data can then become available across multiple Regions. AWS CloudFormation, which you can use to implement an infrastructure as code approach, also helps you define your infrastructure and deploy it consistently across AWS accounts and multiple AWS Regions. Last but not least, Amazon Route 53 and AWS Global Accelerator let you route traffic between multiple Regions. For instance, you may want to always split the traffic in specific proportions between regions, or prefer to route requests based on geo-proximity or based on latency.

That said, operating your workload across multiple Regions will considerably raise the complexity and costs of your solution design. So, make sure to use such a setup only when you really need it.

Automating Recovery for Components Constrained to a Single Location

In some cases, it will not be possible to run components of the workload across multiple AZs. For example, all nodes of an Amazon EMR cluster are launched in the same AZ to improve job performance thanks to reduced latency and thus a higher data access rate. If a component constrained in an AZ is essential for your workload resilience, you need to set up a mechanism to redeploy that component automatically in another AZ whenever needed.

Whenever it is not possible to deploy the workload to multiple AZs due to technological constraints, you must determine an alternate path to resiliency. Don't forget to also automate the recreation of the necessary infrastructure, the application's redeployment, and the data recreation accordingly.

Summary

This chapter covered how you can leverage reliability design principles in designing highly available workloads. Environmental constraints, such as service and account quotas and network topology, were considered first. You then learned how to design workloads to prevent, mitigate, or withstand failure in a distributed environment. You also explored monitoring, leveraging logs, and metrics. Finally, you reviewed how to handle failure by leveraging data backups or multiple AZs or Regions and testing for reliability.

The next chapter will discuss business continuity aspects in detail.

Further Reading

- The Well-Architected Framework reliability pillar: https://packt.link/1zD83
- The *Hybrid Connectivity* whitepaper: https://packt.link/L907A
- *Challenges with distributed systems*: https://packt.link/1yx77
- *Implementing Microservices on AWS*: https://packt.link/Y6jM3
- Reliability and constant work: https://packt.link/jGE1O
- *Timeouts, retries, and backoff with jitter*: https://packt.link/L1b77

7

Ensuring Business Continuity

This chapter will focus on determining a solution design to ensure business continuity. You will look at the different strategies to protect your critical, and less critical, workloads on AWS in case of a disaster.

You will also learn how to design solutions for protecting against a disaster and being able to recover from it, which is paramount to making sure that your business can continue operating. This chapter will guide you through the various possible approaches depending on your needs in terms of business continuity.

The chapter covers the following main topics:

- Disaster recovery versus high availability
- Establishing a business continuity plan
- Disaster recovery options on AWS
- Detecting and testing disaster recovery

Disaster Recovery versus High Availability

Note the essential definitions for the discussion here. A disaster refers to a large-scale event that impacts a broad geographical area. In AWS terms, a disaster may impair an **Availability Zone (AZ)**, multiple AZs, an entire **AWS Region**, or, worse, several Regions.

Disaster recovery (DR) is the process that tackles both the prevention of a disaster and the recovery from a disaster.

High availability (HA) addresses how a workload can keep functioning even though some of its components are impacted by a failure.

How do the two compare with each other? Simply put, HA deals with local failures while DR deals with large-scale failures, so they complete each other.

That said, designing for HA on AWS often brings some form of protection for DR at the same time. Why is that? That is because AWS reliability best practices (see *Chapter 6, Meeting Reliability Requirements*, for more details) recommend measures that also provide some level of protection in case of a disaster. For instance, following AWS best practices, you may have decided to increase the level of resiliency of your workload by distributing it across two or more AZs within a given Region. If your workload serves customers in multiple geographies globally, you may even have designed to run it independently in two or more Regions. In both cases, your design already protects you against some form of disaster – the former against the failure of one AZ, the latter against the failure of one Region.

DR objectives are usually described with two specific KPIs: the **recovery time objective (RTO)** and the **recovery point objective (RPO)**. The RTO describes the amount of downtime allowed to your workload following a disaster before it is back online. The RPO defines the amount of time between the disaster and the latest data recovery point; in other words, how much data your workload is allowed to lose. Both are time measures, typically expressed in minutes, hours, or even days for the least critical workloads. The solution's costs and complexity rise as the RTO and/or RPO values decrease.

You will now explore the processes for preventing and planning for a disaster.

Establishing a Business Continuity Plan

It may sound obvious, but every measure you take to protect your workload against a disaster should be carefully considered and planned accordingly. First, these measures will have a significant impact on your solution design. Second, the cost of your solution will increase with the degree of protection against a disaster. So, you want to keep both aspects under control.

Eventually, you will create a DR plan for your workload. That document will become part of your organization's business continuity plan to make sure the organization can keep operating its business in case of a disaster.

Now, as always, you want to start with your requirements before creating that DR plan. What are your actual business needs in terms of DR protection? The last thing you want is to spend a huge effort and a lot of money on something that will not be useful.

So, when building that DR plan, ensure that every protection you put in place serves a purpose. As an example, suppose that you need to design your workload to survive a major disaster in the AWS Region where it is deployed. You design the solution so that you can recover your workload in a second Region within a reasonable amount of time and start operating again from that Region. Imagine now that a natural disaster, such as a major earthquake across the entire Region, impairs your workload. As expected, thanks to your design you are able to start operating again in another AWS Region. But imagine that the same disaster that impaired the AWS Region also impaired the rest of your organization's business operations. Depending on the business function supported by your workload, it might be useful to have it survive such a large-scale event, but it might also be useless if the rest of your business is down.

The bottom line is, before putting in place a sophisticated DR plan, make sure it is aligned with your organization's business continuity plan, and in particular, make sure to consider the DR plans of other business functions that your workload depends on.

The first thing to do is to conduct a risk assessment. This will help you determine the risk associated with several types of disaster; that is, the impact of a failure of a single AZ, multiple AZs, a single Region, or multiple Regions. Also, remember that AZs are physically separated by many kilometers, so deploying your workload across multiple AZs already provides a fair level of protection against some forms of disaster (for example, local flooding, an earthquake, a power outage, or a lightning strike). Depending on the criticality of your workload, you will examine the diverse options at your disposal with the measures that you can take and the associated costs. Then, you will compare the various options, the associated risks, and the costs of each variant of the solution design to eventually decide which option is the best fit.

Now take a look into the possible protection measures you can take on AWS.

DR Options on AWS

As opposed to traditional infrastructure deployment on-premises, the cloud lets you take a much more dynamic approach to DR. It frees you from the capital investment that you need to make on-premises to provision DR infrastructure. It also makes DR infrastructure readily available for you to speed up recovery; thus, you can focus on designing a solution to meet your RTO and RPO requirements.

Your options to deal with a disaster in AWS can be mapped onto four major DR strategies: **backup and restore**, **pilot light**, **warm standby**, and **active-active**. The idea in each case is to be able to restore or continue operating your workload in a second Region. The cost and complexity of the various strategies range from lower (backup and restore) to higher (active-active), as shown in the following diagram:

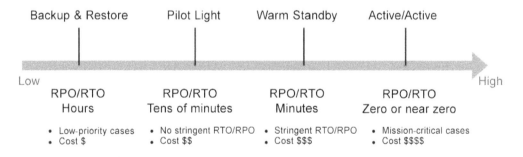

Figure 7.1: DR strategies cost and complexity

The following sections will take you through each strategy in detail.

Backup and Restore

This is the easiest mechanism you can use. It is also the least expensive of the four. The idea is to simply take a backup of your workload where it operates in Region A. Then, you somehow transfer that backup at regular intervals to the location where you intend to restore your workload in case of a disaster. Thus, if you are planning to handle a disaster that would impair your workload in Region A, you should look at restoring it from the most recent backup available in Region B and start operating in Region B. This approach is illustrated in the following diagram:

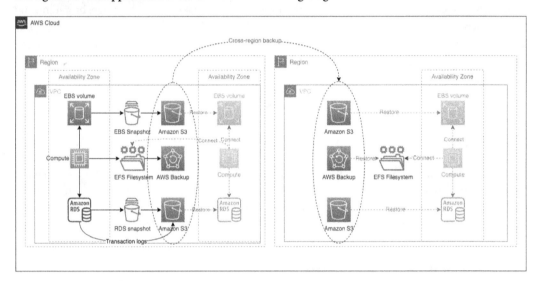

Figure 7.2: Backup and restore approach

Now, the preceding description is overly simplified on purpose. There are many details that will need to be addressed along the way, but if it satisfies your RTO and RPO requirements, this is a good solution and likely your best option.

The following sections describe a few considerations to be noted before defining your backup strategy.

Re-Building or Backing Up Everything

The first consideration to make is about what exactly you should back up. AWS lets you fully automate your workload deployment using a combination of **infrastructure as code** (**IaC**) and automated deployment. This is considered best practice. First, it allows you to gain agility and speed in your development life cycle. Secondly, it reduces your deployment risks and, at the same time, paves the way for deploying your workload anywhere. There are a few caveats before you can really deploy in any Region, but nothing too complex to handle. You will particularly need to pay attention to the resources you refer to in your automation templates and scripts.

For instance, when referring to an **Amazon Machine Image** (**AMI**) for deploying EC2 instances, you need to make sure to use the correct AMI ID for the Region where the automation runs (an AMI copied in multiple Regions will have a separate AMI ID in each).

Once you have automated the creation of your AWS infrastructure and deployment of your workload, then all you have to back up is your data.

Identifying and Backing Up the Necessary Data

On AWS, **Amazon S3** is *the* backup destination of choice. It is a very reliable and highly durable storage option. It is extremely cost-efficient and also offers built-in life cycle management to transition your backup data through various and less expensive storage tiers as it ages.

In addition, whether you opt for the AWS built-in backup capabilities (such as **Amazon EBS** or **Amazon RDS**, for instance) or a centralized backup solution such as **AWS Backup**, they obviously support S3 as a target destination for backups. If you prefer to opt for one of the popular third-party backup software options out there, chances are that they also support S3 out of the box.

Securing and Encrypting Backup

As for all other data stored on AWS, data security and protection best practices apply. Assuming you back up your data on Amazon S3, you have several choices for encrypting your data at rest. Amazon S3 uses server-side encryption and accepts your objects as unencrypted data; it then encrypts them before persisting them. Using client-side encryption for your workload entails encrypting the data before sending it to S3. Both these methods allow you to use the AWS **Key Management Service** (**KMS**) with either an AWS-managed KMS key that is created and managed for you, or a customer-managed KMS key that you create and manage. AWS KMS allows you to set IAM policies to control access to and usage of your data keys. This lets you define who is authorized, or not, to encrypt or decrypt your data.

Performing Data Backup Automatically

As with many other administrative tasks, your objective is to automate as much as possible. Limiting, or even eliminating, manual work is not just a way to reduce operational overhead; it is also a way to reduce your risks since any manual action is error-prone in nature.

So, set up your backups to run automatically, based on a specific schedule. AWS services such as RDS, EBS, DynamoDB, and S3 can all be configured for automatic backup. The same goes for your preferred third-party backup solution. As already mentioned, you can also centralize your backups using AWS Backup, and define backup policies centrally at the AWS Organizations level. That lets you enforce regular automated backups across your entire organization and/or at the level of **organizational units** (**OUs**). Your organization may enforce some common backup policies by default across the entire organization, but let some OUs override certain elements of the backup policies, for instance, the backup frequency. What an OU or an account can override in an existing backup plan or policy depends on whether and how child control operators have been defined in the parent backup policies.

For more details on managing backup policies at the Organizations level, refer to the AWS Organizations documentation (the link is in the *Further Reading* section at the end of this chapter).

Backup Frequency

How can you determine the optimal backup frequency? Provided that backup and restore can satisfy your RTO, the optimal backup frequency is the one that allows you to satisfy your RPO. The RPO will define how much data you are allowed to lose in case of a disaster. So, you need to make sure that the frequency at which you back up your workload will allow you enough time to do the backup and to store it safely where you can use it to recover. If you plan to recover in a separate AWS Region, you need to allow enough time for the backup data to safely reach that second Region. To give a concrete example, if your RPO is 2 hours, you want your backup frequency to be less than that. In fact, you may want to be on the safe side and plan for a potentially missed backup just in case (better safe than sorry). Therefore, you may want to set an hourly backup frequency. This way, if one of the hourly backups fails, you still have time to respond automatically, or even manually in the worst case. You could wait for the next automated hourly backup to be triggered and validate it goes well or otherwise make a backup manually.

Pilot Light

The pilot light approach consists of maintaining a copy of your data in a secondary region and your workload infrastructure in a pre-provisioned but switched-off mode, so all the necessary storage is already present, as well as the data stores (for instance, RDS read replicas), but all the compute is switched off, except when necessary to support the data replication (for instance, again, RDS read replicas). Any other compute is off, so no EC2 instances are running (other than for supporting data stores replication), no Amazon **Elastic Container Service** (**ECS**) cluster tasks are running, and no Amazon **Elastic Kubernetes Service** (**EKS**) cluster Pods are running.

The overall idea behind the pilot light approach is that you are almost ready to handle incoming requests and get your workload up and running in that secondary region, but you maintain a reduced footprint by having no running compute. This approach is illustrated in the following diagram:

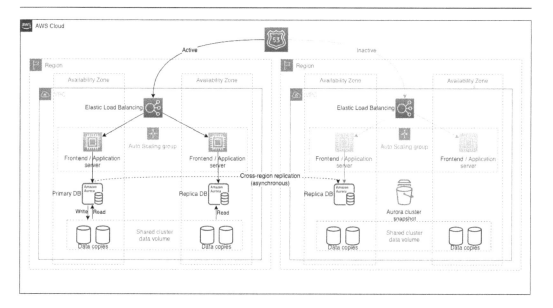

Figure 7.3: Pilot light approach

This approach is valid when your RTO is low enough so that a backup and recovery strategy will not be a good fit, but not so low that you do not get enough time to get everything up and running. This is typically a good fit when your RTO is in the tens of minutes range.

Compared to backup and recovery, because your core infrastructure is already in place and all you need to do is to get the compute capacity up and running, it is easier to test and validate that your DR plan is fully functional.

AWS Services for a Pilot Light Approach

On top of the backup services already mentioned in the backup and recovery approach, you now have to consider services that can offer continuous replication, in particular, if you need to satisfy a lower RPO.

S3 provides automatic cross-region replication natively. Combined with bucket versioning, it also gives you the ability not only to recover in a separate Region but also, if needed, to carry out a point-in-time recovery of specific versions of your objects.

RDS and Aurora also support continuous cross-region replication with Read Replicas and Global Database, respectively. DynamoDB also offers native multi-region support with its global tables.

Promoting a read replica in Region B, upon failure of the master database in Region A, typically takes a few minutes with RDS. On the other hand, promoting a secondary cluster to the role of a primary cluster with Aurora Global Database can take place in under a minute. That period of downtime will constrain the RTO. In both cases, the RPO is typically measured in seconds, conditioned by the lag between the primary and the replicated data stores.

DynamoDB global tables work differently from the RDS and Aurora multi-region mechanisms. Global tables consist of a set of active-active replicas, all part of a single entity (the global table), maintaining the replication across all of the replicas. All regional replicas act both as a primary and secondary. Therefore, a regional disaster will not have any impact at the database level, except that data replication from and to the impacted region will stop for the duration of the disaster.

Now, in case of a failover, in the pilot light scenario, you don't have any compute running in the new region yet. But once your workload comes back up in the second region, you need to update any external reference to your workload, such as DNS domain names. If you use **Amazon Route 53**, you can leverage its health checks mechanism, which will take care of routing the traffic eventually to the healthy endpoints (in the second region). This means that you need to add the new endpoints to your Route 53 configuration once your compute resources become available. There is, however, some time lag, even if you can adjust the DNS record's **time to live** (TTL) before Route 53 actually adjusts to the failover. However, if you rely on another mechanism, **AWS Global Accelerator**, routing failover can happen faster. Global Accelerator associates a set of static IPs to multiple endpoints and essentially routes the traffic to the closest and healthiest endpoint from the AWS edge network through the AWS backbone, and routing failover happens faster because Global Accelerator does not have the same caching mechanism used by DNS services such as Route 53.

In the case of the pilot light strategy, there is one AWS service that can also play a key role, **CloudEndure Disaster Recovery**. That solution does a block-level replication of **virtual machines** (VMs) from a given source environment to the target environment. The source environment can be on AWS but also on-premises. Your target environment in this scenario is your AWS environment in the second Region. CloudEndure Disaster Recovery will continuously replicate the VMs over so that, in case of a disaster in the source region, you are ready to fail over and start the instances in the second region. So, on top of backup, continuous data replication, and IaC practices, this is one more practical tool in your SA toolkit that can be quite handy with a pilot light approach.

At the time of writing, AWS launched **AWS Elastic Disaster Recovery**, which is now the recommended service for pilot light DR, becoming the successor of CloudEndure Disaster Recovery. It does not yet have feature parity with CloudEndure's solution, but it is only a matter of time before it does.

Warm Standby

The warm standby approach goes a step further compared to the pilot light one. It extends the same concept but also maintains a running copy, although scaled down, of your workload. So, your service is already up and running, and the only thing you need is to scale up the compute resources required by your workload. This approach is illustrated in the following diagram:

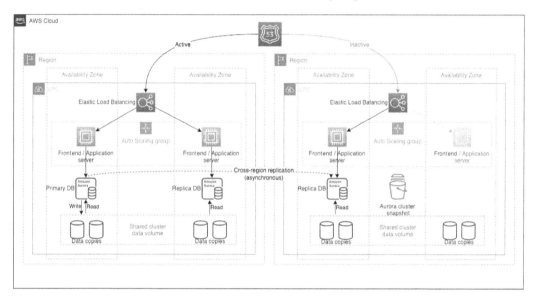

Figure 7.4: Warm standby approach

So, understandably, this targets situations where your RTO is too low for both backup and recovery and pilot light scenarios, but not too low so you have enough time to scale up the environment before you can handle the full production load in the new region. This is typically a good fit when your RTO is in the minutes range.

Compared to pilot light, it is even easier to test and validate that your DR plan is fully functional with this approach because you don't need to take any other action than scaling up to be fully operational.

AWS Services for a Warm Standby Approach

In the warm standby approach, you also use the AWS services already mentioned in the previous two approaches; but, on top, you need to ensure that your workload can rapidly scale up your compute resources to sustain the full production load in the new region. In this case, you're going to rely on **AWS Auto Scaling** to monitor the performances of your compute resources and to adjust the capacity as needed. Auto Scaling works with other AWS services such as EC2, ECS, DynamoDB, and Aurora. EKS uses Kubernetes-specific autoscaling mechanisms, such as the Kubernetes **Cluster Autoscaler** or the recently announced **Karpenter** to scale cluster resources (such as EC2 Nodes) and the Vertical Pod Autoscaler and Horizontal Pod Autoscaler to scale Pods. You would then need to leverage those to ensure that your EKS clusters and Pods are scaled up to the desired capacity.

Active-Active

The multi-region active-active approach is the ultimate DR approach for the most business-critical workloads, for which none of the previous three approaches could satisfy your RTO and RPO. With this approach, your workload is running concurrently in (at least) two separate regions. This is illustrated in the following diagram:

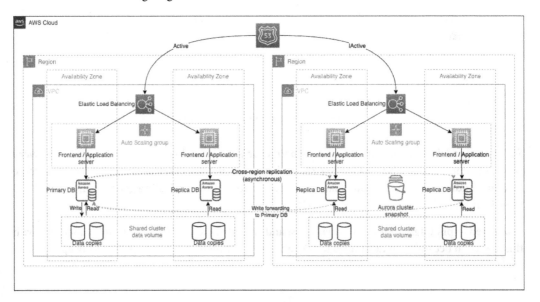

Figure 7.5: Active-active approach

This approach entails scenarios where you need an RTO of zero (no downtime) and an RPO as close as possible to zero. This, however, comes at a cost since you have a fully functional and scaled-up environment to support your workload in multiple regions (at least two).

Compared to warm standby, because you don't need to take any action at all, you are already fully operational in multiple regions, and it is even easier to test and validate that your DR plan is fully functional.

AWS Services for an Active-Active Approach

In this multi-region active-active approach, the same AWS services that were mentioned in the previous three approaches remain useful here. They may only be used slightly differently.

For instance, Route 53 or Global Accelerator would be configured to load balance traffic between both active regions and it is only in the case of a failover that they would redirect all traffic to the remaining healthy region.

Regarding data, all the solutions discussed also remain valid options. Your choice will be based on what you need to achieve in terms of RTO and RPO. Reads are not really an issue, since you can always manage either to redirect the reads to a read replica (such as with RDS or Aurora) or to have the concurrency increased automatically (such as with DynamoDB or S3). On the other hand, writes are often a thorn in your side, but you have a number of options to deal with them as given below:

- You may opt for a "*write global*" approach, as supported, for instance, by Aurora Global Database, where all the writes converge to a single database in a single region. In case of a failover, there is a bit of downtime, but it is limited to a few minutes (RTO) at most (in general, even less than 1 minute) and your RPO can be really close to zero.

- You may prefer to take a "*write local*" approach, such as what you can do with DynamoDB Global Tables. In case of a region failure, there is essentially no downtime since your datastore works in an active-active mode (RTO is zero or very close to zero) and your RPO can also be zero or very close to zero (the missing data from the impaired region will start to be synchronized again when the service comes back up).

- The last option is to take a "*write partitioned*" approach, where you write in a given region based on a partition key. This lets you avoid conflicts when writing data. You could use S3 in this case and configure bi-directional cross-region replication to keep the buckets in the two regions in sync.

Now that you have explored the approaches you can take, you are ready to learn how you can make sure that your DR strategy functions.

Detecting a Disaster and Testing DR

The first step, before you can take any countermeasures, is to detect that a disaster is actually taking place. Your recovery objectives (RTO and RPO) will dictate how much time you actually have to do so. Consider a situation where you have an RTO of 4 hours with an RPO of 1 hour. This implies that you have up to 4 hours to recover in case of a disaster, but you cannot lose more than an hour's worth of data. It also means that, whenever a disaster occurs, you must be able to detect the event rapidly enough to notify the stakeholders, escalate if needed, and trigger the DR response within 1 hour (to meet your RPO).

There are a number of things you can do to make sure to detect disasters on time.

Firstly, AWS offers a general service health dashboard that you can check to get the latest status information about AWS services in near real-time services. You can also subscribe to any of the associated RSS feeds to be notified when a specific AWS service goes down. Secondly, AWS provides the AWS **Personal Health Dashboard** (PHD), which lists the service events that affect your workloads. It presents both the ongoing events as well as the past events with the history of events that occurred in the past 90 days. Now, in some cases that may not be enough. If you have a very stringent RPO and/or RTO, you may need to rely on proactive detection methods, such as health checks, to detect disasters in a timely manner. Going into details on how to meticulously design health checks to assist you effectively is beyond the scope of this book; however, you are encouraged to read the *Implementing health checks* whitepaper referenced in the *Further Reading* section at the end of this chapter. You will have to make sure that the health checks you put in place actually help you detect breaches of your business KPIs and effectively identify disaster conditions early enough to meet your RPO and RTO.

Once your disaster detection is in place, as always in IT, testing is paramount to validate that you can meet your DR objectives.

The above is even more crucial in the case of business continuity as the sustainability of your business relies on its ability to survive a disaster. In this case, what you want to validate is that you can meet your RTO and RPO with the approach you selected.

The recommended approach to this validation is to test your DR strategy on a regular basis. You may even decide to test your strategy at a relatively high frequency, for instance on a weekly or bi-weekly basis, starting from the principle that the things that you repeat often become things that you do well.

In many cases, testing your strategy can be straightforward. In some cases, you may even not have to do anything (think of an active-active scenario). In all active-passive scenarios, testing can also be straightforward, provided that you have automated most of the steps that need to happen in case of a disaster in the primary region.

Even in the case of a backup and recovery approach, if you adopted AWS best practices to leverage IaC and automated release build and deployment, spinning off a test environment should be straightforward. You can then restore your backups there to assess RTO and RPO capabilities and run some checks on data content and integrity.

Summary

In this chapter, you learned the main differences between High Availability and Disaster Recovery. You reviewed how to prepare for a disaster and which major strategies are available in the cloud. You also went through the various AWS services that can support each of these strategies. Finally, this chapter emphasized the importance of testing your DR strategy, as that's eventually the only way to ensure business continuity will be there when you need it most.

The next chapter will discuss meeting the workload performance objectives.

Further Reading

- DR on AWS: `https://packt.link/aqUEE`

- *Backup and recovery approaches on AWS*: `https://packt.link/LNkEt`

- Backup policies in AWS Organizations documentation: `https://packt.link/yGpmx`

- *Implementing health checks*: `https://packt.link/2aRu7`

Meeting Performance Objectives

This chapter will focus on determining a solution design to meet performance objectives. You will look at the best practices and strategies that can be used to design a performant cloud architecture.

Designing solutions capable of meeting performance objectives is essential for the success of your solution and its adoption by end users.

The following topics will be covered in this chapter:

- Performance design principles
- Architecting for performance
- Monitoring performance
- Reviewing and adapting your solution

Performance Design Principles

AWS defines five design principles to enhance performance efficiency in the cloud. The following section will take you through each of them.

Principle #1 – Democratize Advanced Technologies

This first principle is about leveraging the advanced technology services offered by your cloud provider – in this case, AWS. There are two main reasons for that. First, why not let AWS do the heavy lifting of managing the resources supporting a specific technology while you focus on using that technology? Second, there is no benefit in reinventing the wheel. Simply put: if an AWS service matches your needs, use it and don't roll your own. You can thus focus on implementing your solution to bring added value to your business.

Principle #2 – Go Global in Minutes

With AWS, you can easily deploy your resources across the globe, closer to the end users, to improve their experience. It will effectively lower the network latency they encounter when using your solution. Not only can you leverage AWS services that make your content available globally, for instance, through its content delivery network **Amazon CloudFront**, but you can also use automation, in the form of **Infrastructure as Code** (**IaC**), to deploy your solution across multiple geographies in minutes. You can, for instance, use **AWS CloudFormation StackSets** to provision AWS resources across multiple AWS accounts in multiple AWS Regions with a single   operation.

Principle #3 – Use Serverless Architectures

Serverless means that you delegate the management of infrastructure resources (virtual machines or servers) to AWS. Leveraging serverless solutions from AWS lets you focus your entire effort on developing and maintaining your solution, not the infrastructure where it is deployed. AWS provides multiple serverless services, such as **Amazon S3**, which can be used to store your website's static content, such as HTML pages, images, and stylesheets. AWS offers several serverless services, such as data stores (**Amazon DynamoDB**), data streaming services (**Kinesis Data Streams**, **Kinesis Data Firehose**), compute services (**AWS Lambda**, **AWS Fargate**), and many more.

Principle #4 – Experiment More Often

On AWS, you have instant access to all these advanced, serverless, and many other services. Therefore, you should feel encouraged to experiment with them, either to validate a concept or simply to test different configurations (compute, storage, and others) until you find the one that helps your solution deliver the best performance. In most cases, you can use any such service and only pay for the resources you've consumed, which is usually a factor of time and, occasionally, other elements, such as the amount of data (stored or transferred), the number of requests, and so on. So, there should be no barrier to experimentation.

Principle #5 – Consider Mechanical Sympathy

The idea behind this principle is to select the technology that best fits your objectives. For example, before selecting a database technology, make sure to factor in all the use cases involving the database and take into consideration items such as data access patterns, data structure, data lifecycle, and non-functional requirements.

These five principles provide guidelines to drive your rationale when designing the solution for your workload. The next section will cover, step by step, the major building blocks of any solution (compute, storage, database, network) and help you make choices depending on your performance objectives.

Architecting for Performance

As you can probably imagine, there is no single solution design that will provide optimal performance for all kinds of problems. Therefore, you must architect to address the problem at hand and then leverage the AWS services that can provide the best performance for your architecture.

First, you will need to design your solution using industry best practices and reference architectures. Are you building a website to serve end users globally? Are you creating a data pipeline to help data engineers? Are you crafting a platform to collect data in near real time from a fleet of devices? Each of these problems will call for different design patterns and distinct architecture approaches. Suppose after an initial phase of research and analysis of the industry best practices and available reference architectures, if any, you have designed a solution that combines multiple design patterns. Now what? Well, here starts the phase of selecting the right AWS services to best fit your design. But how can you do that?

First, architecting for performance should become part of your solution design process. Make sure to integrate some mechanism to benchmark your solution designs and ensure they offer optimal performance. Optimizing for performance will strongly influence the user experience but also impact the cost of your solution. So, opt for the solution that provides the best cost/performance compromise while meeting your business objectives and providing a great user experience.

Secondly, given the breadth and depth of the AWS portfolio, chances are that you will have more than one option to build your solution. To pick the right building blocks, you need to understand very well how the relevant AWS services could help you optimize your solution for performance. For that, you need to dive deeper into categories that are crucial to every solution: compute, storage, database, and network.

Finally, you can count on AWS resources to help, such as solutions architects or professional services consultants, or on AWS partners to assist with your design. You have the option to reach out to such resources and ask for guidance, or even just a second opinion.

Compute Selection

Start by reviewing the available choices for compute resources on AWS. They fall under three subcategories: virtual instances with **Amazon Elastic Cloud Compute** (**EC2**), containers with **Amazon Elastic Container Service** (**ECS**) or **Amazon Elastic Kubernetes Service** (**EKS**), and functions with AWS Lambda. The following sections discuss each of them.

EC2 Instances

With Amazon EC2, you have access to a broad variety of virtual servers and bare-metal instances. Each of these EC2 instances belongs to a family, and a generation within that family, and possesses unique characteristics. Some of them can satisfy the needs of a large variety of workloads (general-purpose instances such as those from the **M** and **T** families), some are optimized for CPU-intensive workloads (the **C** family), some are optimized for memory-intensive workloads (the **R, X,** and **Z** families), others offer high storage density with high throughput or low latency (the **D** and **I** families), and yet others offer hardware acceleration, such as the **P** and **G** families (GPU-based acceleration) or the **F** family (FPGA-based custom acceleration). There are a few more families that provide specialized chips for one type of task, such as machine learning inference (the **Inf** family), machine learning training (the **Trn** family), or video transcoding (the **VT** family).

On top of their family characteristics, each instance has a specific tee-shirt size that determines the amount of CPU, memory, storage, and network bandwidth available.

Then, some variants also exist within the various families to provide either extra local storage (instance subtypes with a **d**, for instance, **z1d**) or additional network bandwidth (instance subtypes with an **n**, for instance, **C5n**), or sometimes both (for instance, **G4dn**).

Additionally, EC2 instances have supported an increasing variety of CPU processors over time. The latest generations of EC2 instances are now powered by either Intel, AMD, or ARM-based AWS Graviton processors. For instance, the M6 type of EC2 instances, the sixth generation of the M family, has declined in instances with either Intel CPUs (**M6i**), AMD CPUs (**M6a**), or Graviton2 CPUs (**M6g**). AMD-based EC2 instances were first introduced to provide a cost-efficient alternative to Intel-based EC2 instances. Since they both run on the same x86 chipset, they require no change from an application perspective and provide an interesting option if you don't rely on Intel-specific features. Graviton-based EC2 instances were later introduced by AWS to provide an even more appealing alternative to Intel or AMD EC2 instances. They indeed brought significant cost efficiency gains and, with the latest generation of Graviton processors, even added lasting performance improvements over their x86 peers. Nowadays, many enterprises have adopted Graviton-based EC2 instances, citing slashed EC2 costs and boosted performance. They come at a price, though: Graviton-based instances rely on the ARM chipset, so you must first make sure that your workloads can be effectively ported to that chipset. In particular, you will have to procure all the dependencies, such as application libraries, for the ARM platform, and your workload binaries, if any, will require re-compilation.

That said, before picking up an instance type, you must consider the major characteristics of your workload in terms of performance. For that, beyond the initial assumptions you can make, observing and measuring are essential. You should collect metrics showing evidence of those characteristics. Is it mostly bound by the compute power available (number of CPUs or maybe the CPU clock)? Or is it memory-hungry, rapidly saturating all the RAM it can find? Or is it limited by the network bandwidth available? Or is it tied to the disk throughput or the storage latency? You will be able to tell based on observation. Now, which instance type do you start with to make those observations? In the absence of any obvious characteristics, it is recommended to start with one of the general-purpose instances, from the M or T family. T family instances are so-called *burstable* instances, which are useful for workloads that operate most of the time under a moderate baseline but occasionally need to burst above that baseline to meet a punctual spiky demand. Alternatively, if you already have a fair idea of the major characteristics of your workload, you can directly start with an instance from what seems like the most appropriate family (for instance, when you need GPU-based instances).

And then you iterate, based on measurement and observation, either to increase or reduce the size of the instance or to change the family if what you observe invalidates your hypotheses.

As always with AWS, nothing beats experimentation, so don't hesitate to try out multiple families and sizes if you're unsure. If you follow best practices and automate your CI/CD process, testing different EC2 instance families and sizes should be straightforward. When you do, monitor the performance of your workload on each type and size of EC2 instance that you try out. For that, you can leverage EC2 metrics reported by CloudWatch; in particular, as a minimum, pay attention to metrics such as *CPUUtilization*, *MemoryUtilization*, and *NetworkIn*, as well as *NetworkOut*, to understand whether your workload may suffer from a lack of CPU power, or from memory or network bandwidth exhaustion.

Finding the optimal instance type and size for your workload, also known as rightsizing, is a must to optimize your compute resource usage. Remember that the cloud brings elasticity. So, when you have found the ideal instance types to support your workload, remember that you can leverage AWS auto-scaling capabilities to scale out the number of instances used to support the load on the various components of your design. Simply put, it is beneficial most of the time, both from a performance but also a cost standpoint, to use multiple smaller instances that can be scaled out (and then back in), instead of a few large ones. In any case, plan for scalability and put the necessary mechanisms in place to automatically adjust your workload capacity as close as possible to the demand.

Containers

If, instead of deploying your workloads directly on virtual machines, you prefer to deploy them using containers, multiple options are offered to you. First, do you need to manage and have control over the virtual machines running the containers? If you don't, you can opt for **AWS Fargate**, which provides a serverless environment to run containers. If you do, you must pick and manage the EC2 instances that will best support your workload. How do you decide which instance, then? Well, for that, refer to the previous section discussing EC2 instance selection and also cross-check whether there is any incompatibility with your container orchestrator service.

Further, you have to also decide on the container orchestrator service of your choice, either ECS or EKS. So, how do you choose?

Briefly, ECS is an AWS native container orchestrator. It brings the benefits of simplicity of use and integration with other AWS services, such as **IAM** and **AWS Elastic Load Balancing** (**ELB**), to leverage either an **Application Load Balancer** (**ALB**) or a **Network Load Balancer** (**NLB**). Native integration means that no additional layer of abstraction is required to integrate with the specific service at hand.

EKS, on the other hand, brings the flexibility of Kubernetes with its vibrant user community and broad ecosystem. It also integrates well with AWS services such as IAM and ELB, for instance, but it typically relies on a layer of abstraction to integrate with cloud services. If you are familiar with and already using Kubernetes, then EKS will be the natural choice. If that is not the case, especially if you are just starting with containers on AWS, then ECS will definitely be easier to grasp and certainly offers a faster learning curve for the newbie. And, unless you really plan to exploit the flexibility provided by Kubernetes, EKS may well be overkill.

Last but not least, don't forget that AWS offers a serverless container service, Fargate, that supports both ECS and EKS. Unless you need to keep full control over the underlying servers running your containers, Fargate will make your life easier by managing and controlling that part of the infrastructure. That way, you can focus on building, deploying, and managing containers without worrying about the servers behind.

Whichever container service you end up choosing, you will have to configure it to meet your performance requirements. The configuration options will obviously vary per service, but again, exactly like for EC2 instances, you will need to run some experiments to find your optimal configuration. For instance, ECS lets you run your containers inside tasks (a task runs one or more containers). Tasks are allocated CPU and memory to do their job. When you deploy ECS tasks using EC2 instances (EC2 launch type), you have great flexibility when it comes to how you configure allocated CPU and memory to the tasks, and you can even overcommit resources on the same underlying EC2 instance. When you deploy ECS tasks on Fargate (Fargate launch type), you also specify how much CPU and memory are allocated to a task, but you have less flexibility and granular control over things, in particular, because Fargate manages the infrastructure on your behalf (and, for instance, does not overcommit resources). With EKS, containers are deployed on Pods (instead of tasks on ECS), and you control how much memory and CPU are allocated to Pods in a similar fashion. Going any deeper into Kubernetes goes far beyond the scope of this book, so it will suffice to just mention that Kubernetes provides additional flexibility to control and tune the underlying infrastructure configuration (for instance, setting limits not only at the pod level but also at the namespace level). And again, if you are not interested in managing the infrastructure, just let AWS do it for you by leveraging Fargate with EKS.

For monitoring the performance of your containers, AWS provides a solution called **AWS CloudWatch Container Insights** that gathers metrics for containerized workloads running on either ECS or EKS, including Fargate support for both container platforms. As you have already seen for EC2 instances, CloudWatch collects metrics about various resources, including CPU, memory, disk, and network utilization. These metrics typically provide visibility at the ECS or EKS cluster level. For more granular visibility, Container Insights adds insight into container-level resources such as instances (ECS), nodes (EKS), tasks (ECS), and Pods (EKS) regarding CPU, memory, storage, and network utilization. It also provides diagnostic information, such as container restart failures (EKS), to help spot issues. Additionally, you can integrate CloudWatch with an open-source solution such as Prometheus, an immensely popular monitoring toolkit from the **Cloud Native Computing Foundation** (**CNCF**). Integrating Prometheus with CloudWatch allows you to reduce the number of tools being used for monitoring while being able to collect either pre-defined sets of metrics or custom metrics from your container workloads.

Functions

Functions go a step further with the level of abstraction they provide. AWS Lambda functions let you focus on developing the application code while they manage and scale the underlying infrastructure for you. You don't need to package your code in a container, although you can do so if you really want to. Lambda supports several programming languages through runtimes. You can, for instance, develop code in Java, Python, JavaScript, .NET, and more and leverage one of Lambda's built-in runtimes. If your preferred programming language is not supported by any of the built-in Lambda runtimes, you can check whether one of the community-supported Lambda runtimes can help. And, as a last resort, you can also bring your own custom runtime.

Lambda works hand in hand with **Amazon API Gateway** to publish and share your functions' code as services through a set of APIs. API Gateway is also entirely serverless and manages and scales the underlying infrastructure for you.

When deploying a Lambda function, you specify how much memory you want Lambda to allocate at runtime. Lambda automatically allocates an amount of CPU that is proportional to the amount of memory you specified. The more memory (and thus CPU) you allocate, the faster your Lambda function will execute, until it cannot make use of additional memory and CPU power anymore. The optimal performance memory setting is somewhere just before the performance gains start plateauing. Also, since Lambda functions' costs depend on their allocated memory and execution duration time, there is a point where the *cost over performance gain* ratio obtained by adding more memory (and thus CPU) to a Lambda function will start becoming less and less interesting.

An additional consideration for Lambda functions, which long-time Lambda users are familiar with, is cold starts. AWS Lambda functions run on infrastructure managed by AWS. So-called cold starts are experienced when an incoming request requires new infrastructure to be provisioned resulting in an extra delay before executing the function at hand. For a long time, there was no proper solution to that problem. Some tried to keep the infrastructure supporting their Lambda functions warm by generating sufficient synthetic traffic, but that was empirical and far from optimal. Nowadays, although you cannot completely eliminate the issue (in the case of very spiky traffic, for instance), you can greatly reduce its effects by provisioning sufficient capacity for your Lambda functions when you know that cold starts could potentially have a great impact.

As always on AWS, improving performance starts with collecting metrics about your workload. In the case of AWS Lambda, **AWS CloudWatch Lambda Insights** conveniently collects metrics from your Lambda functions on CPU, memory, disk, and network usage. It also provides additional diagnostic information, for instance, about cold starts or worker shutdowns, to help you spot anomalies or issues and fix them more easily.

Storage Selection

Selecting the optimal storage for your solution depends on multiple factors. First, how do you plan to access the storage? Do you require block-level, file-level, or object-level access? Have you identified the storage access patterns your solution requires (random or sequential)? How much throughput do you need? What is the frequency of access (online, offline, archival)? What is the storage update required ? Is it of the **Write-Once-Read-Many** (**WORM**) type, or is it more dynamic? Are there any availability and durability constraints?

There are multiple types of storage solutions available on AWS, so selecting the right storage solution for the right usage is paramount to optimizing performance. And, most likely, your solution will leverage multiple storage solutions according to its needs.

The following presents a brief overview of the various storage options on AWS.

Amazon Elastic Block Store (**EBS**) delivers block-level storage for EC2 instances. It offers storage volumes backed either by **Solid State Drives** (**SSDs**), for very low-latency and **Input/Output Per Second** (**IOPS**)-intensive workloads, or **Hard Disk Drives** (**sHDDs**), for throughput-intensive workloads. EBS volumes can be attached to an EC2 instance and then used to support a filesystem, a database, or anything else that can make use of a block device (think of it as a hard drive). EBS volumes come in different flavors, and it's important to understand their characteristics before picking one up for your workload:

- General Purpose SSD volumes (gp2, gp3) provide a good balance between price and performance for many workloads. They are your best bet to get started when your workload does not have a behavior that directly points to one of the other types mentioned in this list. They can be used for anything really, from filesystems to databases.

- Provisioned IOPS SSD volumes (io1, io2, io2 Block Express) are designed for I/O-intensive workloads. They provide consistent IOPS and can predictably scale to tens of thousands of IOPS per volume.

- Throughput Optimized HDD volumes (st1) provide low-cost magnetic storage for throughput-intensive workloads. This makes st1 volumes ideal for workloads accessing large volumes of data sequentially, such as big data, **Extract, Transform, and Load** (**ETL**), data warehouses, and log processing.

- Cold HDD volumes (sc1) offer even lower-cost magnetic storage for throughput-intensive workloads manipulating cold data. Providing a lower throughput than st1 volumes, sc1 volumes are ideal for workloads requiring infrequent access to large volumes of data sequentially when keeping costs low is the driving factor.

As for all things on AWS, you should monitor the performance of your workload and the components it uses. Leverage EBS metrics such as *VolumeTotalReadTime*, *VolumeTotalWriteTime*, *VolumeReadOps*, and *VolumeWriteOps*, as well as *VolumeQueueLength*, to understand whether your workload suffers from storage latency. Check *VolumeReadOps* and *VolumeWriteOps* to make sure you stay below the IOPS limit for your EBS volume. These metrics should give you an initial understanding of whether you are making the most of your EBS volume performance. If you are too close to your IOPS limit or, even worse, get throttled because you reach it repeatedly, you either need to limit the IOPS of your workload or increase your volume IOPS limit. The latter will require you to either increase the EBS volume size, unless you have already reached the max IOPS supported by your specific EBS volume type, or possibly to change the EBS volume type, for instance, from a gp2 to a gp3 volume type, or even to an io2 volume type depending on the situation.

Likewise, monitor *VolumeReadBytes* and *VolumeWriteBytes* to make sure that you get enough throughput from your EBS volumes, and understand whether you might benefit from either a larger volume or a different type with higher throughput.

Here are some considerations:

- EBS volumes store data redundantly in a specific **Availability Zone (AZ)**. Making the data on an EBS volume available in another AZ (in another Region) requires taking a snapshot of the volume and restoring the snapshot to a new volume in the target AZ (as illustrated in *Figure 7.1*).

- EBS volumes were originally meant to be mounted on a single EC2 instance at a time. Recently, AWS made it possible to attach some EBS volumes to multiple EC2 instances at a time. However, several restrictions apply. For instance, you cannot use a traditional filesystem such as *ntfs* or *ext4*; you need a clustered filesystem that can maintain data consistency. Additionally, multi-attach is supported on provisioned IOPS volume types only (io1, io2, and io2 Block Express).

- io2 volumes provide the highest level of durability.

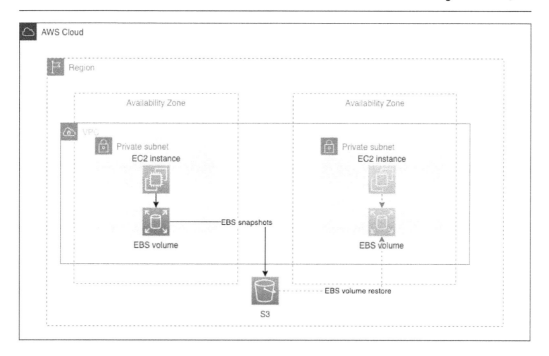

Figure 8.1: EBS volume attach, snapshot, and restore

Amazon Elastic File System (**EFS**) offers a managed **Network File System** (**NFS**) that can scale to petabytes without disruptions. With EFS, typically the amount of IOPS and throughput you get depends on your storage size. The larger your storage, the more IOPS and the larger your throughput. That said, EFS proposes two performance modes, a general-purpose mode and a max I/O mode. The general-purpose mode fits a large palette of use cases with low-latency requirements, such as home directories, general file shares, or content management systems, for instance. The max I/O mode is useful for workloads that need higher throughput and more IOPS. For EBS volumes, when you are not sure which one to pick, the best option is to start with the general-purpose mode and measure your workload performance. In the case of EFS, monitor the *PercentIOLimit* metric reported by **Amazon CloudWatch** to make that decision. If the metric gets close to or reaches 100 percent, it implies that your workload would benefit from the max I/O mode.

EFS also offers two throughput modes, **bursting throughput** and **provisioned throughput**. In bursting mode, EFS has a throughput baseline depending on its storage size. It allows you to hit a throughput above the baseline during a limited amount of time on a daily basis and is ideal when you have a workload requiring a relatively stable storage throughput, with an occasional short peak above the baseline. The provisioned throughput mode makes more sense when you have a workload that requires a sustained throughput higher than the baseline your EFS filesystem could normally offer. In provisioned throughput mode, the throughput does not depend on the filesystem storage size.

Amazon FSx is another AWS service, or a family of services that provide a choice of fully managed filesystems among Windows File Server, NetApp ONTAP, OpenZFS, and Lustre. Thanks to its architecture, FSx can sustain high throughput and low latency.

Amazon FSx for Windows File Server provides a native Windows filesystem and supports the **Server Message Block** (**SMB**) protocol. It comes with either SSD-based or HDD-based storage and can support a large variety of workloads, such as home directories, general file shares, and content management systems, for instance. FSx for Windows File Server leverages an in-memory cache to store your hot data to consistently deliver sub-millisecond latency for frequently accessed data. For data not in the cache, the latency is still very low, going from sub-millisecond for SSDs to single-digit milliseconds for HDDs. You simply adjust the IOPS and throughput settings based on your workload requirements. Additionally, it lets you provision the amounts of storage and throughput independently of each other. On top of that, you can configure multiple filesystems in a single Microsoft **Distributed File System** (**DFS**) namespace to boost performance up to tens of Gbps and millions of IOPS.

Amazon FSx for OpenZFS provides a managed file storage service built on top of the open source OpenZFS filesystem supporting the NFS protocol. FSx for OpenZFS runs on Graviton-chip-based systems and leverages SSD storage to consistently deliver sub-millisecond latency. It also leverages an OpenZFS built-in caching mechanism that can provide even lower latency for frequently accessed data. Throughput capacity grows with storage size up to 12.5 Gbps, while FSx for OpenZFS can deliver up to 1 million IOPS. Note that you can achieve max throughput and IOPS from a single client by using multiple NFS connections in parallel via NFS nconnect.

Amazon FSx for NetApp ONTAP provides a managed file storage service built on top of NetApp's ONTAP filesystem. ONTAP offers filesystem access through the standard file-sharing protocols SMB and NFS, but also supports block storage access via iSCSI. It is worth noting that it supports multi-protocol concurrent access to the same data. So, you could have, for instance, a Linux server accessing a given file share via NFS and a Windows server accessing the same file share via SMB at the same time. FSx for NetApp ONTAP offers two data tiers: primary storage and capacity pool storage. The primary storage tier runs on high-performance SSDs and is typically for your hot or most frequently accessed data. The capacity pool storage tier runs on cost-optimized storage that grows automatically as your data grows and is meant to store the least frequently accessed data. You can increase the size of your primary storage tier at any time if needed. Primary storage delivers sub-millisecond latency, while capacity pool storage offers tens of milliseconds of latency, and both storage tiers can deliver hundreds of thousands of IOPS.

Amazon FSx for Lustre offers a managed Lustre filesystem, an open-source filesystem very popular in the **High-Performance Computing (HPC)** world. FSx for Lustre thus provides a high-performance filesystem for very demanding workloads. Similar to FSx for Windows File Server, FSx for Lustre lets you choose between SSD-based and HDD-based storage, depending on whether your workload is mostly sensitive to latency or throughput. For HDD-based filesystems, you have the ability to add an optional SSD-based cache, representing 20% of your filesystem size, to improve performance in terms of latency. Note that you can also make your content on S3 available to your workload through FSx for Lustre so that objects on S3 are transparently presented as files on your Lustre filesystem.

As you can see, the FSx family offers quite some choice in terms of filesystems. For further details about how they compare to each other in terms of features and performance, it is highly recommended you review the service documentation guidance: `https://packt.link/ieCBn`.

Confronted with all those choices, which storage solution should you pick up, then? Start with your workload requirements. What are the data access patterns? Which access protocols do you need to support? Do you plan to share storage among multiple compute resources? What volume of data and number of read/write requests do you need? What storage growth is expected in the near future? What requirements do you have in terms of fault tolerance and durability? Does the storage layer need to be versatile enough to easily adapt to future access patterns? Which storage metrics are important to consider for your workload?

Answering these questions will take you in the right direction. Then, as already mentioned, if, despite your initial investigation, you hesitate between two solutions and cannot easily make up your mind, simply benchmark one against the other to find out empirically which one best satisfies your needs.

Database Selection

Selecting a database depends on many factors, and you may have to leverage more than one database for your workload to deliver optimal performance. It all very much depends on the type of data you handle, the type of access to support, the querying capability expected, and additional factors such as latency, scalability, consistency, and partition tolerance.

So, the first thing to do is very clearly understand your workload requirements. People tend to stick with the technology they know; nobody is immune to that behavior. If you have been working with relational databases all your career, chances are that they will be a central piece in your solution design. But is it the right choice for this particular workload in the cloud?

First, you do not have to rely on a single technology anymore because AWS presents a plethora of database technologies and because these technologies are available as pay-as-you-go services, including fully managed services. So, you don't have any reason to consider one technology for your entire solution.

Second, your workload components or services will have very distinct requirements from each other. Some may require strong and reliable transactionality, à la **Atomicity, Consistency, Isolation, Durability (ACID)**. Some others may require strong partition tolerance and need to be almost always on. Some others may need consistent ultra-low latency. These needs will make it difficult for you to just use a single technology. If you try to force a single technology, you may have to sacrifice performance to do so or end up with much higher costs ultimately because more resources would be needed to meet those performance needs. The diverse options at hand in AWS in terms of databases are discussed next.

Starting with relational databases, this type of data store is optimal for managing data that follows a pre-established (and stable) structure, or schema, composed of inter-related data entities. It is also very much adapted for workloads relying on ACID transactions, strong data consistency, and referential integrity. AWS offers multiple managed services supporting relational data stores: **Amazon Relational Database Service (RDS)**, **Amazon Aurora (Aurora)**, and **Amazon Redshift (Redshift)**. RDS proposes a managed service implementation of your preferred database engine, choosing between MySQL, PostgreSQL, MariaDB, Microsoft SQL Server, and Oracle. Aurora provides a MySQL-compatible or PostgreSQL-compatible database tailored to AWS, bringing improved performance, scalability, and availability over the plain open-source implementations. Redshift provides analytics capabilities at scale, giving you the ability to run a data warehouse in the cloud but also to launch analytical queries combining data from multiple sources, such as your data lake or other operational data stores.

Further, there are non-relational databases or NoSQL databases. Unlike relational databases, NoSQL databases are better at handling data with a dynamic schema and also better at scaling horizontally (think internet-scale here) but less good at handling random queries. AWS offers several managed services in that space: **Amazon DynamoDB**, **Amazon DocumentDB**, **Amazon Keyspaces**, and **Amazon Neptune**. DynamoDB is a serverless key-value database, meant to support cloud applications at any scale. It also works in a multi-master multi-region mode for applications that need to operate globally. DocumentDB is a document database specialized in storing and querying JSON-like documents. It is also compatible with the Apache 2.0 open-source **MongoDB** APIs. Keyspaces is a serverless wide-column database that offers compatibility with the **Apache Cassandra** CQL API. Neptune is a graph database capable of handling high-throughput and low-latency requirements. It also supports queries using both **Gremlin** and **SPARQL**.

AWS also provides a number of additional managed database services that do not directly fall under the previous two families. Amazon MemoryDB for Redis is a managed Redis-compatible in-memory database, distributed across multi AZs and supporting microsecond read and single-digit millisecond write operations. **Amazon ElastiCache** is an in-memory data store, compatible with either **Redis** or **Memcached**. Although ElastiCache can also be used as a persistence layer, it is typically employed as a caching layer in front of an existing database, such as RDS, for instance, to accelerate datastore performance. ElastiCache delivers sub-millisecond response latency for both read and write operations. **Amazon Timestream** is a time-series database specialized in the collection and storage of time-series data (for instance, telemetry or IoT device/sensor data). Its architecture is optimized to handle fast writes and large analytical queries. Lastly, **Amazon Quantum Ledger Database** (**QLDB**) is a ledger database that stores data in an append-only journal where all changes are cryptographically verifiable.

With such a plethora of choices, how do you pick the right database solution to provide optimal performance to your workload?

Well, you need to start with your workload requirements. What type of data do you need to handle? What volume of data will you need to store? How will the data be queried? How many concurrent requests (on average and max) do you expect? Do you have latency constraints when accessing the data (reads and/or writes)? What are your availability requirements (such as uptime)? In the case of disaster, how fast do you need to be back up and running for your **Recovery Time Objective** (**RTO**), and how much data can you afford to lose for your **Recovery Point Objective** (**RPO**)? What is the lifecycle of your data (for instance, if you have regulatory or operational constraints)? Do you have a predilection for a given technology (for instance, if your team is used to working with a specific database engine)?

Answering those questions will rapidly narrow down your choices. Perhaps you cannot easily isolate a single choice following your initial rationale; say you are left with two alternatives that could potentially both do the job (for instance, Amazon RDS PostgreSQL or Amazon Aurora PostgreSQL). In that case, simply benchmark them for your use cases. Don't take anything for granted: collect metrics using AWS services such as CloudWatch and **AWS X-Ray** to trace end-to-end performance, including your database transactions. You can easily deploy any of the previously mentioned database services to test them out on your most demanding use cases, collect some solid evidence of how they perform under stress, and validate the best choice.

When ensuring good database performance, you typically have two primary areas of focus:

- First, providing fast access to data for reads and/or writes

- Secondly, avoiding resource contention between reads and writes

If you need ultra-low latency, such as sub-millisecond data access, you have to consider storing the data, or part of it, in memory. The choice of solution for that will depend on whether you need it for the entire dataset or just a portion of it, for reads or for writes, without forgetting the previous questions that already narrowed your database choice based on the type of data, the type of queries, and so on.

There are essentially two caching patterns used commonly. The first is cache-aside, or lazy caching; it entails loading the cache when reading data. If the data is not in the cache, it gets pulled out of the data store and then copied into the cache. The other pattern is called write-through and loads the cache in the reverse order; when the data is written to the data store, the cache gets populated. For instance, as illustrated in the following diagram, ElastiCache can be used either in a cache-aside or write-through configuration, or both at the same time, depending on what makes more sense to meet your workload requirements.

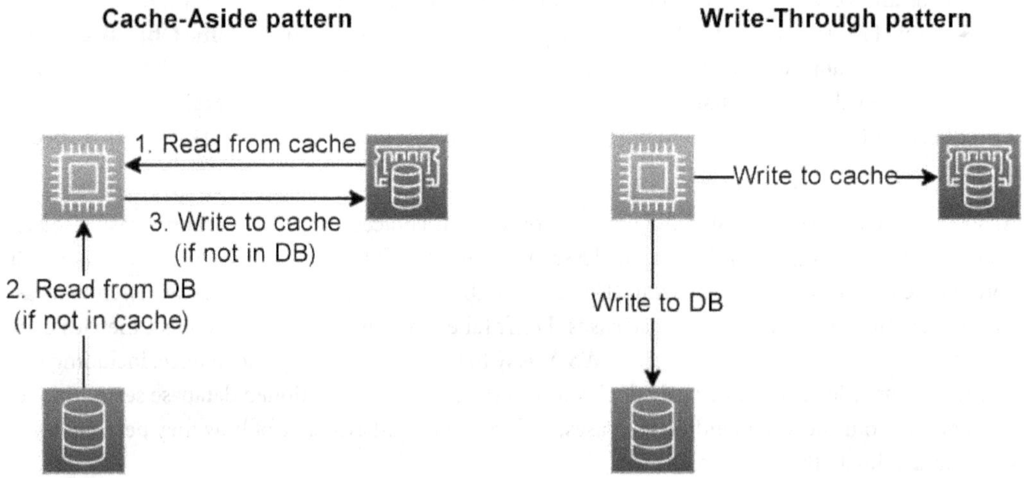

Figure 8.2: Cache patterns: cache-aside and write-through

Some database engines provide native caching mechanisms. For instance, Aurora has a built-in write-through cache to handle fast writes, and DynamoDB offers **DynamoDB Accelerator (DAX)**, which is an in-memory cache add-on that transparently speeds up database requests.

To reduce resource contention between reads and writes, you need to split the database resources serving reads, on the one hand, from those serving writes, on the other hand. Either you have a native mechanism to do it, such as read replicas for RDS or Aurora, for instance, or you need to rely on an external mechanism, such as a caching layer to serve reads.

Don't forget, you don't always need a Ferrari: sometimes a more mundane car with a less powerful engine is good enough for your needs so you can save yourself some money.

Network Selection

In a distributed environment, the performance you get from the network layer strongly influences the overall performance of your workload. It is thus paramount to select the right network architecture. As for the rest, everything starts from your workload requirements.

Consider your infrastructure requirements:

- Do you need connectivity from AWS to one or more on-premises locations?

- Do you need connectivity from AWS to another service provider location (for instance, if your workload depends on a SaaS solution or a service hosted by a third party)?

- Does your workload rely on connectivity to other AWS accounts?

For network connectivity between AWS and any other central location, such as your own on-premises infrastructure or a third-party hosted infrastructure, you have the choice between using the public internet or using a private connection (whether shared or dedicated). You can secure communication over the public internet by using a **VPN** connection on top of it, but if you need large network bandwidth and consistent throughput, **AWS Direct Connect** (**DX**) is the way to go. Whichever option you pick (VPN or DX), make sure to adequately size your network bandwidth to support data transfer in and out of AWS. In the case of VPN, remember that each VPN connection is limited to 1.25 Gbps, so if you need more bandwidth, you should consider using **Equal Cost Multi Path** (**ECMP**) on **AWS Transit Gateway** (**TGW**), setting up multiple parallel VPN connections and adding up their bandwidth. Now, if you plan to interconnect multiple VPCs to each other, use a highly scalable and fully managed service such as TGW to build a scalable and performant AWS network infrastructure. For more details on network connectivity, please refer to *Chapter 2, Designing Networks for Complex Organizations*.

Then, consider your end users:

- Where are the end users located and how do they access your workload?

- Will there be any external access to your workload, and, if so, will it be public access through the internet or private access through a corporate network only?

- How are the end users spread geographically? Are they located in one country or a limited geographical area, or are they all over the globe?

If your workload delivers content through a web frontend to end users across multiple geographies, consider leveraging CloudFront. Firstly, it will lower the latency they experience when accessing the content by caching it at the edge; secondly, it will also reduce the requests to your origin servers to get that content. If your end users are not geographically dispersed, they could still benefit from the reduced latency unless they are located in a country where AWS owns Regions, AZs, or Local Zones.

When your users are widespread but, for some reason, you prefer not to distribute your workload across the globe and instead want to run it in a single region, for instance, you have the option to leverage a service such as **AWS Global Accelerator** to optimize the network path. As illustrated in the following figure, Global Accelerator reduces the number of network hops needed on the internet by entering the AWS backbone at the closest edge location and then navigating to the Region where your workload is deployed and which is the closest to the end user location. It provides a set of static IP addresses (and yes, you can bring your own if you want to) that serve as entry points for your workload and are announced by multiple edge locations at the same time. Within your AWS environment, you can associate those IP addresses with your resources or endpoints deployed in any AWS Region, such as elastic load balancers (classic, ALB, or NLB), EC2 instances, or **Elastic IP Addresses (EIPs)**. Then, either you let Global Accelerator determine the optimal network path from the edge to a healthy resource, or you define your own custom routes, which can be useful if you need to maintain connectivity between a set of end users and the same group of AWS resources over multiple requests, for instance, for gaming or **Voice over IP (VoIP)** sessions.

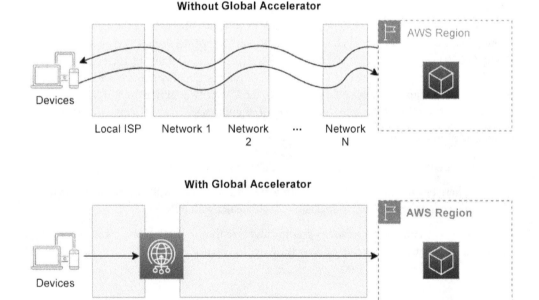

Figure 8.3: Side-by-side visualization: with and without AWS Global Accelerator

Similarly, Amazon Route 53, an AWS DNS service, can also help reduce latency with its built-in latency-based routing mechanism. When you deploy your workload in multiple Regions, latency-based routing is an efficient way to route traffic to your AWS resources, presenting the shortest network round-trip time from the end user's location. Route 53 uses its data on latency to reach the various Regions and then routes the end user's request to your AWS resources, such as load balancers (classic, ELB, or NLB), EC2 instances, or EIPs, in the region offering the lowest latency.

Now, if one of the features of your workload offers the ability to your end users to upload content to Amazon S3, they may benefit from the same acceleration provided to CloudFront customers using **Amazon S3 Transfer Acceleration** (**S3TA**). Similar to CloudFront, the further away your end users are from your S3 buckets, for instance, for end users far away from your buckets Region, the more S3TA will help. And the good thing is that you only pay for the transfers to S3 that are actually accelerated.

Remember the direct requirements your workload may have as provided below:

- What sort of processing does your workload perform (real-time, batch, or a combination)?

- What data throughput does your workload expect to handle (for ingesting and/or sending data)?

- Do you have any latency requirements to meet your end users' expectations (and provide a good user experience)?

If you plan to use EC2 instances for your workload, make sure to pick up instances that support the network throughput you expect. The recent generations of EC2 instances (from the fifth generation, for instance, M5) provide enhanced networking leveraging the AWS custom **Nitro card** and an **Elastic Network Adapter** (**ENA**). That allows some of the instances from the N series, such as M5n, to deliver up to 100 Gbps of network throughput. If your workload heavily relies on inter-node communication, **Elastic Fabric Adapter** (**EFA**) is a network interface available on some EC2 instances, for example on M5n, that allows low-latency inter-node communication. EFA enables HPC applications to efficiently run at a scale of hundreds or thousands of interconnected EC2 instances.

Similarly, if you rely on EBS volumes to store data, leverage EBS-optimized EC2 instances to reduce resource contention between EBS I/O and other network traffic. EBS optimization is enabled by default on recent EC2 instance types, since the fourth generation of instances (such as M4). It is also available on older instance types, although it is not enabled by default. If you want to make sure your instance is properly sized for your EBS I/O traffic, monitor the **EBSIOBalance%** and **EBSByteBalance%** metrics available in CloudWatch. If these metrics consistently stay low, you should consider increasing your instance size, because if you don't, EBS I/O traffic could be throttled. Inversely, if these metrics stay at 100% all the time, you may want to consider downsizing your instance. Note that this is only regarding the EBS I/O aspect; before downsizing an instance, you should also look at the overall picture, considering other factors such as CPU usage, memory usage, and network traffic to make sure you will still have enough of those resources after the size reduction.

In recent years, AWS has also added a number of options so that you can now deploy workloads supporting ultra-low latency on AWS. Local Zones were mentioned earlier. Local Zones are located close to areas with a large population or industry and let you run workloads closer to your end users on an infrastructure managed by AWS. Unlike AZs, Local Zones offer a reduced set of AWS services, including EC2, EBS, ECS, EKS, ALB, EMR, and RDS. A Local Zone is connected to a Region and lets you expand your VPCs in that Region and have subnets in the Local Zone as you do in the Region's AZs. You simply rely on the attached Region to access AWS services not supported in Local Zones. For more details on network connectivity, please refer to *Chapter 2, Designing Networks for Complex Organizations*.

However, you might need to consider a situation in which your end users do not have a Local Zone nearby or if you require additional services not supported yet by Local Zones? One option is using **AWS Outposts** to actually run AWS services and infrastructure on-premises. Outposts comes in various form factors: either an industry-standard 42U rack or 1U or 2U servers. The servers can be deployed standalone or mounted on industry-standard racks. Similar to Local Zones, Outposts lets you expand your VPCs on-premises and provides a subset of AWS services. You simply rely on the attached Region to access AWS services not supported on Outposts. It is a useful option when you need ultra-low latency for some applications, for instance, to provide your AWS resources with single-digit millisecond latency access to your systems on-premises. That gives you an extra opportunity to efficiently process data locally using AWS services but without actually moving that data into an AWS Region. It is also a valid option in other cases unrelated to performance, for instance, to enforce data residency in locations where you might be constrained by the local legislation or your industry regulators.

You have just been through the major building blocks constituting the foundations of any solution built on AWS. In particular, you have examined for each layer the rationale that should guide you to select the best fit for your workload among those components. You are now ready to learn how you can ensure you eventually reach your performance objectives.

Monitoring Performance

Part of the overall architecting for performance approach is to monitor how your resources actually perform. The idea is to do it not just once at the beginning, when you lay out your solution design, but continuously, to monitor how your workload performs over time and detect any deviation.

First, identify the metrics that are important for you to monitor. For instance, say that you set up an e-commerce application; it is essential for you to know the transaction throughput and its variation over time, but also to spot any I/O bottlenecks, the evolution of the request latency, and so on. This obviously varies from workload to workload and depends on the performance criteria that your workload must meet. Once you have identified these metrics, make sure that you collect and record them. Leverage a monitoring tool to assist you in this task: either native services from AWS such as CloudWatch, or any other third-party tool you prefer.

Then, you could create a dashboard with the relevant metrics. Define some **Key Performance Indicators (KPIs)** calculated using these metrics, then add them to your dashboard to help you understand how well your workload performs compared to its objectives. In the case of an e-commerce application, one such KPI could be the volume of sales, for instance. Then, create some alarms and notify the relevant people when an alarm is triggered. In some cases, you might be able to remediate automatically when an alarm is triggered; for instance, if an alarm identifies a need for more compute resources, you may be able to handle the situation by scaling said compute resources horizontally with their built-in autoscaling mechanisms.

Before testing your workload for performance, make sure to have all the aforementioned elements in place. Then, lay out your test plan to simulate real-life scenarios you expect to encounter. After that, go through your test plan and adjust your solution design based on your findings.

Remember to include tests that validate the end user experience when your workload runs under stress, performing real-user monitoring to spot issues that may affect that experience. You can again leverage CloudWatch RUM and other CloudWatch digital experience monitoring capabilities for that.

Then, routinely review the metrics and KPIs collected during normal functioning or during or after an event or incident occurs. These reviews will help you confirm the metrics that were essential to spot the issue and help you identify additional metrics that you should be monitoring to obtain a fuller picture and hopefully prevent a similar event or incident from taking place again.

So, at this stage, you have created your solution design and tested it against your workload performance objectives, your monitoring is in place, and you are actively collecting metrics and measuring KPIs. Is the task done, then? Well, almost, but not quite.

The following section discusses the last stage of designing for performance.

Reviewing and Adapting Your Solution

The last stage of the overall process consists of conducting regular performance reviews of your workload's behavior. For that, you simply need to apply the best practices that were covered in this and previous chapters.

First, you want to automate the creation of your AWS resources to speed up your release process. So, define your **Infrastructure as Code** (**IaC**), using CloudFormation, CDK, Terraform, or any other templating engine – whichever makes more sense in your case. Once most, if not all, of your AWS resources have been coded taking an IaC approach, testing variations of your solution design becomes almost effortless to the point that it becomes an incentive to try out new things to improve your workload performance.

Make sure that you collect all relevant metrics to establish good judgment of whether your current design is optimal or not. This inevitably requires factoring cost metrics in. Anyone can increase the performance of a workload without any consideration of costs, but as the frugality **Leadership Principle (LP)** of Amazon says, *"there are no extra points for growing budget size."* So, your objective is actually to do more with less, to perform better, and possibly optimize your costs at the same time. Therefore, make sure to factor cost metrics into your performance review and take these into account in your solution design and re-design exercises.

One additional element that needs to be taken into account is the launch of new services and features by AWS. Technology evolves fast and cloud technology evolves even faster. So, chances are that between the moment you produce your first solution design and the moment it gets released in production, AWS will have launched dozens of features, including some that may more efficiently address your workload requirements. How should you deal with AWS technology evolution? This is something important to include in your regular solution performance reviews. As an AWS solution architect, an essential part of your role is to keep up to date with AWS' new services and features. These updates should then feed your performance reviews with fresh ideas to proactively improve the status quo. If you don't do it, you will eventually build up technical debt in your workload. So, it is paramount to keep it in mind.

This review phase will greatly help you, especially if your workload performs poorly and misses its target.

Summary

This chapter started with the introduction of the AWS Well-Architected Framework's performance design principles. You looked at what it actually means to architect for performance and what considerations you should make. Some of the essential services and design patterns that should be part of your toolkit as a solution architect were discussed. The chapter then concluded by focusing on the importance of monitoring, reviewing, and actually adapting your solution.

The next chapter will dive deeper into workload deployment.

Further Reading

- The *Well-Architected Framework performance efficiency pillar*: https://packt.link/vxKIp

- Tips for right-sizing: https://packt.link/DE8af

- Amazon ECS vs Amazon EKS: making sense of container services: https://packt.link/dhPcj

- *Best Practices Design Patterns: Optimizing Amazon S3 Performance*: https://packt.link/i1sbP

- *Performance at Scale with Amazon ElastiCache*: https://packt.link/mVnbi

9

Establishing a Deployment Strategy

This chapter will focus on determining a deployment strategy to meet your business requirements. We will look at the various options offered by AWS for deploying and updating your solution.

Choosing the right deployment strategy is paramount to successfully manage your solution stack and meet your business requirements.

This chapter will cover the following topics:

- Deployment strategies
- AWS deployment services

Deployment Strategies

You might be wondering why you need a deployment strategy in the first place. A deployment strategy ensures that you can efficiently deploy new workloads and roll out new features or fixes for existing workloads, which is essential to any enterprise's change management process. The deployment strategy is the key element in your change management process, because deployment is the element that most affects the speed, agility, and efficiency of your overall change management process.

Nowadays, numerous enterprises have adopted a DevOps culture to advance their change management process. Their objectives are to deliver software faster to production, reduce time to market, and improve software quality. Furthermore, as part of these objectives, they emphasize achieving a high degree of automation for deployment. Deployment is part of an end-to-end software lifecycle management process that, in an ideal DevOps world, consists of **Continuous Integration (CI)** and **Continuous Delivery (CD)**, aka CI/CD.

CI involves improving software quality and delivering software updates faster. In this phase of the software lifecycle management, developers commit their code to a code repository, which then triggers a software build and a series of tests to verify that the expected quality is met. If the tests fail, the developer must address the identified issues. When the tests are passed, the process continues to the next stage, CD, which involves delivering new software releases into production faster and safer. Once the software has been built and tested during the CI stage, it enters the CD stage to be packaged and deployed into the relevant environments (for instance, Integration, **User Acceptance Testing (UAT)**, and Production environments). This is where deployment takes place.

Before you deploy a new software release, you must decide how you're going to proceed to actually roll it out in the target environment(s). First of all, do you intend to reuse any existing infrastructure for your workload, or do you consider your infrastructure to be immutable? DevOps advocates treating infrastructure as code (that is, taking an **Infrastructure as Code (IaC)** approach), so that you can manage infrastructure using the exact same tooling you used for software development, including development environments, source code repositories, testing tools, and so on.

Then, how do you want new releases of your application to be deployed? Do you want the new release to replace the pre-existing release *all at once* and *in-place*? Or do you want to progressively roll out new releases, updating your infrastructure in batches (aka *blue/green deployment*)? Alternatively, when you progressively roll out your new release, do you want to keep the pre-existing release up and running until you have verified that the new release works as expected (either *canary deployment* or *linear deployment*)?

How you answer the preceding questions will inevitably be influenced by your very own software development culture, which will vary from one organization to the next. In any case, answering these questions is the first step in determining your deployment strategy. For instance, you may opt to use immutable infrastructure with an IaC approach and then enforce canary deployment for every new software release in Production.

Once you have decided on a strategy, the next stage consists of deciding on the tooling to actually implement it.

AWS Deployment Services

Multiple AWS services can be involved in deployment, depending on the technology used and the degree of operational control you want to have.

The following subsections will walk you through the various solutions available with regard to controlling the deployment stage of the software lifecycle.

AWS OpsWorks

Let's start with **AWS OpsWorks**. OpsWorks is an application configuration management service that comes in three flavors: **AWS OpsWorks for Chef Automate**, **AWS OpsWorks for Puppet Enterprise**, and **AWS OpsWorks Stacks**. The first two OpsWorks flavors provide either a Chef Automate or Puppet Enterprise environment, managing the underlying infrastructure to run Chef Automate or Puppet Enterprise resources on your behalf, including patching, updating, and backing up the servers. For those of you who do not require the premium features of Chef Automate or Puppet Enterprise, OpsWorks Stacks offers a simplified application configuration management system that helps to simplify the configuration management of the AWS resources supporting your workload, including infrastructure such as servers, load balancers, databases, and so on. Both OpsWorks for Chef Automate and OpsWorks Stacks let you leverage Chef recipes. The former manages a fully fledged Chef server on your behalf and provides the premium features of Chef Automate. The latter does not deliver a Chef server environment but performs some of the work of a Chef server on your behalf.

OpsWorks Stacks is the legacy OpsWorks solution. It lets you define your workload in terms of stacks. A stack is meant to represent a group of resources that serve a common purpose and can group various types of resources. For instance, a stack could represent a web application, including the load balancers, the application servers, and the database server. You could also define multiple stacks for the same web application: one stack corresponding to the integration environment, another one for the user acceptance environment, and a separate stack for the production environment. Stacks are, in turn, composed of layers. A stack contains at least one layer. A layer represents a type of component within a stack. Therefore, coming back to the web application example, you could have one layer for the load balancers, one layer for the application servers, and one layer for the database server. OpsWorks Stacks can contain layers of compute resources running on AWS, such as **Amazon Elastic Container Service** (**ECS**) clusters and **Amazon Elastic Compute Cloud** (**EC2**) instances, Linux, or Windows. It can also handle layers with compute resources running outside of AWS, provided that they use one of the Linux distributions OpsWorks Stacks supports. OpsWorks Stacks has, however, some constraints. For instance, at the time of writing, it does not support the most recent Linux or Windows Server releases, which might be okay if you deal with legacy systems but less so if you prefer to build your workload on the most recent operating systems' releases. Essentially, nowadays, it is preferable to work with the more recent flavors of OpsWorks, using either Chef Automate or Puppet Enterprise depending on what your preferred configuration management tool is. Both services allow you to configure, deploy, and manage the lifecycle of your software solutions. Both of them offer a starter kit so that you can take your workload for a test drive using a set of sample (*Chef*) recipes or (*Puppet*) modules.

But the OpsWorks family is only one of several options for deploying software on AWS.

AWS Elastic Beanstalk

Now you are ready to take a look at another commonly used option, especially for newcomers on AWS, which is **AWS Elastic Beanstalk**. Beanstalk is an application management platform that targets web applications; it is commonly referred to as a **Platform-as-a-Service** (**PaaS**) solution. It is a great option, especially for newcomers, because it hides a lot of the underlying complexity of setting up a complete environment for a web application and, as such, it doesn't require you to have a deep knowledge of cloud computing. At the same time, it supports a broad variety of runtime environments, such as Java, .NET, Python, Node.js, PHP, Ruby, Go, and Docker. So, if your application is developed in a language for which Beanstalk does not directly offer a runtime environment, you can package it with all its dependencies in Docker containers, which Beanstalk can handle.

Now, even if Beanstalk makes your life easier by abstracting away much of the complexity of the underlying infrastructure, it still allows you as much control as you need over your infrastructure, should you prefer to manage parts of it yourself. Beanstalk relies on other AWS services, such as EC2, **AWS Auto Scaling**, **Elastic Load Balancing** (**ELB**), **Amazon Relational Database Service** (**RDS**), and **Amazon S3**, to provision an environment that supports your application.

To illustrate how it works, start with a simple Hello World sample application. You can grab the application from GitHub, for instance, using the following **Command-Line Interface** (**CLI**) in a shell terminal window:

```
$ git clone https://github.com/aws-samples/aws-elastic-beanstalk-
express-js-sample.git
```

> **Note**
>
> Please note that these commands will work in a Bash shell terminal on macOS or Linux, so you will need to adapt them if you use a different shell or even a different OS.

Then change to the newly created directory where your sample application code now resides:

```
$ cd aws-elastic-beanstalk-express-js-sample/
```

The following steps make the assumption that you have already installed the Beanstalk CLI. If you haven't, please refer to the Beanstalk CLI documentation at `https://packt.link/EcLMT`.

The immediate next step is to initialize the Beanstalk environment for this application. You can either provide the required parameters or use interactive mode to feed them to Beanstalk. For the latter, you can simply run the following command:

```
$ eb init
```

The Beanstalk CLI will then guide you by asking questions to define the environment where your application will run and will provide default values for when you're not sure what to choose. You will be asked to specify things such as these:

- The AWS Region where you want to deploy your application

- The application you want to use (for a new application, as in our case, you want to select `Create new application` and provide a name for it)

- The runtime engine Beanstalk should use (in our present case, the latest release of Node.js will do just fine)

- The **AWS CodeCommit** repository that Beanstalk will use to manage your application code lifecycle

Once the `eb init` command completes, your application is ready to be created. You can proceed to the next stage, which is the actual creation of the environment on Beanstalk and the deployment of your application to the newly created Beanstalk environment.

First, make sure you have an EC2 instance profile that can be used for your Beanstalk environment. If you're wondering what that is or how to create an EC2 instance profile, please refer to *Chapter 5, Determining Security Requirements and Controls*. The sample application simply displays `Hello World` on a web page, so you do not need to provide any specific permission to the instance role assigned to your instance profile. Now, after your instance profile has been defined, you can proceed to create your Beanstalk environment and deploy your application. Carry on with this next step:

```
$ eb create --instance_profile aws-elasticbeanstalk-sample-role
```

This command instructs Beanstalk to do the following:

1. Create the necessary resources to support your application.
2. Deploy your application once those resources are in place.

Beanstalk will again ask you a couple of questions to clarify what sort of infrastructure you want to create. You will provide additional information such as the following:

- The name for your Beanstalk environment, so that you can look it up later on (for instance, in the AWS Management Console)

- A DNS CNAME that Beanstalk will use to create the necessary DNS records with **Amazon Route53** so that your web application can be reached

- The type of ELB load balancer that you would like to use (**Application Load Balancer** or **Network Load Balancer**)

Beanstalk will then create all the relevant resources on your behalf: the ELB load balancer, the EC2 instance, the Auto Scaling group (the mechanism for scaling your infrastructure), and security groups (virtual firewalls to protect your resources and control the traffic that can reach them). Once the environment creation is complete, Beanstalk will install and configure the engine runtime you specified and deploy your application on top of it.

And that's it! From this point on, your application is up and running. As you can see, it's pretty straightforward and doesn't require any prior knowledge of AWS besides a high-level understanding and the ability to look up the AWS IAM documentation to create an EC2 instance profile. You could even have a simpler experience if you use the AWS Management Console (you then don't need to worry about the EC2 instance profile) to do everything we just did via the command line.

When you need to make further developments and iterate over multiple versions of your application, Beanstalk offers multiple options to release updates for your application. These options are discussed in detail here:

- **All-at-once deployment**: This option consists of replacing your existing application with the new version on all the currently deployed EC2 instances.

- **Rolling deployment**: With this second option, Beanstalk does a rolling update by progressively deploying your new version on a subset of your environment, updating a batch of EC2 instances at a time. With this option, the old and new releases of your application co-exist for the time the rolling update takes, so a portion of the incoming requests will still be served by your existing application. Before updating a batch of EC2 instances with the new release of your application, Beanstalk will detach them from the load balancer. You can enable connection draining before this happens to avoid breaking in-flight transactions. After deploying the new release on a batch of EC2 instances, if the release passes health checks successfully within the allocated time, Beanstalk carries on with the next batch of instances. In the case of health check failures or timeouts, and unless you specified ignoring health checks, Beanstalk will declare the deployment as failed and stop rolling out the new application. This may have the undesired effect of leaving your environment partly running the new release of your application and partly still running the old release. It is then up to you to carry out a new deployment to fix this situation in whichever way you see fit, for instance, by re-deploying the old release.

- **Rolling deployment with an additional batch**: As you've just seen, during a rolling deployment, some instances are taken offline to proceed with the new release's deployment, which means that your workload stops operating at full capacity. This might be an issue depending on the incoming traffic hitting it at that moment in time. This is why Beanstalk made a third option available, namely, `Rolling deployment with an additional batch`. This option lets you spin up a batch of fresh instances before taking any of your existing instances offline for updating. Once the deployment is completed, Beanstalk will tear down the extra capacity.

- **Immutable deployment**: With this option, Beanstalk adds brand-new EC2 instances in a separate Auto Scaling group where it deploys the new release of your application. Your old release, including all its infrastructure resources, is left intact until your new release successfully passes health checks. Only then does Beanstalk stop the old release and tear down all its infrastructure resources (such as EC2 instances and their Auto Scaling group). If the health checks fail, then the old release of your application and all of its infrastructure remain in place as if nothing happened, and the freshly created resources supporting the new release are terminated.

- **Traffic-splitting deployment**: Beanstalk provides yet a fifth deployment option, namely, `Traffic-splitting deployment`, which, similar to immutable deployment, launches a set of new instances wherein you can deploy the new release. Once the new release is deployed, only a portion of the incoming traffic is diverted to these new instances for the evaluation period that you specify. If, at the end of the evaluation period, the new instances remain healthy, Beanstalk will send all traffic to them and terminate the old instances. If they're unhealthy, all traffic moves back to the old instances and the new instances are terminated. This deployment option lets you avoid any traffic interruption.

If you don't explicitly specify a deployment strategy, Beanstalk defaults to an all-at-once strategy. However, if you deploy your application via the CLI, as we just did, Beanstalk defaults to a rolling update strategy.

AWS App Runner

AWS App Runner is an AWS service that simplifies the deployment of web applications and APIs on AWS. App Runner goes a step further than Elastic Beanstalk by offering a software deployment environment fully managed by AWS. It does not just manage the EC2 instances supporting your application but also manages the application language runtimes, the load balancers and auto-scaling capability, and the deployment pipeline. It also provides observability from the get-go without further effort being required from your DevOps teams. The objective of App Runner is to make it easier for developers to get their web application or API deployed on AWS without having to worry about the plumbing. Furthermore, because App Runner abstracts a lot of the underlying infrastructure away, it surfaces a limited set of options to pick from to customize your application environment. For instance, you can let App Runner manage your CI/CD pipeline for you (that's the automatic deployment option) or you can manage it yourself (that's the manual deployment option). It is a much more limited choice compared to the multiple deployment options we had with Elastic Beanstalk (all-at-once deployment, rolling deployment, immutable deployment, and a few more).

So, when should you use App Runner and when should you use Elastic Beanstalk?

Well, it mostly boils down to the degree of control you'd like to have over your application resources. Elastic Beanstalk currently supports various types of applications, including not only web applications but also other types of worker applications, while App Runner only supports web applications at the time of writing.

Elastic Beanstalk also lets you customize more of the supporting infrastructure since those resources run in your account and you have full control over them (VMs, load balancers, etc.). App Runner completely hides away the infrastructure resources, which are fully managed by AWS, leveraging serverless technology.

Additionally, with Elastic Beanstalk you pay separately for the various AWS resources your application uses (compute, storage, etc.). With App Runner, you also pay for the resources used by your application, but the billing is integrated within the service. So, you can more easily visualize the exact costs associated with your applications deployed using App Runner.

AWS CodeDeploy

Another option to deploy applications is to use **AWS CodeDeploy**. CodeDeploy lets you automate the deployment of your applications, whether they are deployed using EC2 instances, on-premises servers, containers, or **AWS Lambda** functions. It can collect application artifacts from S3 buckets or directly from your source code repositories. It currently supports code repositories on AWS Code Commit, GitHub, and Bitbucket. CodeDeploy is often used in combination with the other members of the AWS Code family of services (**AWS CodeCommit**, **AWS CodeBuild**, and **AWS CodePipeline**) to form an end-to-end CI/CD solution. It also integrates with configuration management tools such as Ansible, Chef, Puppet, and Salt.

Deployment Groups

CodeDeploy is meant to be used with existing compute resources, using which it can deploy application artifacts. Note that it does not create compute resources on your behalf. With that in mind, CodeDeploy organizes compute resources in deployment groups. When deploying application components to ECS, Lambda, and EC2/on-premises servers, deployment groups specify the various elements required by the target compute service as well as how a new version of the application components should be rolled out. For instance, a deployment group for ECS would specify when to re-route traffic to replacement tasks and when to terminate the original tasks after the successful deployment of an ECS application release. When deploying to Lambda, a deployment group would specify how to re-route traffic to the new version of a Lambda function. Optionally, deployment groups may also specify alarms (using **Amazon CloudWatch**), notification triggers (delivered by **Amazon SNS**), and automatic rollbacks. For more details on this, please consult the CodeDeploy documentation at `https://packt. link/h36GV`.

Deployment Configurations

CodeDeploy also needs to know about the deployment mechanism you want to follow and how to route traffic to the newly deployed application release. This is defined through the deployment configuration. This deployment configuration varies based on the compute resources you want to use.

When deploying to EC2/on-premises instances, you can either choose between the three pre-defined deployment configurations (all-at-once, half-at-a-time, and one-at-a-time) or specify your own custom configuration. If you don't specify any deployment configuration, CodeDeploy defaults to the one-at-a-time deployment configuration. Now, for each of the pre-defined configuration options, you can also opt between an in-place deployment or a blue-green deployment. It is important to select the right configuration option for your needs as the success or failure of the deployment is defined differently according to different configuration options, as discussed here:

- In an *all-at-once deployment* configuration, CodeDeploy deploys your new application release to all EC2/on-premises instances simultaneously. In this case, the deployment is considered successful if the new application release is deployed successfully to all of the instances and is considered failed otherwise.

- In the case of an *in-place deployment*, the deployment is considered successful if the new application release is deployed successfully to at least one instance and is considered failed otherwise.

- In the case of a *blue-green deployment*, the deployment is considered successful if traffic can be re-routed to at least one instance in the replacement environment.

- In the *half-at-a-time configuration*, CodeDeploy deploys your new application release to half of the EC2/on-premises instances simultaneously. In the case of half-at-a-time deployment, the deployment is considered successful if the new application release is deployed successfully to at least half of the instances and is considered failed otherwise.

- Finally, in a *one-at-a-time configuration*, CodeDeploy deploys your new application release to one EC2/on-premises instance at a time. A one-at-a-time deployment is deemed successful if the new application release is successfully deployed to each individual instance, one after the other.

Note that there is an exception when deploying to multiple instances: CodeDeploy considers that the deployment is still successful if the new application release fails to deploy to the last instance after having been successfully deployed to all the other instances before. When using a custom deployment configuration, you define how deployment takes place and what success looks like. The following command-line instruction shows an example of what you can do:

```
$ aws deploy create-deployment-config --deployment-config-
name EightyPercentHealthy --minimum-healthy-hosts type=FLEET_
PERCENT,value=80
```

This example creates a custom deployment configuration called `EightyPercentHealthy` requiring that at least 80% percent of the target EC2/on-premises instances stay healthy upon deployment.

When deploying to ECS, CodeDeploy always uses a blue-green deployment approach (unlike EC2/on-premises, it does not offer in-place deployment). Your deployment configuration then determines how traffic gets shifted to the new application release. You essentially choose between a set of pre-defined deployment configurations (either canary, linear, or all-at-once) or specify your own custom canary or linear configuration. When using one of the pre-defined canary deployment configurations, the traffic gets shifted in two increments: 10 percent of the traffic is initially sent to your new application release, and the remaining 90 percent is then shifted over after either 5 or 15 minutes. There are also two pre-defined variants of the linear deployment configuration: 10 percent of the traffic is shifted to the new application release either every minute or every 5 minutes (until all traffic has been directed to the new release).

When deploying to Lambda, CodeDeploy also always uses a blue-green deployment approach (similarly to ECS, it does not offer in-place deployment). The deployment configuration options are also similar to ECS: CodeDeploy deploys to Lambda following either a canary, linear, or all-at-once deployment. However, CodeDeploy offers many more pre-defined configuration options for Lambda. For canary deployment, CodeDeploy preset deployment configurations shift 10 percent of the traffic in a first increment and then the remaining 90 percent after either 5, 10, 15, or 30 minutes. For linear deployment, CodeDeploy preset deployment configurations shift 10 percent of the traffic every 1, 2, 3, or 10 minutes.

Application Specification

So far, you have specified the AWS resources supporting the deployment (deployment group) and also instructed CodeDeploy on how to deploy the new application release and how to shift traffic to it (deployment configuration). There is one major element left to define, and that's the application specification to let CodeDeploy know what to deploy. This information is passed to CodeDeploy through an application specification file (or AppSpec file), written either in **Yet Another Markup Language** (**YAML**) for deployments on EC2/on-premises, ECS, and Lambda, or **JavaScript Object Notation** (**JSON**) for ECS and Lambda deployments only.

The AppSpec file contains details about the application to be deployed. In the case of deployments on EC2/on-premises instances, it provides a mapping between source files in the new application release and their destination on the target instances. It also defines permissions to apply to the deployed files. Furthermore, it can additionally specify a set of scripts to perform validation tests during the deployment process.

In the case of ECS and Lambda, the AppSpec file provides, respectively, the name of the ECS service as well as the container name and port used to direct traffic to the new task set, and the Lambda function version to deploy. It can also specify a set of Lambda functions to be used to perform validation tests at various stages of the deployment process.

The validation scripts or functions, which are defined in the `Hooks` section of the AppSpec file, will be executed when the corresponding deployment lifecycle event is triggered, such as "before install" or "after install" events. Lifecycle events vary sensibly according to the target compute platform. Although ECS and Lambda share the same set of deployment lifecycle events, deployment on EC2/on-premises instances provides a totally different set. For more details on the available lifecycle events, please consult the CodeDeploy documentation at `https://packt.link/NJhbu`.

Here is an example of an AppSpec file to deploy a Lambda function:

```
version: 0.0
Resources:
  - myAwsomeLambdaFunction:
      Type: AWS::Lambda::Function
      Properties:
        Name: "myAwsomeLambdaFunction"
        Alias: "myAwsomeLambdaFunctionAlias"
        CurrentVersion: "1"
        TargetVersion: "2"
Hooks:
  - BeforeAllowTraffic: "LambdaFunctionToValidateBeforeTrafficShift"
  - AfterAllowTraffic: "LambdaFunctionToValidateAfterTrafficShift"
```

As you can see, the content of the file is self-explanatory. It is composed of three sections: version, resources, and hooks. The `version` section is reserved for CodeDeploy for future use and must be set to `0.0` for now. The `Resources` section describes the Lambda function to be deployed, while the `Hooks` section lists the optional lifecycle events to handle and the associated test functions.

CodeDeploy really shines when used in combination with other members of the AWS Code family (such as CodePipeline) and/or third-party solutions, such as source code repositories or configuration management tools. It also handles automatic rollback on your behalf if a deployment fails or if a specific monitoring alarm is triggered (provided that you have defined such an alarm with CloudWatch). So, it is an extremely helpful tool for your delivery process when you deploy software to EC2/on-premises instances, ECS, or Lambda.

AWS CloudFormation

Another option for deployment is to use **AWS CloudFormation**. CloudFormation is a service that lets you provision all sorts of AWS resources, offering an IaC approach. You define your resources in one or multiple templates, written in either YAML or JSON. YAML is typically easier and less verbose, compared to JSON, for humans to read and write. CloudFormation can be used to deploy infrastructure as well as application resources. Therefore, you can create infrastructure resources, such as VPCs, subnets, route tables, IAM roles, security groups, and more, and also deploy EC2 instances, create S3 buckets, instantiate RDS databases, and work with many other resources. For instance, you can also deploy a Lambda function or an application on Elastic Beanstalk using CloudFormation templates. For the complete list of resources supported by CloudFormation, please consult the AWS documentation at `https://packt.link/6cofc`. Without getting into too much detail and the complete anatomy of a CloudFormation template, we will simply illustrate how resources are actually declared through a concrete example. Take the example of a simple application on Elastic Beanstalk, similar to the sample application we previously created.

Your CloudFormation template first needs to contain the declaration of your Beanstalk environment, for instance, something like the following:

```
SampleEnvironment:
  Type: AWS::ElasticBeanstalk::Environment
  Properties:
    Description: AWS Elastic Beanstalk Environment running a sample
Node.js Application
    ApplicationName:
      Ref: SampleApplication
    TemplateName:
      Ref: SampleConfigurationTemplate
    VersionLabel:
      Ref: SampleApplicationVersion
```

In the preceding YAML snippet, you can see that the Beanstalk environment definition references additional resources, highlighted in the code. Your CloudFormation template then also needs to include the definition of those referenced resources, that is, the Beanstalk application, the application version, and the application configuration.

The definitions of these resources would look something like this:

```
SampleApplication:
  Type: AWS::ElasticBeanstalk::Application
  Properties:
    Description: AWS Elastic Beanstalk Sample Node.js Application
SampleApplicationVersion:
  Type: AWS::ElasticBeanstalk::ApplicationVersion
  Properties:
```

```
      Description: Version 1.0
      ApplicationName:
        Ref: SampleApplication
      SourceBundle:
        S3Bucket:
          Fn::Join:
          - "-"
          - - elasticbeanstalk-samples
            - Ref: AWS::Region
        S3Key: nodejs-sample.zip
  SampleConfigurationTemplate:
    Type: AWS::ElasticBeanstalk::ConfigurationTemplate
    Properties:
      ApplicationName:
        Ref: SampleApplication
      Description: SSH access to Node.JS Application
      SolutionStackName: 64bit Amazon Linux 2018.03 v4.7.1 running
  Node.js
      OptionSettings:
      - Namespace: aws:autoscaling:launchconfiguration
        OptionName: EC2KeyName
        Value:
          Ref: KeyName
      - Namespace: aws:autoscaling:launchconfiguration
        OptionName: IamInstanceProfile
        Value:
          Ref: WebServerInstanceProfile
```

This is essentially all you need to define your simple Beanstalk application using CloudFormation.

Note that some additional resource definitions have been omitted from the preceding YAML template for the sake of brevity. For a complete and fully functional template, please check out the various examples included in the CloudFormation documentation at https://packt.link/QLYFJ.

CloudFormation templates are deployed in your AWS environment as a stack, using either the AWS Management Console, the AWS CLI, or CloudFormation APIs. Once a CloudFormation stack is deployed, you manage the lifecycle of its resources by updating the stack. Before committing any changes to the stack, you can generate a change set that lets you assess the impact that your changes will have on existing resources. You can then decide whether to carry on with the stack update or amend your changes as needed.

CloudFormation is a vast topic and diving deep into details goes beyond the scope of this book. However, this book will focus on a few points of importance for a solutions architect to make informed design decisions.

Organizing Your Stacks

First of all, for complex environments, it is recommended to split the definition of your environment resources across multiple CloudFormation templates. But how should you split them? Best practice is to organize your stacks according to the lifecycle and the ownership of the AWS resources they define. For instance, suppose you need to manage multiple layers of resources such as infrastructure, platform, and/or application layers. Depending on your organization's governance model, these resources may be owned and managed by different teams. An infra team may be responsible for infrastructure resources while a DevOps team may be in charge of application resources. Because of this responsibility split, chances are that the lifecycles of infrastructure and application resources will differ. It would then be wise to use separate CloudFormation templates between infrastructure resources on the one hand and application resources on the other. For instance, you could have one template managing the infrastructure resources, such as the VPCs and subnets, network routing tables, and so on that are handled by the infrastructure team, and another template defining application resources, such as the S3 buckets, databases, Lambda functions, APIs, and so on that are handled by the DevOps team. Feel free to refine your model and split the templates in a more granular manner depending on your own organization. For instance, application resources might be split further down the line; for instance, perhaps databases might be managed by a database administrators team while APIs and Lambda functions might be handled by an application development team.

Does this sound familiar to you? Well, probably because it follows the approach of a layered architecture, where multiple layers of components are stacked one on top of the other to compose your solution architecture. Think of a **Service-Oriented Architecture (SOA)** or **micro-service architecture** where each service has its own lifecycle and could represent a stack, that is, a CloudFormation template.

In this regard, CloudFormation offers features to let you reference resources from other stacks in your own stack. The first mechanism consists of exporting output values in your CloudFormation template (think, for instance, of a VPC or a subnet ID), using the `Export` field in the `Output` section of the template. Other templates can then import these exported output values, using the `Fn::ImportValue` function, and use them as if those values were their own resources. Please note that this imposes a constraint on the stack exporting the output values: that stack cannot modify or delete the exported output values until all the imports of those values have been removed.

The second mechanism relies on nested stacks. A nested stack is a stack that is created (and referenced) inside another stack, creating a parent-child relationship between the two stacks. The output values of a nested stack can be directly accessed by the parent stack and indirectly accessed by other nested stacks (the parent stack can pass the output values of one nested stack to other nested stacks using stack parameters).

The difference between nested stacks and completely independent stacks is that nested stacks are typically managed and deployed from the parent stack. So, using one mechanism or the other will be dictated by the relation and lifecycle of the stacks.

> **Organize Your Stacks**
>
> Don't put everything in a single CloudFormation template to manage your AWS resources. This is especially true for large and complex environments. Split your templates and organize them by ownership (one owner = one template, at minimum) and lifecycle (for instance, one SOA-like service = one template).

Reusing Common Patterns

As your environment expands on AWS, you may identify a need to reuse the same resource configurations over and over from one stack to the next across teams or projects. For instance, you may want to implement best practices across various projects by enforcing some specific guidelines in their infrastructure resource configurations, whether they are for network, storage, or compute resources. CloudFormation modules let you do just that. You can package specific resource configurations reflecting your best practices into modules and have other CloudFormation stacks reference these modules. This is quite useful for reusability as well as enforcing standards and best practices across multiple teams or projects. Moreover, it is also beneficial to improve traceability as CloudFormation will retain knowledge of which resources in a given stack come from a particular module. This approach also helps maintain current best practices over time since you register modules in the CloudFormation registry and maintain their lifecycle, which includes versioning. You also decide the AWS Regions and accounts in which each module is made available. Check out the *Further Reading* section at the end of this chapter if you're interested in finding out more about how to reuse patterns at scale within your organization.

Maintaining IaC Resources

As has been emphasized many times in this book, infrastructure is code—well, at least in the cloud. So, your CloudFormation templates, modules, and all other artifacts that you use to define resources using IaC should be maintained exactly like you maintain application code. This means storing IaC code in a code repository (CodeCommit, GitHub, etc.) to properly manage its lifecycle as well as handling the rest of the lifecycle using CI/CD best practices.

Furthermore, it's also essential to manage your stack resources through CloudFormation. Avoid, as much as possible, making changes to stack resources deployed outside of CloudFormation. After all, if you make the effort to codify your best practices and automate your AWS resources deployment using IaC, it's not to see them spoiled with potentially harmful manual changes. If such manual changes occur, you will inevitably end up with a configuration drift between what your stack says should be deployed and what is actually deployed in your AWS environment, and reconciling the two can be quite grueling. Therefore, it is advised that you avoid making manual updates on stack resources. What is the best practice to update such resources, then? Simply use CloudFormation. First, you should make the changes in the template(s) or module(s) behind your stack and then create a change set to evaluate and verify the actual changes that will be made to your stack if you ask CloudFormation to commit those changes.

Never directly run an update to an existing CloudFormation stack—remember to use change sets. You may sometimes be surprised by the impact a template change has on your deployed stack resources. For instance, making what looks like an innocuous name change to an RDS database may lead to unrecoverable data loss if you forget to make a database backup just before rolling out the change; this is because CloudFormation will simply create a new database and tear down the old one. Once you're happy with the changes documented in your change set, you can apply the change set to your stack, which will roll out the changes to your resources.

> **Use change sets...Really do**
>
> Don't deploy CloudFormation updates directly without first using change sets, unless you're okay with potential data losses and breaking changes to your stack resources. It's better to be safe than sorry: remember to use change sets.

Scaling with CloudFormation StackSets

Imagine you have created your AWS resources using CloudFormation templates only to realize some time later that you actually need to roll out and manage the same stack in multiple accounts across multiple AWS Regions.

If the stacks were to be managed independently of each other, creating them separately one by one would make sense. However, since in this case you want to manage them centrally, it doesn't. This is where StackSets comes in handy. StackSets allows you to manage and roll out the same CloudFormation template across multiple accounts and in multiple Regions. It also lets you specify how many accounts the deployment should be applied to concurrently. Therefore, you can decide whether you prefer to maximize the speed of deployment or to be more conservative and roll out one at a time, possibly starting with the lowest-impact region, if any, to measure impact first.

Another important aspect of StackSets is security. Deploying a stack set means having the necessary permissions to roll out a CloudFormation template in the target accounts across the target Regions. You can naturally handle this yourself with self-managed permissions, creating an IAM role with enough permissions to let CloudFormation operate. However, it becomes particularly interesting when you manage accounts in your organization using AWS Organizations. CloudFormation is integrated with AWS Organizations, which allows you to leverage service-managed permissions, provided that you have enabled all features in AWS Organizations (and not consolidated billing only). When all features are enabled, CloudFormation creates the necessary roles for you in your organization's accounts so you can more easily roll out stack sets across your organization.

CloudFormation is a vast topic and covering it in detail is beyond the scope of this book. Please refer to the resources listed in the *Further Reading* section of this chapter to learn more. You've looked at the general concepts around cloud deployment, along with some of the ideas you should be concerned with when designing a deployment. Now we will look at how we can use AWS to define IaC using the AWS Cloud Development Kit.

The AWS Cloud Development Kit

The **AWS Cloud Development Kit** (**CDK**) is an open source framework created by AWS to allow cloud application developers to leverage widespread programming languages, such as Java, Python, TypeScript, and more, to create AWS resources using an IaC approach. The AWS CDK uses CloudFormation behind the scenes and translates code written in Java, Python, TypeScript, and so on into CloudFormation templates.

The AWS CDK provides a higher-level programming language interface, abstracting away CloudFormation YAML and JSON templates. Its basic building block is called a **construct**. A construct defines an AWS component that maps to one or more AWS resource(s). The AWS CDK comes up with a library of constructs, called the AWS Construct Library, which provides constructs for many AWS resources out of the box. You will, for instance, find constructs to define an S3 bucket, a DynamoDB table, and many more. Constructs can also be more complex and can group together multiple AWS resources, for instance, an S3 bucket and an EC2 instance, to provide a higher level of abstraction for resources frequently used by your applications and to simplify your deployment process.

Construct Hub, which can be found at `https://packt.link/M5DwN`, provides a centralized repository for CDK constructs offered by AWS, various providers, and the CDK community of users.

Take a look at the following simple example that highlights the benefits of using the CDK. It illustrates the creation of an S3 bucket using the AWS CDK in TypeScript:

```
import { App, Stack, StackProps } from 'aws-cdk-lib';
import * as s3 from 'aws-cdk-lib/aws-s3';
class MyFirstCdkStack extends Stack {
  constructor(scope: App, id: string, props?: StackProps) {
    super(scope, id, props);
    new s3.Bucket(this, 'MyFirstBucket', {
      versioned: true
    });
  }
}
const app = new App();
new MyFirstCdkStack(app, "MyFirstCdkStack");
```

> **Note**
>
> To run this code, it is a prerequisite to have the CDK CLI installed on your terminal.

First, as illustrated in the preceding example, creating resources with the CDK is usually concise and much less verbose than with CloudFormation and its templates in YAML or JSON. This is due to the fact that the CDK assumes many of the default parameters that you normally have to specify when using CloudFormation. This is something that you can easily verify by letting the CDK generate the supporting CloudFormation template—that's the synthesization step (`cdk synth` in the CLI). The CDK will generate a CloudFormation JSON template that will be used to deploy your resources. As you will see in this case, it is not unusual for the resulting CloudFormation template to be 10 times as big as the piece of code that you actually wrote.

Second, and most important for them, developers can now create AWS resources using a programming language they're familiar with. This makes their life so much easier compared to writing CloudFormation templates using either YAML or JSON.

Additionally, for those familiar with or already using **HashiCorp Terraform**, there also exists a **Cloud Development Kit for Terraform** (**CDKTF**). The CDKTF lets you access the Terraform ecosystem without having to learn **HashiCorp Configuration Language** (**HCL**).

Amazon Elastic Container Service (ECS)

Amazon ECS is AWS's native and fully managed container orchestration service. ECS integrates with core AWS services such as CloudWatch, Auto Scaling, and IAM to provide a seamless experience for monitoring, scaling, and load balancing your containerized applications. It lets you deploy Docker containers without having to worry about installing, operating, or scaling the container management infrastructure, that is, the control plane. It supports three types of compute engines, or *launch types*, for deploying containers: EC2 instances, **AWS Fargate**, and external.

EC2 allows you to run and manage Docker containers on a cluster of Amazon EC2 instances that you manage. With this option, you have control over the underlying infrastructure, giving you the flexibility to choose the instance types, manage the cluster's scaling policies, and configure the networking.

Fargate offers a serverless experience to container developers by managing the underlying AWS infrastructure on their behalf. The external launch type lets you deploy containerized applications on ECS on infrastructure that you run on an on-premises server or **Virtual Machine** (**VM**).

ECS groups containers in tasks. A task is simply a unit of deployment for containers and can consist of multiple different containers (up to 10 at the time of writing). You would typically combine containers that are closely related into a single task. However, it is good practice to split an application into multiple tasks, each representing a component of your application.

In terms of infrastructure, when working with ECS, you start by creating a cluster, that is, a logical grouping of your application components. You then deploy your application components onto the cluster using tasks. ECS also has two important additional concepts:

- **Service**: ECS services ensure that tasks run as expected and are scheduled, stopped, and restarted when needed.

- **Capacity provider**: ECS capacity providers manage the infrastructure supporting the tasks and, in particular, handle scaling and manage placement constraints, if any.

When you want to actually deploy your application components onto ECS, you have several options, called deployment types, to roll out new releases of your ECS services. ECS currently offers three deployment types—rolling update, blue/green, and external.

In the case of the rolling update deployment type, the ECS service scheduler replaces the currently running tasks with new tasks. You control how many tasks ECS adds or removes through the deployment configuration. What happens behind the scenes is that the ECS service scheduler launches enough new tasks to reach the desired count that you specified in your deployment configuration (or when the service was defined). It is worth noting that the rolling update deployment type also allows you to optionally use circuit breaker logic on the service deployment; this causes the deployment to transition to a failed state if it can't reach a steady state. This functionality can be used to trigger ECS to roll back to the last completed deployment upon a deployment failure.

In the case of the blue/green deployment type, ECS leverages the blue/green deployment model controlled by CodeDeploy. Traffic essentially shifts to the new release of the service in one of three ways:

- **Canary**: Traffic is shifted in two increments. The first portion of traffic shifts in the first increment while the rest of the traffic is sent over, after the specified amount of time, in the second increment.

- **Linear**: Traffic is shifted in equal increments with an equal time between each increment.

- **All-at-once**: All traffic is shifted from the original tasks to the updated tasks all at once.

We have already covered blue/green deployment to ECS using CodeDeploy in the section of this chapter dedicated to CodeDeploy; please refer to that section for more details.

Finally, in the case of the external deployment type, you can hand over the control of the deployment to a third-party solution. The external deployment solution then makes use of the ECS APIs for managing services and tasks to pilot the deployment.

Amazon Elastic Kubernetes Service (EKS)

Amazon EKS is a fully managed **Kubernetes** service that makes it easier to build, operate, and manage Kubernetes clusters on AWS. EKS offers a native Kubernetes experience to the developer, where the control plane is managed by AWS. EKS is integrated with core AWS services such as CloudWatch, Auto Scaling, and IAM to offer a seamless experience for monitoring, scaling, and load balancing containerized applications. It also brings rich observability, traffic controls, and security features to applications through its integration with **AWS App Mesh**.

Similar to ECS, with EKS you decide whether to provision the underlying compute resources for your EKS cluster with either EC2 instances, AWS Fargate, or externally. EKS lets you run your Kubernetes Pods on EC2 instances that you either manage yourself (*self-managed nodes*) or have AWS manage on your behalf (*managed node groups*). The benefit of using managed node groups over self-managed nodes is that AWS provisions and manages the configuration of the EC2 instances for you. You can alternatively run your Pods on Fargate, letting AWS fully manage the underlying infrastructure. Finally, you can also decide to run your EKS cluster on AWS Outposts or on your own compute resources, for instance, on-premises.

Regarding application deployment on EKS, you rely on Kubernetes application deployment mechanisms. Among the various possibilities, Kubernetes has the specific concept of the **Deployment**, which lets you define the desired state of a rollout, which you can then leverage to deploy your applications. Deployments on Kubernetes use the same concepts that we have already explored with CodeDeploy, ECS, and others: you do either rolling updates, blue-green deployments, or canary deployments. The Kubernetes ecosystem is now very broad and provides a wealth of solutions to assist you with application deployment. Please refer to the references provided in the *Further Reading* section and to the Kubernetes ecosystem to find out more on this topic. One example is **CDK for Kubernetes** (**CDK8s**). Initially developed by AWS, the CDK8s project has been open sourced and taken over by the **Cloud Native Computing Foundation** (**CNCF**), the organization also responsible for Kubernetes. CDK8s provides you the ability to leverage the CDK to deploy Kubernetes applications. For more details, please consult the project's website at `https://packt.link/LmrBi`.

AWS Copilot

AWS Copilot is an open source CLI for developers to quickly launch and easily manage containerized applications on AWS. You can use Copilot as your CLI to operate such applications across App Runner, ECS, and Fargate.

Copilot makes it easier for you to get started with deploying your first containerized applications on AWS. It helps you with deploying your application across multiple environments, such as test, staging, and production environments. It also helps you set up the necessary resources, such as networking, load balancing, and DNS records.

Additionally, Copilot assists with the lifecycle management of your application. It will let you create and configure CI/CD pipelines to build and package your application and deploy new releases of your application automatically. It does so by leveraging AWS CodePipeline and the relevant AWS Code services, along with your preferred source code repository (GitHub, Bitbucket, or CodeCommit). Have a look at Copilot's documentation at `https://packt.link/Q3JT9` for more details.

AWS Proton

AWS Proton is a service that lets you coordinate the work of infrastructure and platform teams for them to define prescriptive AWS environments that developers can use to deploy their applications.

Using Proton, infrastructure and platform teams can define environment and service templates describing the combination of resources that developers can then use to deploy their applications. An environment template represents the shared resources that Proton services will use. Think of resources such as VPC or load balancers, for instance. A service template defines the entire set of resources composing the infrastructure and platform that developers will be pushing their application to. This can also be applicable to resources such as, for instance, a backend service or an SNS topic to publish alarm notifications.

Proton environment and service templates can be authored either in AWS CloudFormation YAML or Terraform HCL format. The AWS Proton team maintains a library with some curated sample Proton templates, for both CloudFormation and Terraform. These sample templates address common cloud application design patterns, such as event-driven architectures (using Lambda), or internet-facing or backend container services (using Fargate or EC2-based ECS). Additionally, if none of the provided sample templates matches your use case, you will have to roll your own templates, but they can at least serve as an example for some extra guidance.

Proton is an opinionated service in how it delivers support for application lifecycle management. When your infrastructure is managed by Proton, the service automatically provisions a CI/CD pipeline that manages the lifecycle of your application, from storing the code in a Git repository, through building and packaging your application, to the actual deployment of your application on top of the AWS resources defined in your Proton environment and service templates.

Moreover, Proton also created the notion of components as a mechanism for developers to customize service templates. Components provide developers with some flexibility to extend the infrastructure templates. A development team may define their own application-specific components, such as DynamoDB tables, for instance. So, which resources should be part of the infrastructure layer (environment and service) and which should fit in the application layer (component)? Suppose everything in an application design that constitutes a moving target and will likely change frequently is a good fit for components.

Infrastructure and platform teams, owing to Proton, can enforce their standards and guidelines into codified and maintained templates that directly address the specific needs of one or more development teams. Proton also lets them manage their template's lifecycle, so that the infrastructure resources they provide can adapt to technology evolution or changes made to their own standards.

For developers, Proton ensures that they directly focus on writing application code without having to worry about the underlying infrastructure or the CI/CD pipeline to support the development lifecycle. Even if Proton is prescriptive in many ways, developers still have some degree of freedom and flexibility to design and create their applications and focus exclusively on the business logic and not the underlying plumbing.

So far, you have reviewed the plethora of options available to you for deploying resources on AWS, including some of the best practices to follow. The next aspect to cover, one of the key best practices in general for any AWS resource or event, is monitoring.

Tracking Deployment

Tracking and monitoring your deployment is of particular importance since the last thing you want is to end up with a broken release in production without even knowing about it until your users or customers start complaining. You want to make sure that whatever strategy and solution you put in place to deploy your workloads on AWS works as expected in the actual deployment. If it doesn't, you want to be told, preferably sooner rather than later, especially when it comes to releases in production environments. The AWS services previously mentioned in this chapter are integrated with core AWS services, such as CloudWatch and CloudTrail, which are great tools to follow up on the status and progress of things over time. However, be aware that all these services may also behave differently from one another when they encounter failure conditions. For instance, don't expect that Beanstalk or AppRunner will behave exactly like OpsWorks in such circumstances.

Remember that it is ultimately your responsibility to monitor the progress of your deployment process and to see that it remains effective over time. To do this, you must create alarms in CloudWatch to be notified when any of your deployments go wrong for any reason. Thereafter, you want to take the necessary remediation actions either automatically, if that's an option, or manually as needed. For instance, upon failure, your deployment process may have left your application environment in an unstable state with a mix of your previous and new application releases. How do you roll back to the last known successful deployment? Your options will vary based on the deployment solution you use. For more details, please refer to the documentation and best practices of the relevant AWS services.

Summary

In this chapter, you examined various deployment strategies and AWS services that can help you determine the most appropriate deployment strategy to meet your business requirements.

Chapter 10, Designing for Cost Efficiency, explores the diverse pricing models available through AWS and provides insights into how you can select the most suitable option based on your specific needs and limitations.

Further Reading

- Overview of deployment options on AWS: `https://packt.link/YLKZe`

- Introduction to DevOps on AWS: `https://packt.link/ksioB`

- AWS DevOps blog: `https://packt.link/eTs4k`

- Architecture best practices for containers on AWS: `https://packt.link/g5CnZ`

- CloudFormation best practices: `https://packt.link/AdAQk`

- Share reusable infrastructure as code by using AWS CloudFormation modules and StackSets: `https://packt.link/l6kqN`

- EKS best practices: `https://packt.link/iHnMW`

- What is AWS Proton? `https://packt.link/XXNh4`

10
Designing for Cost Efficiency

In the previous section on migration planning, you explored and learned about the various avenues with which solutions architects such as yourself can help guide customers with their migration and modernization journey to the AWS cloud.

In this chapter, we continue along the same lines and explore how to effectively evaluate, design for, and manage costs on the AWS cloud. Throughout this chapter, we will be leveraging the same sample web application as before, exploring how costs are evaluated for each of the application's components over time. You will also explore opportunities to right-size your workloads and, finally, dive deep into the AWS pricing models with the help of a few simple cost evaluation exercises.

The following main topics will be covered in this chapter:

- Understanding AWS pricing models
- Evaluating costs
- Right-sizing workloads

The next section begins with a detailed review of the various pricing models offered by AWS services and how you can leverage them based on your needs.

Understanding AWS Pricing Models

When it comes to diving deep into the 200+ AWS services available, it is really important to take a moment and understand some of the underlying pricing options that accompany these services. The majority of the pricing can be classified into three main categories—on-demand, spot, and reserved; however, there are a few additional pricing options that all Solutions Architects should keep in mind when recommending a particular service for a particular task. In this section of the chapter, you will explore these pricing options in a bit more detail, starting with the compute layer.

Compute

Almost all organizations that leverage the cloud in some form or another also leverage one or more of its underlying compute capabilities to host and run workloads. In the case of AWS, these capabilities are provided by services such as Amazon EC2 instances, AWS Fargate, and AWS Lambda. For such compute services, AWS provides three primary pricing models to choose from. They are discussed in the following subsections.

On-Demand

This is by far the most commonly used pricing model and the easiest to implement. The on-demand mode simply allows customers to pay as per their resource consumption; the more you use, the more you pay, and if no resources are used or deployed, you essentially pay nothing. There are no long-term commitments or contractual obligations with this pricing model. Customers pay at a fixed rate of consumption, based on the resources that they have requisitioned on either an hourly or per-second basis. For example, an **m5.large** instance is priced at $0.102/hour in the Stockholm Region, which equates to $74.46/month.

> **Note**
>
> The prices mentioned anywhere in this chapter reflect the pricing at the time of writing, and AWS might change them at any time.

The on-demand pricing model is ideal for workloads that are required for a short duration or have an unpredictable, spiky workload that can require additional capacity at any given moment in time. One example is commerce websites during sales events. Retailers can experience unpredictable spikes in traffic during a sales event such as Black Friday or Cyber Monday. On-demand resources allow them to handle the increased load without over-provisioning throughout the year. Once the sale ends, they can scale down to normal levels. Similarly, streaming platforms that host sports events or conferences face varying demand, based on live events. An on-demand infrastructure allows them to scale up rapidly during live broadcasts and scale down afterward, optimizing costs.

Reserved

As the name implies, with this model, customers can choose to essentially reserve a fixed capacity for their workloads, based on either a one-year or a three-year commitment. Reserving capacity is an ideal scenario when you have a steady workload that will essentially run 24x7, 365 days without much interruption. AWS offered (and still offers to existing customers) the Reserved Instances pricing model up until the introduction of a more flexible pricing plan, called Savings Plans.

Although the Reserved Instances pricing model provided up to 75% savings from the standard on-demand instances, there was an inherent flaw in this model – you could not easily allocate or apply the pricing benefits to other EC2 instance family members, apart from the one that was selected for the reservation. Additionally, the reservations were tied to a particular Region and in many cases to individual **Availability Zones (AZs)** too, which made transferring the benefits of the pricing model to other Regions even more difficult.

In the case of Savings Plans, you can not only apply the savings (up to 72% on on-demand instances) to any EC2 instance in your workload irrespective of its family, OS, or even tenancy; you can also apply the benefits across your Amazon EC2 instances, AWS Lambda, and AWS Fargate workloads, running across multiple Regions. For customers who only wish to apply the accrued savings to EC2 instances, they can alternatively opt for the EC2 instance Savings Plans as well.

Considering the same **m5.large** instance for reserved, Savings Plans, and EC2 instance Savings Plans for the Stockholm Region, a customer would end up with the options shown in *Table 10.1* to choose from:

		No upfront	**Partial upfront**	**Full upfront**
Standard Reserved	1-year	$560.64	$539.56	$525
	3-year	$1,156.32	$1,061	$1,008
Compute Savings	1-year	$718.32	$683.28	$674.52
	3-year	$1,524.24	$1,419.12	$1,392.84
EC2 Instance Savings Plans	1-year	$560.64	$534.42	$525.60
	3-year	$1156.32	$1077.3	$998.64

Table 10.1: Compute pricing plans for an m5.large instance (the Stockholm Region)

For example, imagine you're running a web application that serves user requests. You have a consistent workload throughout the year. By purchasing a three-year Reserved Instance for the specific instance type and Region used by your web servers, you can significantly reduce costs. The Reserved Instance discount applies to the instances that match its attributes.

Many organizations also have periodic batch jobs such as data processing or report generation. With AWS, you can reserve instances for specific batch-processing tasks. During peak hours, the RI discount applies. This gives the benefit of savings during high-demand periods without over-provisioning, and predictable savings over the long term, especially if your application runs continuously.

Spot

Perhaps the most cost-effective pricing model out there, Spot Instances can provide up to 90% cost savings compared to the on-demand pricing model. However, this drastic price reduction does come with some caveats. Spot pricing is adjusted based on a particular instance's supply and demand in a particular Region, as well as an AZ. If the demand for a particular instance family is low, the instance price is significantly reduced and made available to all to use. However, if the demand for the instance is high across the AZ or Region, its pricing can be more or less similar to an on-demand instance's price as well.

So, when is the right time to use a Spot Instance, and for which workloads? Well, Spot Instances and pricing are best suited for workloads that are stateless and extremely flexible regarding when they are started or stopped. This is because if the demand for a particular instance type is high within an AZ, AWS can reclaim this instance back with a two-minute termination warning. As a result, such types of instances are best suited for batch processing workloads, or even stateless container workloads that are designed to handle flexible start and stop times.

Imagine that you are running a data analysis pipeline that processes large datasets periodically. Instead of using expensive on-demand instances, you can launch Spot Instances during data processing. These instances handle compute-intensive tasks, such as running machine learning models, statistical analyses, or **Extract, Transform, Load** (ETL) jobs. This can mean significant cost savings, especially for non-urgent tasks where interruptions are acceptable.

Although the three pricing models discussed apply to Amazon EC2, AWS Fargate, and AWS Lambda, there is still a difference in the way costs are calculated across the three services. For example, in the case of an Amazon EC2 instance, the pricing is pretty straightforward and based on the instance type that you select. Each instance type comes with its own set of resources, such as CPU, memory, and OS flavor. However, this pricing mechanism changes a bit with both AWS Fargate and AWS Lambda functions. For example, AWS Lambda charges customers based on the following three parameters:

- The time taken to execute the Lambda function, rounded off to the nearest millisecond
- The number of requests made to the function
- The amount of memory allocated to the function

Similarly, for containers running on AWS Fargate, customers pay for the underlying CPU, memory, and storage for the duration of the container's lifetime. Now that we have discussed compute pricing models, it is time to look at another important part of your enterprise infrastructure: storage.

Storage

Storage services are priced depending on the type of service being used. For most workloads, storage services can be broadly split into two main categories:

- **Block storage**: When it comes to block storage, AWS provides three main storage devices to choose from – SSD volumes, HDD volumes, and finally, magnetic volumes. SSD volumes are designed to run the most common workloads, including general-purpose workloads as well as workloads that require frequent reads and writes, and **input/output operations per second (IOPs)**. HDD volumes can be used for generic workloads that require high throughput, and magnetic volumes are used primarily as backup or tape drives.

  Amazon EBS volumes are priced based on three factors:

  - **The storage volume (in GB) provisioned for the EBS volume**

  - **The snapshot storage (in GB)**: Each snapshot is stored in an Amazon S3 bucket and can be used to restore the state of an instance in case of a failure

  - **The data transferred out from the EBS volume**: This can be data that is transferred out from the EC2 instance, or it can also include cross-regional data transfers

- **Object storage**: Similar to the block storage options, object storage is also charged based on the amount of storage that you consume. In this case, Amazon S3 charges customers based on the following factors:

  - **The amount of storage consumed, based on the objects stored within a bucket.**

  - **The storage class used to store the objects**: This can be in the form of S3 Standard, a general-purpose storage ideal for frequently accessed data, or even tiered storage options such as S3 Standard-**Infrequent Access (IA)**, which is designed for less frequent data access patterns. S3 Standard-IA comes at a slightly cheaper rate compared to the Standard tier. Customers can also further reduce their object storage costs by automatically transitioning objects through various storage tiers, including a deep storage archive layer called S3 Glacier, using one or more S3 Lifecycle policies.

  - **The amount of data transferred to the outside world as well as across AWS Regions.**

  - **The number of retrieval requests made for the data and objects**: Different operations may have different costs associated with them. For example, a simple GET request to retrieve an object might have one price, while other types of data access, such as PUT, or management operations may have different prices. Some storage classes, such as S3 Glacier, also have retrieval charges.

> **Note**
> You can estimate your costs using AWS Pricing Calculator. However, remember that AWS Pricing Calculator does not take into account any taxes that might apply to your workloads.

While block storage plays an important part in general data management in your enterprise infrastructure, databases, both SQL and NoSQL, are another common, more specialized storage framework, which we will discuss in the next section.

Databases

We have talked a bit about the various database options provided by AWS in the chapters discussing migration and modernization. For simplicity, let us consider standard relational and **non-relational** (**NoSQL**) database types:

- **Relational databases**: With Amazon RDS, customers can spin up a fully managed version of their preferred open source and proprietary database variants, such as MySQL, MariaDB, PostgreSQL, MS SQL Server, and Oracle, in a matter of minutes by simply selecting the underlying compute (EC2 instance) and storage (EBS volume) associated with it. Accordingly, customers can then choose to run the RDS instances based on on-demand prices, or even save up to 60% by switching to either reserved or Savings Plans as well.

  The database instance costs can also be influenced by other secondary functionalities, such as the number of read replicas created, database volume backups, as well as deployment options that provide disaster recovery (multi-AZ).

- **Non-relational databases**: Conversely, non-relational databases such as Amazon DynamoDB use a different pricing model compared to Amazon RDS. The service offers two capacity modes that can be leveraged by customers, depending on their database usage patterns. Each of these capacity models comes with its own pricing model, as explained here:

  - **On-demand capacity mode**: Very similar to the compute on-demand model, here you simply pay for the read and write requests your application makes to the DynamoDB tables, without having to set or configure any throughput. This mode allows customers to scale their workloads for any random performance peaks, without having to configure additional settings. The actual scaling is taken care of by DynamoDB itself. This is ideal for applications with unpredictable workloads or new tables with unknown workloads.

  - **Provisioned capacity mode**: With this model, customers can essentially configure the number of reads and writes that an application can perform on the database each second. This form of capacity selection is ideal if you understand your application's read-and-write requirements and wish to save costs in the long run.

In addition to compute and storage, it's also important to consider the network infrastructure that an enterprise will run on. The next section will discuss the best ways to consider costs for services such as Amazon VPC.

Network

AWS also offers various networking services that help customers set up basic networking constructs, such as subnets and VPNs, as well as more advanced services that provide content delivery as a service.

It is often a misconception that Amazon VPC does not cost anything to run. While true, VPC itself relies on other services that do have a cost associated with them. For example, leveraging a NAT gateway incurs charges per NAT gateway, as well as the price per GB processed by the gateway. Similarly, if you leverage a VPC's advanced features, such as network analysis using traffic mirroring, you will incur an hourly charge for each **Elastic Network Interface** (**ENI**) that is enabled for traffic mirroring.

For other advanced networking services, such as Amazon CloudFront, the pricing model is based on the volumes of data transferred to either the internet or back to the origin, and on the number of HTTP/HTTPS requests made.

Keeping the preceding discussion in mind, it is evident that one pricing model does not match each of the customer's workload requirements. In an ideal world, customers can end up leveraging a mix of almost all of the pricing models, depending on the workload type. Consider a .NET application migration. If you were to take the lift-and-shift version of the application and host it on AWS, you would end up with the cost breakdown shown in *Table 10.2*:

	Option	Description	Pricing model
EC2 instance type	M5.large	General-purpose instance, providing two vCPUs and 8 GB of memory	Instance cost/hour, which includes OS (Windows), CPU, and memory
Utilization	100%	Web servers running 24 hours a day	Mix of on-demand and Spot pricing (assuming web servers are stateless or they can handle abrupt start/ stop times)
Instance storage	EBS volume (GP2) 30 GB	General-purpose EBS volumes mounted to each of the instances for persistent storage	GB/month
Minimum instance count	3	One EC2 instance in each of the available AZs	

	Option	Description	Pricing model
Maximum instance count	6	Based on workload demand, instances can scale to accommodate the application's load	
Instance backup	Daily EBS volume snapshots	Backups made using AWS Backup on a daily basis	GB/month
RDS instance type	db.r5.4xlarge	Memory-intensive instance providing 16 vCPUs and 128 GB of memory	Instance cost/hour, which includes SQL Server license, CPU, and memory
Utilization	100%	Database server running 24 hours a day	Reserved Instance
DB instance storage	Provisioned IOPs EBS volume (IO2) 200 GB	Provisioned IOPs EBS volume capable of providing up to 16,000 IOPs each	GB/month
DB Configuration	Multi-AZ	Standby database replica running in a different AZ	Cost baked into the service

Table 10.2: A cost breakdown sample of a typical web application hosted on Amazon EC2 and Amazon RDS

Now that you understand the nuances of how different AWS services can be priced, as well as what pricing models are available to customers, you can examine some of the methods using which a customer can evaluate the costs of their workloads on AWS.

Evaluating Costs

Understanding and evaluating costs for your applications is an important aspect of migrating workloads to the cloud, and AWS provides numerous tools as well as services that customers can use to forecast and evaluate the runtime costs of resources. This section discusses three such commonly used tools, starting with AWS Pricing Calculator.

AWS Pricing Calculator

AWS Pricing Calculator is a simple web-based tool that allows you to easily calculate and estimate the runtime costs of your AWS resources. The tool is exceptionally useful when it comes to planning and estimating resources during a migration activity, as well as when it comes to building a new application on AWS.

The tool provides mechanisms with which customers can create numerous estimates depending on their workloads, group them as per their requirements, and share them across their organizations as well.

> **Note**
>
> At the time of writing, AWS Pricing Calculator supports around 150 AWS services.

The following steps will enable you to use the pricing calculator to make accurate estimates of the services you might need:

1. Visit `https://packt.link/PT4DC` to launch AWS Pricing Calculator.

2. Once launched, select the `Create estimate` option.

3. This will bring you to the `Select service` page, where you can browse and select the AWS service that you want to work with. This example uses the Amazon EC2 service for demonstration purposes, as shown in *Figure 10.1*. Select the `Configure` option to continue:

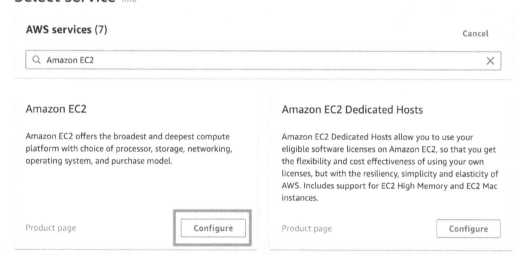

Figure 10.1: Service selection using AWS Pricing Calculator

4. On the `Configure Amazon EC2` page, provide a description of your workload and select the location (whether an AWS Region, Local Zone, or Wavelength zone).

5. Toggle between `Quick estimate` and `Advanced estimate` to prefill the form with default values. Alternatively, you can configure each of the fields manually, based on your exact requirements.

6. Once you have completed the form, click on `Save and add another service` for your workload.

7. Once all the services are filled in, you will end up with a dashboard as shown in *Figure 10.2*:

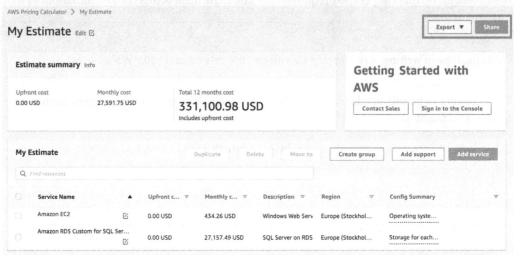

Figure 10.2: The estimates dashboard

8. You can choose to either export the calculations into a PDF or CSV file or simply share the estimates using a web URL as well. Do note that the URLs are valid for a period of three years from the date of creation.

> **Important Note**
> AWS Pricing Calculator does not take into account any taxes that might apply to your workloads.

AWS Pricing Calculator is a useful tool that helps you determine the cost of running your workloads on AWS. However, it can be helpful to granularly analyze, forecast, and plan costs based on actual usage; the next section will look at a tool that can help you do just that – AWS Cost Explorer.

AWS Cost Explorer

AWS Cost Explorer is used to help customers view and analyze costs up to a 12-month period from the given date. It also forecasts for an additional 12-month period based on existing usage patterns. The tool also provides reports and recommendations based on monthly costs incurred by AWS services, Savings Plans reports, and Reserved Instances reports.

To get started with AWS Cost Explorer, you first need to enable it using the AWS Management Console. To do so, navigate to `https://packt.link/O1dNR` and select the `Cost Explorer` option from the navigation pane on the left-hand side of the web page. If this is the first time you are visiting this page, you will be prompted to launch Cost Explorer on the welcome page. Select this option to view the Cost Explorer dashboard, as shown in *Figure 10.3*:

Figure 10.3: The AWS Cost Explorer dashboard

With Cost Explorer now enabled, you can use the dashboard to do the following tasks:

- View the estimated cost of your AWS resources to date
- Forecast costs for the same month
- View your five top cost trends as well as any recently viewed reports

Unlike AWS Pricing Calculator, AWS Cost Explorer provides you with a programmatic API that can be used to obtain fine-grained cost and usage metrics for your workloads.

AWS Cost and Usage Reports

The third option available to all customers for evaluating their AWS costs is AWS **Cost and Usage Reports (CUR)**. CUR provides by far the most detailed and granular reports that a customer can have at their disposal to gain a comprehensive understanding of their overall AWS resource consumption and pricing. These reports are essentially the AWS Billing reports, which provide the cost breakup of resources by the hour, day, week, or month, by AWS service or product, by tags, and so on that a customer may define. Once a report is created, it is made available inside the customer-owned S3 bucket in the CSV format. You can then view the report and analyze it using your compatible toolsets.

Setting up CUR is a two-step process that first begins with the creation of a simple S3 bucket in your own AWS account. Once the bucket is created, you then apply a bucket policy that grants the AWS Billing service permission to both list the contents of the bucket and add items to it. The following is a sample policy for your reference:

```
{
    "Statement": [
        {
            "Effect": "Allow",
            "Principal": {
                "Service": "billingreports.amazonaws.com"
            },
            "Action": [
                "s3:GetBucketAcl",
                "s3:GetBucketPolicy"
            ],
            "Resource":"arn:aws:s3:::EXAMPLE-BUCKET",
            "Condition": {
                "StringEquals": {
                    "aws:SourceArn": "arn:aws:cur:us-east-
1:${AccountId}:definition/*",
                    "aws:SourceAccount": "${AccountId}"
                }
            }
        },
        {
            "Sid": "Stmt123456789",
            "Effect": "Allow",
            "Principal": {
                "Service": "billingreports.amazonaws.com"
            },
```

```
            "Action": "s3:PutObject",
            "Resource": "arn:aws:s3:::EXAMPLE-BUCKET/*",
            "Condition": {
                "StringEquals": {
                    "aws:SourceArn": "arn:aws:cur:us-east-
1:${AccountId}:definition/*",
                    "aws:SourceAccount": "${AccountId}"
                }
            }
        }
    ]
}
```

Once the two aforementioned steps are completed, perform the following steps from the AWS Management Console to create a new CUR report:

1. Sign in to the AWS Billing and Cost Management console by navigating to https://packt. link/Szv6s.

2. Once logged in, select the Costs and usage reports option from the navigation pane on the left-hand side of the page.

3. The first step to perform on the AWS CUR page is to enable the service. To do so, select the Enable option from the dashboard. Remember that it will take approximately 24 hours to populate the spend data for the CUR report.

4. Next, select the Create report option.

5. On the Report content page, provide a suitable name for the CUR report, followed by optionally selecting Include resource IDs as well as the Automatically refresh your Cost & Usage Report when charges are detected for previous months with closed bills option. Click Next to continue.

6. On the Delivery options page, select the S3 bucket you created and configure it using the Configure option.

7. You can configure the Time granularity, Report versioning, and Data integration options as per your requirements. Click on Next to continue.

8. On the Review page, review your CUR report's configuration, and select the Review and complete option to complete the process.

Your new report should be created and ready for analysis from the `AWS Costs and Usage Reports` dashboard, as depicted in *Figure 10.4*:

Figure 10.4: AWS CUR

With this, we have reached the end of yet another section. In the next and final section of this chapter, you will explore some tools as well as tips to keep in mind when right-sizing AWS resources.

Right-Sizing Workloads

One of the key benefits as well as the driver for cloud adoption is the reduction in operational as well as resource costs, and one of the key factors that contributes to this is called right-sizing. In this section of the chapter, you will learn what right-sizing is all about and explore some tools and essential tips to keep in mind when right-sizing your own workloads.

Right-sizing, as the name implies, refers to the process of providing workloads with the right number of resources required to run. This becomes increasingly essential in cloud environments where resources are billed based on the quantity or volume of usage, so any unnecessary resources that you assign to your workloads will incur additional costs.

But if right-sizing is so important, why isn't it talked about that often? Well, most organizations that adopt the cloud journey through a lift-and-shift migration size their workloads based on the assumption that the number of resources required to run a workload on the cloud will be the same as that required in on-premises architecture. However, this isn't entirely true, as the cloud offers way more advanced and purpose-built hardware on which applications can run, unlike your common off-the-shelf servers and racks found in a data center. This means that your application may be over-provisioned and cost you more on the cloud compared to on-premises. However, this isn't the only reason right-sizing is often overlooked. Most lift-and-shift migrations are characterized by the need to migrate fast and ensure the best-of-breed performance while doing so. As a result, costs are often given less priority, and before you know it, your workloads now cost more to run on the cloud than on-premises.

Now that you have understood the need to right-size, you might be wondering how and when to do so. The answers to both these questions are simple – you do right-sizing as a continuous process. You do it before your migration activity has started and then after you have migrated workloads in the cloud as well. Before migrating your workloads, you can also obtain a performance reading as well as average resource utilization of your workloads, using monitoring tools and metrics provided by common virtualization software such as VMware vSphere and Microsoft Hyper-V. Using this data, you can map your workloads to the right instance family in AWS (e.g., compute, general purpose, memory, I/O, etc.) and select the right instance size (e.g., small, medium, large, extra-large, etc.) for your workloads in AWS. Once your workload is successfully migrated to AWS, you can utilize AWS services such as Amazon CloudWatch, AWS Cost Explorer, and AWS Trusted Advisor to understand current consumptions and plan for future requirements as well. Further, in all these cases, you as the customer are responsible for building up the dashboards, monitoring the usage of resources, as well as analyzing the various performance patterns. But what if all of this could be automated and used as a service? Fortunately, AWS ensures this automation through AWS Compute Optimizer.

AWS Compute Optimizer is a machine learning-based right-sizing service that helps customers reduce the creation of over-provisioned as well as under-provisioned resources. AWS Compute Optimizer does this by analyzing utilization metrics and data points over a certain period of time, and then it uses machine learning models to provide recommendations and best practices regarding resource usage.

AWS Compute Optimizer supports the following five resources at the time of writing:

- Amazon EC2 instances
- Amazon EC2 Auto Scaling groups
- Amazon EBS volumes
- AWS Lambda functions
- Amazon ECS services on AWS Fargate

In order to get started with AWS Compute Optimizer, you first need to enable the service using either the AWS Management Console, the AWS CLI, or the AWS SDK. The following steps show you how to set up AWS Compute Optimizer using the Management Console:

1. Log in to the AWS Management Console and select the AWS Compute Optimizer service using the following link: `https://packt.link/iuXpP`.

2. Once logged in, you may need to enable the service if this is the first time you are using it. To do so, simply select the `Get started` option on AWS Compute Optimizer's landing page.

3. Next, in the `Setting up your accounts` section, select the appropriate scope you wish to provide for AWS Compute Optimizer; either only enable it for this particular account or choose to enable it for all the accounts under the existing AWS organization.

4. Once completed, simply select the Opt in option, and that's it! You are now ready to use Compute Optimizer!

> **Note**
>
> Compute Optimizer recommendations may take up to 12 hours to be generated from the time of opt-in.

With the service now enabled, you can use Compute Optimizer's dashboard to view recommendations, insights, as well as findings, as illustrated in *Figure 10.5*:

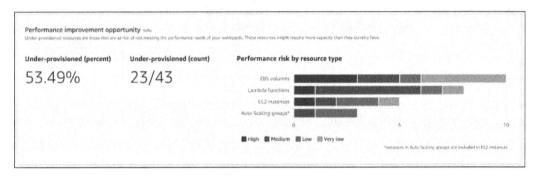

Figure 10.5: Over-provisioned and under-provisioned resource findings

AWS Compute Optimizer also allows you to export your findings and recommendations to an Amazon S3 bucket of your choice in either a CSV or JSON format.

Based on the discussion so far, the following are some handy tips and recommendations that solution architects need to be aware of when right-sizing resources on AWS:

- **Select the right instance family for the right job**: AWS provides over 400 different instance types that you can choose from. However, most of these instances can be categorized into families based on their characteristics, such as general-purpose, compute-intensive, memory-intensive, network-intensive, and I/O-intensive. Make sure to select the most appropriate instance family for your workload, based on its usage patterns and characteristics.

- **Analyze performance at regular intervals and for sustained periods**: By leveraging services such as Amazon CloudWatch and AWS Compute Optimizer, you can always monitor and analyze the resource consumption of your workloads and plan for optimizations accordingly. Remember to not rely on stale data that may be too old to use or data points that are not collected over a period of time – for example, a month or two.

- **Understand application usage patterns**: Selecting the right type of instance as well as the right pricing strategy for your workloads starts with having a good understanding of your workloads first. Applications that have a steady state of consumption are ideal for pricing models, such as Reserved Instances or Compute Savings/EC2 Savings Plans, where the resources can be reserved for a period of 1–3 years at a fraction of the on-demand costs. Conversely, applications that have a variable resource requirement or are too spiky in nature can require on-demand or Spot Instances, based on a workload's design and setup.

- **Don't keep what you don't need**: Unlike the traditional on-premises data center where resources once turned on are almost never switched off, resources created in the cloud can be powered off or, better yet, terminated when not in use. This helps save costs and also minimizes the management of unwanted resources in the long run.

You have now reached the end of yet another chapter. Before moving on to the next, let's quickly review all the things you learned with a chapter summary.

Summary

This chapter started off by discussing the various AWS pricing models and how each of these models varies from service to service. You also learned about the importance of evaluating costs and explored a few tools and services that help customers understand their spending on the cloud, both before and after a migration. You also learned about the importance of right-sizing resources and how you can automate right-sizing recommendations, using AWS Compute Optimizer. Finally, the chapter presented a few key tips and recommendations to keep in mind when right-sizing resources on AWS.

In *Chapter 11, Improving Operational Excellence*, you will focus on operational aspects to identify areas of improvements.

Further Reading

- AWS Compute Optimizer user guide: `https://packt.link/HdVF1`

- Right-sizing best practices: `https://packt.link/FyDzY`

11

Improving Operational Excellence

Being able to keep workloads operating efficiently over time is essential to meet your business objectives. As part of the continuous improvements that you want to make to an existing solution on AWS, improving operational excellence should be relatively high on your to-do list. This is of particular importance if you have deployed workloads in production whose health could be critical to your business. For those workloads, it is paramount to ensure that operations run smoothly to avoid any mishaps.

This chapter will discuss the specifics of operating workloads on AWS. You will look at the following topics and learn how you can follow operational best practices:

- Design principles
- Improving the organizational fit
- Identifying operational gaps
- Evolving operations

Design Principles

The operational excellence pillar of the **AWS Well-Architected Framework** is the primary reference for defining and improving your operations on AWS. It includes guidelines for you to support your business objectives, run workloads more effectively, gain insight into your operations, and continuously improve supporting processes and procedures to deliver business value.

The operational excellence pillar provides several principles to keep in mind when designing solutions for the cloud to operate workloads effectively. They are discussed in the following subsections.

Principle #1 – Perform Operations as Code

In the cloud, it is best practice to apply the same engineering discipline used for application development to your entire cloud environment. Your entire workload, including not only applications but also the infrastructure supporting them, can be defined as code, and that code life cycle management can follow the same principles you use to maintain application code. All your operation procedures can also be scripted and entirely automated – for instance, by triggering them in response to specific events. Performing operations as code allows you to contain human error and allows you to provide consistent responses to events.

Principle #2 – Make Frequent, Small, Reversible Changes

Make sure to design your workload in such a manner that its components can be updated frequently to let your workload benefit from improvements over time. Ensure that you make changes in small increments so that they can be more easily reversed, should they fail. This approach will also help you identify and solve issues introduced in your environment while reducing the impact on your workload's end users.

Principle #3 – Refine Operations Procedures Frequently

Look for opportunities to improve your operations procedures. As your workload evolves, so should your operational procedures. Make sure to review and validate that these procedures are effective and that the various teams operating in your cloud environment are familiar with them. Leverage game days regularly to that end.

Principle #4 – Anticipate Failure

It is essential to identify potential sources of failure to either eliminate them or, at the minimum, put some mitigation measures in place. You must do two things – first, test your failure scenarios to make sure you understand their impact, and second, test your response procedures to ensure their effectiveness. Organize game days to test workload and team responses to various simulated events.

Principle #5 – Learn from All Operational Failures

Run a post-mortem analysis on all operational events and failures to draw lessons to drive improvements. Make sure to share the lessons learned with your entire organization so that the relevant teams are well aware of what went wrong, why, and the countermeasures you have put in place or plan to take.

Principle #6 – Use Managed Services

Rely on cloud services provided by your cloud provider (in this case, AWS) rather than managing everything yourself. Using managed services reduces the operational burden on your team, thus allowing you to focus on your core business logic.

Principle #7 – Implement Observability for Actionable Insights

Observability refers to the ability to gain insights into your workload's behavior, performance, reliability, cost, and health. It enables proactive identification of issues before they impact users and should give you actionable insights, allowing you to make informed decisions.

Improving the Organizational Fit

In simple terms, organizational fit refers to the alignment between an employee's values, beliefs, and working style and those of their organization. The starting point of improving the organizational fit is the business objectives defined by your organization's leadership team. Your workload must deliver some business outcomes aligned with these objectives. It is up to you to make sure that the workload is properly instrumented and monitored so that you can validate, at any time, whether the expected results are delivered and take appropriate steps when some abnormal behavior is observed. You must also ensure the fluidity of the application development life cycle, automating as much as possible its various elements, such as build, testing, and deployment. The life cycle must also be instrumented and monitored, making sure it works as expected and addressing issues in a timely manner.

As you monitor your cloud environment, you will collect various types of metrics. These metrics will allow you to assess whether your workload has attained or breached its business outcomes. Similarly, you should define some operational metrics to assess the performance of your operations in the cloud, ensuring that they effectively enable your workloads to reach their business outcomes and drive potential operational improvements.

An essential part of this is the organizational fit. Every organization has its own characteristics, and grasping them fully is key to successful operations. You must understand how your organization works, its culture, its priorities, and how the various teams interact with each other. All of these elements will contribute to workloads operating effectively.

Organization Priorities

Starting from the context in which your workload will operate, such as your organization's and your team's cultures or the objectives at stake, you will have to establish a set of priorities to determine which aspects to focus on to maximize the business value of your efforts. For instance, you may realize that you need to enhance your team's skills as well as improve their performance and reduce costs. This may lead you to focus on training people, then improving how you monitor workloads, and ultimately, on automating runbooks.

Then, as your team and your organization naturally evolve, you will need to regularly re-visit these priorities over time and modify them to meet the new conditions. Here is a list of attention points to help drive the definition of your priorities:

- **Evaluating customer needs**: Your workload is meant to deliver a business outcome to your customers. Whether your customers are internal or external to your organization, you must clearly understand their needs to determine what is necessary in terms of operations to deliver the expected business outcome.

- **Evaluating governance requirements**: Your organization may already have some guidelines or policies in place, defining things such as IT standards for software development or application deployment. Make sure to evaluate the impact of such governance requirements on your workload and monitor their evolution.

- **Evaluating compliance requirements**: Your organization is also most likely bound by a set of regulations that apply to the industry and/or the geographies within which it operates. Make sure to determine the impact of such external factors on your workload. When in doubt, refer to your compliance team and the **AWS Compliance** page at `https://packt.link/DgAwv` for additional insights and guidelines.

- **Evaluating the threat landscape**: A number of threats may endanger your organization, such as business or operational risks, legal liabilities, and IT security threats. It's important for you to be aware of their impact on your workload. Depending on this assessment, you may, for instance, emphasize securing your workload environment or raise awareness about specific threats within your team. The Well-Architected Framework and, in particular, conducting a Well-Architected review can help you detect potential risks and determine the best approach to address them.

- **Evaluating trade-offs**: When making architectural decisions, you will be confronted with choices, presenting various trade-offs between benefits and drawbacks. For instance, you may need to choose between delivering some additional features, offering extra business value to your customers, and optimizing your workload to reduce costs, bringing higher profitability to the business. Assess the trade-off of each alternative and make the best decision for the business, considering your organization's objectives and priorities.

- **Managing benefits and risks**: Finally, you have to balance the benefits and risks, making sure that the risks you take to increase the potential business benefits are well mitigated and acceptable for your organization. Risk management is very tightly coupled with your organization's culture. If it is risk-averse, then you will likely emphasize reducing all possible risks; if, conversely, your organization is more risk-tolerant, then you might focus on increasing the business benefits.

For all of the aforementioned items, you want to take advantage of the wealth of resources available online, such as AWS documentation and the best practices they suggest. For instance, the AWS Well-Architected Framework and **Amazon Builders' Library**, **AWS Training**, and **AWS Support** can assist you.

Operating Models

Once your organization's priorities have been clearly established, the various teams must understand clearly the role they have to play to deliver the business outcome for your workload. They also need to understand their roles relative to other teams' success, as well as the role of other teams in their own success. For instance, the DevOps team of an e-commerce company will need to align its approach to feature delivery with the insights provided into customer behavior and needs by the business analysts. A well-defined set of responsibilities will reduce conflicts or redundant efforts. An effective alignment between business, development, and operations teams will make business outcomes more easily achievable.

Relationship and Ownership

It is very important to have a clear understanding of the relationship and ownership of the entities at stake. The following points discuss this in depth.

Relationships, Ownership, and Responsibility

An operating model should define the relationships between the various teams involved and identify ownership and responsibility for the resources at stake. If responsibility or ownership remains unclear or, worse, unknown, your organization becomes at risk of not undertaking the necessary initiatives promptly and making redundant or conflicting efforts toward those activities.

Resources Have Identified Owners, and So Do Processes and Procedures

You must have a clear, unambiguous understanding of who owns each workload, application, platform, and infrastructure component and the associated business value they provide. Understanding how these components contribute to business outcomes sheds light on the processes and procedures in which they're involved.

Similarly, it is equally important to understand who is responsible for creating, documenting, and maintaining operational processes and procedures and what they are used for. Having a thorough understanding of the reasons behind specific processes and procedures will let you identify improvement opportunities.

Operations Activities Have Identified Owners Responsible for Their Performance

It is essential for you to be aware of the person responsible for specific operational activities in your workload. Having this understanding helps to establish agreements on how tasks are to be executed and by whom, ensuring that there's a clear line of accountability. It also ensures that feedback can be provided to the owner of the activity as and when required.

Team Members Should Know What They Are Responsible For

If team members have a clear understanding of their role within your organization, they will be able to prioritize their tasks and act appropriately.

Mechanisms Exist to Identify Responsibility and Ownership (and Any Changes Made to Them)

Pre-defined escalation paths help the relevant person reach out to someone with the necessary authority to assign ownership or address the issue when ownership is missing.

Responsibilities between Teams are Predefined or Negotiated

There usually are pre-existing arrangements between various teams that describe their interactions, such as service-level agreements. Understanding the impact that your team's work has on business outcomes and on other teams will guide the prioritization of your tasks accordingly.

Choosing an Operating Model

Which operating model should you adopt then?

There isn't a single model that fits every team's needs. These needs are shaped by the industry in which the team operates, the organization, the composition of the team, and the characteristics of the workload at stake. Therefore, you cannot expect a single operating model to support all types of teams and their workloads across organizations and industries. For instance, consider an organization that operates in a highly regulated market. It is bound by strong compliance constraints, most of which are probably totally irrelevant to another organization operating in a completely different and, maybe, deregulated industry. They might eventually adopt the same operating model in some cases (for instance, for business-critical workloads), but they are also likely to adopt very different operating models for most of their workloads and teams.

Similarly, within a given organization, multiple operating models may well co-exist simply because they need to serve multiple teams with different priorities and objectives in possibly very distinct business contexts. Therefore, your organization may end up using various operating models, and there's nothing wrong with that.

However, this doesn't mean that a team can use any operating model they want. It's good to establish a set of operating models that your organization supports and to have teams pick one or the other, depending on their needs, priorities, and objectives.

One thing well worth mentioning here is that adopting IT standards can simplify your operations and limit the support burden on your operating model, by prescribing common rules and guidelines to follow. The more teams that adopt IT standards within your organization, the more benefits you will reap in terms of operations. For instance, the right IT standards can help teams mitigate risks by standardizing the best practices to follow for security, data management, and other critical IT areas. However, standards must not represent a brake to innovation, so it's also essential to leave room for changes to include new standards or evolve existing ones.

The following section presents some examples of operating models to clarify this topic further.

Operating Model Examples

This example uses a graphical representation to make things clearer and groups activities along two axes:

- On the first axis, workloads are split into applications (any code implementing the business logic, either custom development or commercial off-the-shelf software) and platforms (all the resources and tooling supporting applications, including compute, storage, networking, data, security, operations, and support)

- On the second axis, workload life cycle management activities are split between engineering (responsible for the workload development life cycle) and operations (in charge of deploying, operating, and managing workloads)

This will give us a 2x2 graphical representation with four individual quadrants, as shown in *Figure 11.1*:

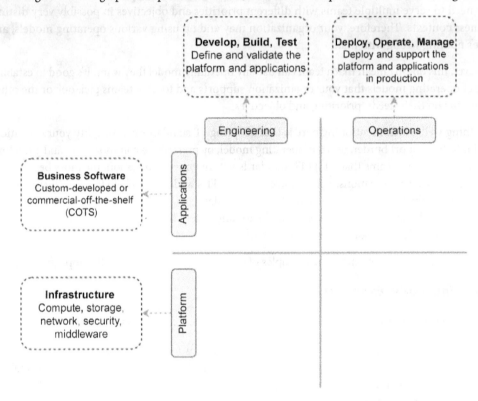

Figure 11.1: A 2x2 template representing an operating model graphical representation

Using this graphical template, a few examples of possible operating models on AWS can be illustrated.

Fully Separated Operating Model

First, there is the traditional IT operating model that has been around for decades and is still present in many organizations. It is a fully separated operating model, where activities in each quadrant are led by separate teams, as illustrated in *Figure 11.2*:

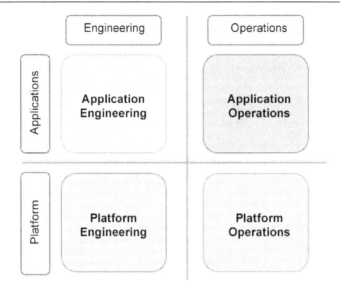

Figure 11.2: A fully separated operating model

In this operating model, each team carries its work independently of other teams but still depends on the other teams' activities (for instance, the development team depends on the platform team to make the infrastructure available to host their application environment, and on the operations team to deploy their application).

This model has the advantage that each team can be highly specialized, with very sharp skills in their limited perimeter of activity. For the interaction between teams to work smoothly, requests for work or work items must be passed across teams through some mechanism, such as a queueing or ticketing system or an **IT service management** (**ITSM**) tool. Activity handover across teams is typically highly structured and codified. However, this transition across teams adds complexity, constitutes a bottleneck, and eventually increases the risks of delays (with each team having its own priorities and agenda). Further, with each team working in isolation, there is also a high risk of misalignment between all of them.

Separated AEO and PEO with Centralized Governance

In modern-day software development, many organizations have started to adopt or, for some of them, already fully adopted a "*you build it, you run it*" policy. According to this policy, teams that develop an application are also in charge of operating it after it is deployed. This approach merges engineering and operations activities into the same team, so you end up with **application engineering and operations** (**AEO**) teams on one side and **platform engineering and operations** (**PEO**) teams on the other side.

This *"you build it, you run it"* approach is illustrated in *Figure 11.3*.

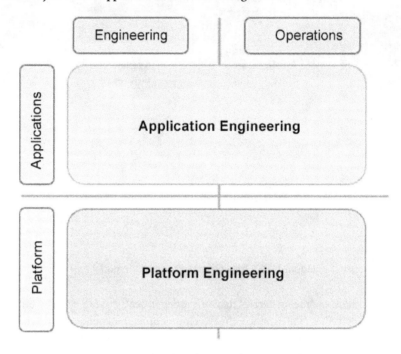

Figure 11.3: Separated AEO and PEO with centralized governance

In this operating model, platform teams are responsible for providing all the necessary supporting elements (infrastructure, data, and tooling), while application teams are in charge of developing the software and running it once deployed. Considering centralized governance in this operating model, IT standards and best practices typically flow from the platform teams to the various application teams, which have little to no room to maneuver to customize the platforms.

In this model, platform teams have to provide all the necessary visibility into the underlying infrastructure so that application teams can detect issues and clearly identify where the problem lies (whether at the application or the infrastructure level). Platform teams are also welcome to provide assistance and guidance to application teams with the configuration and setup of their platform.

Since application engineering teams have little to no flexibility to customize the underlying platforms, it is essential to have a feedback loop mechanism for them to request changes, improvements, or exceptions.

AWS services such as AWS Organizations and AWS Control Tower are great help for platform teams in such a context. They allow you to enforce IT standards and governance centrally within your AWS environment, providing AWS accounts with specific configurations and access to allow-listed AWS services. Additionally, if the platform engineering teams want to accelerate platform delivery to application teams, they can also leverage AWS Service Catalog to deliver self-service products, implementing architecture blueprints that reflect the IT standards and best practices they want to enforce.

Separated AEO and PEO with Decentralized Governance

This model follows the same *"you build it, you run it"* approach, except that this time, we want to establish decentralized governance and leave more room for application teams with regard to platform capabilities. The idea behind this model is to leave more room for application teams to innovate and remove many of the previously identified issues, such as bottlenecks or delays inherent in a more centralized governance model.

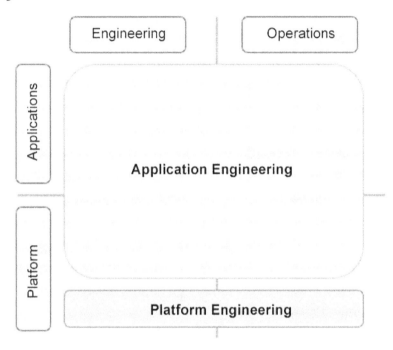

Figure 11.4: Separated AEO and PEO with decentralized governance

As illustrated in *Figure 11.4*, application teams now face fewer constraints and have more room for maneuver to customize and extend the platform they use for their application. As much as this model can bring application teams benefits, it also puts more responsibility on their shoulder, since they are now responsible for a part of the platform themselves.

Decentralized governance does not mean the absence of control. Platform teams still have a role to play in enforcing some IT standards (for instance, on security or networking aspects). Apart from the fact that you may still want to enforce some standards globally across all teams, it might also become counter-productive to leave architecture decisions open to application teams, since each team is likely to reinvent the wheel again and again on common architectural patterns.

Again, AWS services such as AWS Organizations and AWS Control Tower can be of great help to platform teams in such a context. Similar to the centralized governance model previously described, it is important to have a feedback loop mechanism in place for application teams to request improvements or exceptions to the supported platforms in this model.

The examples cited here are by no means exhaustive. Their purpose is to illustrate a few possible operating models serving organizations with different governance approaches.

The Role of Organizational Culture

In this section, you will examine the importance of organizational culture – in particular, some cultural aspects that are essential to improve your organization's operational excellence. The following subsections discuss these aspects in depth.

Executive Sponsorship

Senior leaders are responsible for setting expectations for the organization and evaluating success. They should sponsor, advocate, and drive the adoption of best practices and the evolution of the organization.

Team Members Should Be Empowered to Take Action When Outcomes Are at Risk

There should be clear guidance in place for team members to know what to do when an unexpected event negatively impacts the workload. Further, they should have all the resources they need at their disposal when they are performing the necessary remediation.

Escalation Should Be Encouraged

There should be mechanisms in place to let team members escalate issues early and often if they believe that the workload business outcome is at risk. Team members should feel comfortable escalating concerns to decision makers and stakeholders in such a situation. Timely escalations help identify and prevent serious incidents.

Communication Should Be Timely, Clear, and Actionable

Communication within an organization is essential, even more so when it is to notify team members of known risks and planned events in a timely manner. Any communication pertaining to operations should include the necessary contextual information, with enough details to determine which action must be undertaken, if any. Examples include notifying the team responsible about security vulnerabilities spotted within an application so that patching can be expedited, or providing notice of planned business events (such as a marketing campaign) so that a change freeze can be triggered to avoid the risk of service disruption.

AWS Systems Manager provides a number of useful features for operations to record events (**Change Calendar**) or schedule maintenance activities (**Maintenance Windows**).

Experimentation Should Be Encouraged

Experiential learning is encouraged to accelerate learning and keep people engaged. Through experimentation, team members can find ways to unlock innovation and turn ideas into business value. If they stumble upon a path that does not lead to success, they are not punished but encouraged to carry on experimenting.

Team Members Should Be Enabled and Encouraged to Maintain and Grow Their Skill Sets

A distrusting manager might think along the lines of, "*What if we train our people and then they leave?*" A counterargument could be, "*Well, what if we don't and they stay?*" It should be a no-brainer. It is absolutely essential for teams to grow their skills and to adopt new technologies, simply to support the changes in demand and responsibilities regarding your workloads. This will also bring satisfaction among team members, help retain talent, and support innovation as well. Similarly, team members should feel encouraged to pursue and maintain industry certifications as a great way to acknowledge their growing skills. Also, remember to promote knowledge transfer between team members to reduce the risk of impact if, despite your efforts, you happen to lose skilled and experienced people.

AWS provides tons of online and offline resources that can be leveraged to build up your skills, in the form of guidance and best practices, such as the AWS Well-Architected Framework or the Amazon Builders' Library blogs, videos, community forums, and trainings.

Resource Teams Appropriately

Avoid overloading team members with too many activities to reduce the risk of incidents resulting from human error. Leveraging tools and resources (for example, to automate frequently executed tasks) helps to scale your team, freeing them from more mundane tasks to support additional activities.

Diverse Opinions Should Be Encouraged and Sought Within and Across Teams

Cross-organizational diversity is recommended to ensure various unique perspectives are considered. Leverage diversity and inclusion to increase innovation, challenge your assumptions, and reduce the risk of confirmation bias. Organizational culture has a direct impact on team member job satisfaction and retention. Hence, your team members' capabilities and engagement have a strong influence on the success of your business.

Identifying Operational Gaps

When preparing for operational excellence, the first thing to do is to understand your workloads and their expected behaviors. You will then be able to adapt their design to retrieve information on their state and create the necessary operational procedures to support them. In order to understand workloads, you need to be able to analyze data, and to be able to do so, you first need to collect it. Telemetry refers to the automated process of collecting data from remote or inaccessible systems and transmitting it to a receiving system for monitoring and analysis. This is covered in the next section.

Designing Telemetry

When operating any workload, it is paramount to be able to tell at any moment in time whether that workload is working as expected. Systems should also be in place to otherwise be notified, typically by triggering alarms, when any unexpected event or abnormal behavior occurs. This is what monitoring entails.

Monitoring can only be efficient if it can rely on the right level of information, and whether it comes from the environment in which the workload runs (for instance, a measurement of CPU, memory, or disk usage when that makes sense) or from the application layer itself (for instance, metrics from application logs). The more useful the information that your applications log, the more informed your decisions can be when operating your workload.

It's particularly important for development teams to instrument their applications for operations to be effective. Instrumentation will feed data into operations monitoring dashboards and let you issue alarms when needed, ensuring that the operations team can take the relevant action in a timely manner. The goal of instrumentation is really to enable situation awareness for whoever needs it. Therefore, development teams should make sure to share enough information in application logs for them to be useful and allow complete visibility of your workload.

Application logs include error messages as well as informational data that presents a clear picture of an entire workload, an application, or even just a component of health. Once this instrumentation is in place, you can leverage the information to decide when and how to respond to any given situation. Consider an operations team that monitors whether their application can handle the increased load on a Black Friday sale. The informational data available to them might consist of the depth of a queue, some data store read/write latency, and the end-to-end duration of a transaction. They can utilize this information to obtain a clear picture of what's happening, and they can identify potential issues before they impact the user experience and take proactive measures, such as adding more resources or optimizing certain processes.

As much as possible, development teams should standardize application logs, structuring them consistently across the various application components to allow end-to-end application flow traceability and monitoring. You can, for instance, structure your logs using JSON so that the contents can be more easily processed by operational procedures.

The following example of logging data without much structure illustrates the preceding point:

```
2022-08-09T09:30:00.001Z      INFO      component-xyz: Hello world,
foo=bar
```

Instead, you would publish a structured log message that becomes easier to read, search, or parse, such as the following:

```
2022-08-09T09:30:00.001Z      component-xyz :: result ::
{
    "requestId": "123456-abcdef",
    "message": "Hello World!",
    "level": "INFO",
    "foo": "bar"
}
```

Additionally, you should aim to centralize your logs to make it easier for anyone to search and analyze them. Use built-in mechanisms such as the unified CloudWatch agent, CloudWatch API, or one of the Amazon SDKs to send logs to CloudWatch. Multiple AWS services also allow you to publish logs to CloudWatch – for instance, **Amazon API Gateway**, **Amazon Cognito**, **Amazon Elastic Container Service** (**ECS**), and **AWS Lambda** (for the complete list, please refer to the CloudWatch documentation at `https://packt.link/4xm6F`). When using these services, you only need to make sure that the proper configuration is in place (such as the necessary IAM permissions) for the AWS service to publish your logs to CloudWatch.

Designing for Operations

Designing for operations entails enforcing best practices across your development teams to streamline operational processes, enabling quality feedback, code refactoring, and bug fixing. When you do this, you raise code quality and limit the number and impact of issues deployed in production, allowing quicker issue identification and remediation in the development life cycle. The following are best practices that you should keep in mind when designing for operations:

- Define infrastructure as code, and manage it using the same principles and tooling that you use to manage application code (code versioning, testing, automatic building and deployment, etc.).

Infrastructure is code

In the cloud, code is not limited to your application anymore, as your entire workload can be defined as code, including infrastructure components.

- Leverage version control for application and infrastructure code using your preferred tool – for instance, **AWS CodeCommit**, **GitHub**, **GitLab**, and **Bitbucket**. Consistently managing versions of your code, whether it's for infrastructure or an application, lets you track the changes made by your teams and the various releases of your workload.

- Implement frequent, small reversible changes as much as possible to limit their impact on your workload. It also makes troubleshooting easier and facilitates remediation.

- As mentioned in previous chapters, automate the build, testing, integration, and deployment phases in your workload life cycle, taking a **continuous integration/continuous deployment (CI/CD)** approach to manage the releases of your workload. Use build and deployment tooling to help you with that (for instance, **AWS CodeBuild** and **AWS CodeDeploy**) or your preferred third-party equivalent.

- Make sure to also test and validate changes to identify as many quality issues and defects as possible before they hit your production environment. Automate testing as part of your CI/CD pipelines. Leverage the benefits of the cloud for that. In the cloud, in general as well as when you specifically define your entire workload as code, it is effortless to create new environments to perform tests as needed before deploying a new release into production. Therefore, it's good practice to use various environments to validate changes (such as testing, integration, and user acceptance).

- Finally, when you maintain your own infrastructure resources (such as EC2 instances and ECS or EKS clusters), don't forget to patch your AWS environments regularly with the latest security fixes or OS updates. AWS Systems Manager provides capabilities that can assist you with that, such as **AWS Systems Manager Patch Manager** and **AWS Systems Manager Maintenance Windows**.

Mitigating Deployment Risks

The previous section emphasized the importance of implementing best practices when designing your workload to bake in telemetry, as well as smoothen operations. Just before operations kick in, your workload must be deployed. You have a number of means to limit deployment risks to the bare minimum.

The first thing to do is to make sure you have a plan B. Whenever you do a release into production, you want to be sure that you can roll it back to the last working state should things go wrong.

You can also deploy frequent and small reversible changes to production for easier troubleshooting and to facilitate remediation if there are issues, with the option to roll back immediately.

Another best practice is to use parallel environments to deploy new releases, where you can also enforce additional validation – for instance, using blue/green or canary deployments to prevent a new release from causing severe issues in production. We've extensively covered the various options you have on AWS to do so in *Chapter 9, Establishing a Deployment Strategy*, so refer to that chapter for more details on deployment strategies.

Operational Readiness and Change Management

One last review step that you must take before actually operating your workload is to assess the readiness of your operations and change management process. This also includes assessing the actual readiness of your teams, that is, assessing whether your personnel is trained and aware of the various processes to support your workload once in production and whether your teams are large enough to cover all activities, including those outside of business hours if needed. AWS provides extensive education and training material in this regard that is available online, including the AWS Well-Architected Framework, AWS blogs, AWS events and webinars, and the Amazon Builders' Library. Further, AWS Training provides either self-paced or instructor-led training sessions. You can leverage these resources to make sure you and your team are ready, especially if you need to support a business-critical workload in production. Additionally, you must also regularly review your operational readiness. Procedures, teams, and workloads evolve over time. Being ready on day one doesn't mean you'll be ready down the road after several releases or be ready to support a totally different workload.

Another important step is to have a clear validation process to manage changes in your AWS environments. Chances are that changes involving low versus high complexity, risk, or impact on your workload will need to be evaluated appropriately within your organization. In particular, you may have to follow a totally different approval flow for high-risk, high-impact changes (such as a data model change) compared to a low-risk, mundane code update of limited scope. When such a variable approval flow is required by an organization, it is not uncommon to find at least three different types of flows – one for emergency changes (e.g., an urgent bug fix to respond to an incident), one for normal changes (e.g., infrastructure upgrade), and one for standard changes (e.g., pre-approved ones requiring a reduced set of tests). You may also define some additional categories for change to cater to high-risk or high-impact changes, for which some extra care is needed. Also, as you build up expertise in operating your workload and making changes, the border between normal and standard changes is expected to shift –changes that used to fit into the normal category become part of the standard category after a while when the change management process and the teams have both matured. Note that some tools out there can very well help to manage change (for instance, **AWS Systems Manager Change Manager** or an equivalent third-party solution) if you prefer. Change Manager lets you define how you want changes to be requested, approved, and implemented within your organization. It is integrated with **AWS Organizations** and allows you to manage changes across multiple accounts in multiple AWS Regions. You can, for instance, define change templates to automate your change processes (for instance, for pre-approved changes) to avoid having to make manual operational changes.

There are two very useful instruments that should become the bread and butter of your operations – runbooks and playbooks. You can use **runbooks** to perform procedures meant to achieve specific outcomes, including routine activities. You can, for instance, leverage runbooks to enable consistent and timely responses to expected events (such as an increase in traffic for a web application) by documenting what needs to be done in such cases. You could also implement runbooks as code (for instance, using Change Manager) and trigger the execution of such runbooks in response to events where appropriate, reducing errors caused by manual processing. Then, use playbooks to guide you in issue resolution. Playbooks are clearly documented processes that teams can use to investigate issues so that when the unexpected happens, your teams don't have to improvise. Playbooks facilitate consistency and speed when responding to failure scenarios. You can also implement playbooks as code and initiate playbook execution in response to events where appropriate, allowing you to reduce errors caused by manual processing. You could leverage **AWS Systems Manager Run Command**, **AWS Systems Manager Automation**, **AWS Step Functions**, AWS Lambda, or any other tools to script the necessary remediation actions and leverage **Amazon EventBridge** to intercept events to trigger those scripts.

Evolving Your Operations

You should keep improving your operations over time. Ideally, you want to introduce small incremental changes based on the lessons learned in your operation activities. Additionally, you also want to measure the success of such changes to determine whether they brought the expected benefits.

First, you must develop a plan to allocate time to analyze your operations activities and failures, experiment with new ideas, and make improvements. You should evaluate and prioritize opportunities for improvement on a regular basis and focus your efforts on where they can provide the greatest benefits. Note that there might be opportunities to improve in all your environments (not only production but also maybe development, testing, and user acceptance). It is also paramount to learn from failures and to share the lessons learned with the engineering community in your organization, as part of the continuous improvement process.

Second, it is important to perform post-incident analysis and review any events that had an impact on your customers to identify their causes. You can then leverage this information to mitigate the impact of such events or to prevent them from happening again. When possible, develop procedures that will be used to respond in a timely manner, possibly automatically. Here, as well, it is important to communicate the outcome of your analysis with the engineering community within your organization to spread the knowledge.

In order to achieve these different types of analysis, AWS offers a host of tools. Amazon CloudWatch gives you the ability to aggregate all your logs in a central location. On top of that, Amazon CloudTrail gives you visibility of all the API calls (through the AWS Management Console, the CLI, SDKs, and APIs) occurring in your AWS environment and across all of your accounts. So, you could, for instance, track deployment activities with CloudTrail and get insights into whether they brought any benefit using the information in CloudWatch logs (for instance, measuring improved response times or any other significant metric). You can throw in additional AWS services to help you make this analysis over time, allowing you to compare results across teams and workloads. CloudWatch also gives you the ability to export logs to **Amazon S3** (which is a best practice both for long-term storage of your logs and to lower your CloudWatch costs). Once your logs are in S3, you can leverage additional services such as AWS Glue, Amazon Athena, and Amazon QuickSight to sift through the logs, make whatever necessary data correlation, and visualize the results.

Then, once you have conducted such an analysis, make sure to review your results and discuss the responses that you plan to take with cross-functional teams and business owners. Such reviews are paramount to corroborate a common understanding, uncover additional impacts, and define countermeasures.

It is also valuable to perform a retrospective analysis of incidents and operation metrics with cross-team participants, including your leadership team and people from different areas of the business. These reviews are an effective means to identify opportunities for improvement and potential remediation, and to share the knowledge gained through lessons learned with other teams.

> **Sharing is caring**
>
> Spreading knowledge is paramount to building effective operations. Make sure to document and share the lessons you learned within your team with the rest of the engineering community in your organization (and even beyond).

The final consideration is that improvements won't happen by magic without any investment. You will have to dedicate the time and necessary resources to make continuous improvements possible in your operations. This includes setting up extra development, testing, or staging environments where you can experiment with new features, configurations, or services without affecting your live operations. Thereafter, you validate the outcomes of your tests and determine whether they meet your objectives. If the new elements prove to be successful in your extra environments, you test and develop them further to ensure that they are ready for your production environment.

Summary

This chapter first discussed the importance of culture and organizational priorities for your operating model. You then examined how to gain a better understanding of your workloads' expected behaviors by implementing telemetry, configuration and change management, deployment release management, and by preparing runbooks and playbooks. Finally, you learned about the importance of having a thorough understanding of your operations' health and how to improve them over time.

In the next chapter, you will look into bringing improvements to the reliability of your workload.

Further Reading

- The Well-Architected Framework Operational Excellence Pillar: `https://packt.link/CZbeA`

- The **Operational Readiness Review (ORR)**: `https://packt.link/53IfO`

- Instrumenting distributed systems for operational visibility: `https://packt.link/lgV1C`

12
Improving Reliability

This chapter covers the specifics of reliability engineering in AWS. The following four areas are important to ensure that you are leveraging reliability best practices in AWS: checking the foundations, assessing the architecture, adapting to change, and adapting to failure.

Being able to continuously improve the reliability of existing workloads in the cloud and adapt to an ever-changing environment is key to meeting your business objectives.

Now, prior to reading the current chapter, it will be beneficial if you go through *Chapter 6, Meeting Reliability Requirements*, first. The key concepts from *Chapter 6* are important to consider when designing your workloads for reliability. Therefore, this chapter will make constant references to *Chapter 6* without explicitly repeating the concepts.

Before you start examining the best practices for ensuring reliability, here is a quick refresher on the five design principles emphasized by AWS in the **AWS Well-Architected Framework's** reliability pillar:

- Automatically recover from failure
- Test recovery procedures
- Scale horizontally to increase aggregate workload availability
- Stop guessing capacity
- Manage change through automation

These principles are meant to guide your overall approach to improving reliability. For more details on these principles, refer to *Chapter 6, Meeting Reliability Requirements*.

Checking the Foundations

The term foundations here refers to the elements of your AWS environment that are not specific to a single workload or project and that can impact the reliability of any of your AWS workloads. In this context, your foundations consist of, on the one hand, your AWS resource constraints and, on the other hand, your overall AWS network topology.

Resource Constraints

Resource constraints can be divided into categories: service quotas and environmental constraints. The following subsections discuss these in detail.

Service Quotas

Service quotas are default pre-defined values on each AWS account; they are meant to protect you from over-provisioning AWS resources and to protect the AWS cloud from abuse. Different quotas apply to each service and could represent very different items and quantities. Service quotas are defined per AWS account and typically apply at the Region level, unless otherwise specified. Some of these quotas are adjustable and represent soft limits, while others cannot be changed and represent hard limits. When assessing your solution design, it is essential to take these quotas into account and be aware of the hard and soft limits. If you haven't done so, put a mechanism in place to monitor your usage of the AWS services to detect whenever you're getting close to any relevant quota limit. When you get close to a soft limit, you can request a service quota increase by submitting a request via the **Service Quotas console** or API at any time. However, when this happens for a hard limit, your choices will be more constrained. If you reach a service quota that is a hard limit, you will likely need to adapt your solution design somehow. Consider the options you have to amend your design and their impact on your workload. Consider whether you require an in-depth re-design of your solution. What will be the impact? What will be the effort needed? How urgently do you need the fix? Essentially, you must assess how likely you are to reach that hard limit in the near future. Answers to these questions will help you pick the right option for your workload.

But how can you effectively monitor your usage in the first place and also be alerted if you're getting too close to some service quotas? Well, that's where **AWS Trusted Advisor** comes in. It's a recommended best practice to systematically activate Trusted Advisor on your AWS accounts since it makes useful recommendations to optimize your AWS environment across several pillars of the WellArchitected Framework, such as security, cost optimization, performance, and fault tolerance, based on its observations of your AWS environment. Trusted Advisor also checks how you're currently doing against AWS services quotas. The check results provide both a summary and detailed information about your environment with regard to AWS services quotas. Further, if you work with **AWS Organizations** (for more details, see *Chapter 3, Design a Multi-Account AWS Environment for Complex Organizations*), Trusted Advisor can also report on your entire organization or a subset of it from a central location. You can enable this through the Trusted Advisor console or API. For instance, using the CLI, you can find out which checks Trusted Advisor performs, as shown in the following snippet:

```
aws support describe-trusted-advisor-checks --language "en" --region
us-east-1 | jq '.checks[] | select(.category == "service_limits")'
```

The output will consist of a succession of items, describing the various service quotas checked by Trusted Advisor, and will look something like this:

```
{
  "id": "jL7PP017J9",
  "name": "VPC",
  "description": "Checks for usage that is more than 80% of the VPC
Limit. Values are based on a snapshot, so your current usage might
differ. Limit and usage data can take up to 24 hours to reflect any
changes. In cases where limits have been recently increased, you
may temporarily see utilization that exceeds the limit.<br>\n<br>\
n<b>Alert Criteria</b><br>\nYellow: 80% of limit reached.<br>\nRed:
100% of limit reached.<br>\nBlue: Trusted Advisor was unable to
retrieve utilization or limits in one or more regions.<br>\n<br>\
n<b>Recommended Action</b><br>\nIf you expect to exceed a service
limit, request an increase directly from the <a href=\"https://
us-east-1.console.aws.amazon.com/servicequotas/home?region=us-east-1\"
target=\"_blank\">Service Quotas</a> console. If Service Quotas
doesn't support your service yet, you can create a support case in
<a href=\"https://aws.amazon.com/support/createCase?type=service_
limit_increase\" target=\"_blank\">Support Center</a>.<br>\n<br> \
n<b>Additional Resources</b><br>\n<a href=\"https://docs.aws.amazon.
com/AmazonVPC/latest/UserGuide/VPC_Appendix_Limits.html#vpc-limits-
vpcs-subnets\" target=\"_blank\">VPC Limits</a>",
  "category": "service_limits",
  "metadata": [
    "Region",
    "Service",
    "Limit Name",
    "Limit Amount",
    "Current Usage",
    "Status"
  ]
}
```

> **Note**
>
> If your account does not have an active support plan, you may get an error notice with the preceding code.

The description of the check regarding the VPC quota informs us that it will change the status from Green to Yellow when 80% of the allowed quota is reached, and to Red when the quota is reached. Now, you can check whether you are close to reaching your VPC quota in any Region by filtering out all check results that don't have a Green status:

```
aws support describe-trusted-advisor-check-result --check-id
jL7PP017J9 --language "en" --region us-east-1 | jq '.result.
flaggedResources[] | select(.metadata[5] != "Green")'
```

And if you are close to reaching your VPC quota in any Region, the previous command will produce something like the following output:

```
{
  "status": "warning",
  "region": "eu-west-1",
  "resourceId": "avvKEBoqypBNOebybZ18aSj_txWurcQZ9qb3KtDMgAQ",
  "isSuppressed": false,
  "metadata": [
    "eu-west-1",
    "VPC",
    "VPCs",
    "15",
    "12",
    "Yellow"
  ]
}
```

The output tells you that you are close to reaching your VPC quota in the eu-west-1 Region where you currently have a Yellow status for this check since the 12 VPCs currently deployed on your account represent 80% or more of the allowed quota of 15 VPCs.

This is just an example of a one-off analysis; it is recommended to put a mechanism in place to monitor service quotas systemically and at scale. You have multiple options to do so, depending on your preferences. The simplest approach is to register with Trusted Advisor to receive weekly notifications by email. A second, more elaborate approach is to leverage **Amazon EventBridge**. When the status of a Trusted Advisor check changes, a corresponding event is emitted to EventBridge. It gives you multiple choices to take further action, such as sending a notification using **Amazon SNS**, dropping a message in a queue using **Amazon SQS**, pushing the event to a data stream with **Amazon Kinesis Data Streams**, and triggering an **AWS Lambda** function or an Amazon CloudWatch alarm.

You could also directly configure CloudWatch alarms for specific quotas that you are particularly interested in monitoring more closely. When configuring the alarm, you can also decide on any relevant action to take based on the type of quota and the threshold level breached. The resulting action could consist of sending a notification using SNS, triggering any further processing down the line or creating an incident with **AWS Systems Manager**; that could initiate the relevant response plan.

The next subsection circles back to the second type of constraint that was mentioned previously: environmental constraints.

Environmental Constraints

Environmental constraints refer to the constraints imposed by the physical resources supporting the AWS infrastructure. For instance, it could be the amount of storage available on a physical disk used for Amazon EC2 instances or the network bandwidth available between your AWS environment and your on-premises environment. Such environmental constraints might impact your workload and need to be considered. For instance, if you need to retrieve data from an operational data store located in your on-premises environment, the bandwidth and latency of the network connection between your on-premises and AWS environments will naturally constrain the possible use cases.

There are a number of recommended best practices concerning network topology on AWS. The next subsection briefly discusses the most important network topology aspects; for greater details, please refer to *Chapter 6, Meeting Reliability Requirements*.

Using Highly Available Network Connectivity for Your Public Endpoints

If your workload has internet-facing endpoints that must be highly available, you must make sure that whichever component you lay on their path is also highly available, whether it is a DNS service, a **content delivery network** (**CDN**), a load balancer, or a gateway.

The major networking-building blocks for internet-facing endpoints are as follows:

- **Amazon Route 53**, an AWS DNS service, is both scalable and highly available out of the box. It provides domain name resolution, registration, and health checks. Route 53 provides consistent and reliable DNS service, independently of local or regional network conditions, routing traffic to your AWS resources and to resources outside of AWS if needed.

- **Amazon CloudFront** is an AWS CDN service. It can significantly reduce network latency by delivering content closer to the end users, improve the availability of your content owing to its distributed nature, and also limit access to your origin servers owing to edge and regional caches.

- The AWS **Elastic Load Balancing** (**ELB**) service provides various types of load balancers: **Classic Load Balancer** (**CLB**), **Application Load Balancer** (**ALB**), **Network Load Balancer** (**NLB**), and **Gateway Load Balancer** (**GWLB**). The first three types (CLB, ALB, and NLB) offer a **service-level agreement** (**SLA**) of 99.99% availability. GWLB's availability depends on how you implement your preferred third-party appliances.

- **AWS Global Accelerator** offers a way to deliver a service through a set of static IP addresses and from the optimal endpoint based on your user's location. It is also well suited for cross-Region failover scenarios (single-Region failover scenarios are usually better served by ELB load balancers). Further, unlike CloudFront, Global Accelerator can also process non-HTTP requests.

Depending on the AWS services or third-party components that you implement, make sure that the terms of their SLAs and their deployment models match the requirements of your network connectivity.

You are now ready to examine the private networking aspects at hand when building for resiliency.

Provisioning Redundant Connectivity between Your AWS and On-Premises Environments

When connecting your on-premises environments to your AWS environment, it is highly recommended to make the connections redundant to sustain the failure of any single one of them. You essentially have two private connectivity options with AWS—**AWS-managed VPN** and AWS **Direct Connect** (**DX**).

When using an AWS-managed VPN, it is highly recommended to connect your **Virtual Gateway** (**VGW**) or AWS **Transit Gateway** (**TGW**) to two separate **Customer Gateways** (**CGWs**) on your end. By doing so, you establish two separate VPN connections, and if one of your on-premises devices fails, all traffic will be automatically redirected to the second VPN. Further, when using AWS DX, it is recommended to have at least two separate connections at two different DX locations. It will provide resiliency against connectivity failure due to a device failure, a network cable cut, or an entire location failure. You can increase the level of reliability of your DX setup further by adding more connections to additional DX locations.

It is also possible to combine both technologies. By combining both, you can benefit from the consistency offered by DX most of the time and also, in case of DX failure at your DX location, fail over from DX to VPN. It can also be a cost-effective option compared to multiple DX connections to multiple DX locations.

Once secure connectivity is in place between your on-premises and AWS environments, you should be able to securely access AWS services or your AWS resources from your on-premises environment, or vice versa. The key mechanism here is **VPC endpoints**. VPC endpoints are used to provide redundant entry points for any traffic targeting a supported AWS service or a VPC endpoint service. They materialize through **Elastic Network Interfaces** (**ENIs**) deployed in your VPC subnets. Additionally, deploying VPC endpoints in multiple **Availability Zones** (**AZs**) is a good practice to reinforce resilience in case of failure in one or more AZs.

For more details about the various connectivity options, please refer to *Chapter 2, Designing Networks for Complex Organizations*.

Ensuring IPv4 Subnet Allocation Accounts for Expansion and Availability

The number of IP addresses needed by your solution might have been overlooked initially. Review these requirements and don't forget to include all the resources that will be deployed in a VPC, such as elastic load balancers, **Amazon EC2** instances, **Amazon RDS** databases, **Amazon Redshift** clusters, or **Amazon EMR** clusters. Make sure to accommodate address space for multiple VPCs in a Region and within each VPC for multiple subnets spanning several AZs. Also, make sure to leave room for expansion in the future by maintaining unused CIDR block space. As a general rule of thumb, it is better to size VPCs and subnet CIDR blocks by excess than the other way around. This is especially true if you plan to accommodate highly variable workloads that will need enough room to scale up. Thus, it is recommended to deploy large VPC CIDR blocks. Note that the largest IPv4 CIDR block you can create on AWS is a /16 block, providing 65,536 IP addresses.

Using Hub-and-Spoke Topologies instead of a Many-to-Many Mesh

Hub-and-spoke topology is highly recommended if you plan to connect either multiple VPCs to each other or multiple VPCs to an on-premises environment. This technology is meant to limit the number of connections, which can otherwise rapidly become unmanageable when you expect tens, hundreds, or even thousands of different address spaces (including VPCs and on-premises environments). In a many-to-many mesh, each node of the network will have a connection to every other node, exponentially increasing complexity as the network grows. With hub-and-spoke, there is a central node (the hub) acting as a TGW through which all other nodes (spokes) are connected.

With a central node, it becomes easier to attach new VPCs to the network and monitor traffic. Security and compliance can also be centralized through this mode. Although it does create a single point of failure, it overall increases operational efficiency, lowers costs, and simplifies scaling.

AWS TGW provides a managed hub-and-spoke solution that is highly available, by default, across multiple AZs.

Enforcing Non-Overlapping Private IPv4 Address Ranges where Private Networks Are Interconnected

It is best to avoid overlapping address spaces to avoid running into trouble when interconnecting multiple address spaces across VPCs and on-premises environments. If an overlapping issue ever occurs, you'll have to rely on NATing or switch to IPv6, if that's an option, or use VPC endpoints based on **AWS PrivateLink** or any other workaround to sort it out.

Thus, it is best to avoid running into the issue if you can and plan accordingly. Leverage Amazon VPC **IP Address Manager** (**IPAM**) or a third-party equivalent to help you with that. IPAM is a VPC feature that lets you plan, track, and monitor IP addresses for your AWS workloads.

The next section will guide you through the process of evaluating your current workload architecture. It will discuss the importance of conducting a Well-Architected Review and how this review can help identify areas for improvement in your architecture.

Assessing the Architecture

When assessing your workload architecture to determine what could be improved, the best way to get on the right path, especially if you're unsure where to start, is to conduct a Well-Architected Review. Before you proceed with it, make sure you have a clear view of the current status of your application.

Understanding Application Growth and Usage Trends

A particularly important aspect of improving reliability is to have a clear picture of your workload usage over time. First, it will help you understand whether its current usage meets your expectations. When you designed the solution initially, you made some architecture decisions based on what you knew at the time about the expected traffic your workload would need to sustain. Now, you should have the actual data to evaluate how good those assumptions were. Second, having a view on the usage growth until now gives you some initial data points to forecast the usage growth for the foreseeable future. For a more accurate usage prediction, make sure to also take into account any additional information regarding future traffic expected to hit your workload. For instance, there might be a marketing campaign planned for a specific period of time or the business might be expecting to onboard many additional clients on your workload in the following few months. Finally, putting your workload usage growth in perspective with how it actually handled that growth so far will help you determine whether your initial solution design was well suited.

Evaluating the Existing Architecture to Determine Areas that Are Not Sufficiently Reliable

In order to review your design along a specific pillar of the Well-Architected Framework, it is best to conduct a review focusing exclusively on that pillar. So, in this case, you would go through the set of questions related to the reliability pillar in the Well-Architected Review to reveal potential issues and get some advice from AWS on how to address them. The easiest way to do this in practice is to use the **AWS Well-Architected Tool**, which is accessible from the **AWS Management Console**.

The Well-Architected Review's reliability section currently takes you through the following questions:

- REL1 – how do you manage service quotas and constraints?
- REL2 – how do you plan your network topology?
- REL3 – how do you design your workload service architecture?
- REL4 – how do you design interactions in a distributed system to prevent failures?
- REL5 – how do you design interactions in a distributed system to mitigate or withstand failures?
- REL6 – how do you monitor workload resources?
- REL7 – how do you design your workload to adapt to changes in demand?
- REL8 – how do you implement change?
- REL9 – how do you back up data?
- REL10 – how do you use fault isolation to protect your workload?

- REL11 – how do you design your workload to withstand component failures?

- REL12 – how do you test reliability?

- REL13 – how do you plan for disaster recovery?

> **Note**
>
> The Well-Architected Framework is a living set of best practices. Its constituents (pillars and lenses) naturally evolve over time along with AWS services and best practices. So, don't be surprised if the next time you conduct a Well-Architected Review, you find an updated set of questions to answer, as it is expected to happen over time.

Answering the preceding questions, or at least those relevant to your workload, is a good starting point. Upon completion of the review, the Well-Architected Tool will identify potential issues and provide recommendations that follow AWS best practices. Often, you may have already conducted a Well-Architected Review of your workload in the past, and it might be time to do another run instead. In that case, create a milestone to store the results and status of the previous review and conduct a brand-new review. Comparing the results of the new review with your previous milestone will highlight the areas of improvement yet to be tackled.

As an example, consider that you've just conducted a review of your workload focusing on the reliability pillar. The report generated by the Well-Architected Tool then clearly identifies all the high-risk issues to be tackled first, before you can address the medium-risk ones, as shown in *Figure 12.1*:

Improvement plan

Improvement item summary

High risk: 10

Medium risk: 3

Pillar	High risk	Medium risk
Reliability	10	3
Security	0	0
Operational Excellence	0	0
Performance Efficiency	0	0
Cost Optimization	0	0
Sustainability	0	0

Figure 12.1: Example of a Well-Architected Review results overview

Then, you simply go through each identified high-risk issue to find out what the issue is and what recommendations AWS makes to address it. *Figure 12.2* shows an example of the assessment for one of the 10 high-risk issues identified:

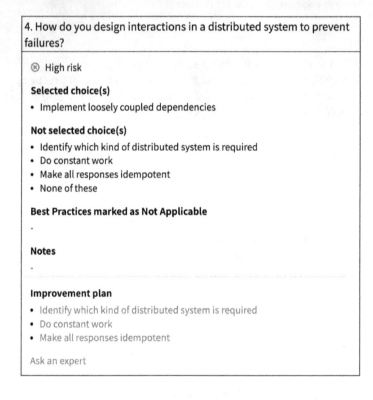

4. How do you design interactions in a distributed system to prevent failures?

⊗ High risk

Selected choice(s)
• Implement loosely coupled dependencies

Not selected choice(s)
• Identify which kind of distributed system is required
• Do constant work
• Make all responses idempotent
• None of these

Best Practices marked as Not Applicable
-

Notes
-

Improvement plan
• Identify which kind of distributed system is required
• Do constant work
• Make all responses idempotent

Ask an expert

Figure 12.2: Example of a high-risk issue

This example documents that we implemented loose coupling, when it made sense, between the components of our solution design to prevent failures. As a result, the Well-Architected Tool detected a potential high-risk issue since we fell short of implementing a set of best practices considered essential by AWS to actually prevent failures. Now, the outcome of the Well-Architected Tool is to be taken with a pinch of salt. Did we purposely ignore the recommended best practices or were we simply unaware of them? In both cases, the outcome of the tool is the same, except that, in the former case, we may already have a mitigation in place, while the latter requires us to re-visit our design in light of the newly discovered best practices.

After you've gone through the entire report from the Well-Architected Tool, you end up with a relatively exhaustive list of improvement points, each with a link pointing to the Well-Architected Tool questions documentation, which shares some extra information about possible remediations. You will notice that the questions documentation is relatively succinct, but it contains various links to other AWS documentation, whitepapers, blogs, and videos that provide more elaborate guidance to address the issue at stake.

While it is strongly recommended that you review the suggested remediations for all the identified high-risk issues, you don't need to implement every single one of them. Again, take the review results with a pinch of salt. The remediation recommendations are generic and you must evaluate whether they make sense in the context of your own workload. Ultimately, you must have a good enough understanding of them to decide which ones should be implemented, in which order of priority, and why you would drop some of them, if any. To make these decisions, evaluate how the recommendations can help meet your business objectives or fulfill the SLAs, if any, that your workload must adhere to. You may identify some quick wins, that is, recommendations that are relatively easy to put in place but would bring a significant improvement to your workload reliability. These are very likely the recommendations you want to start with before considering those that might require a profound re-design of your solution and much more effort to implement.

The Well-Architected Framework is a powerful tool to drastically improve your workload solution design but it is not the only one. You can also rely on your own experience or the experience of other team members to analyze the existing workload. In particular, you can analyze the data collected on the workload so far, such as workload profile information, data related to failure events, if any, and data collected from either stress tests or reliability tests. Such a dataset can deliver a wealth of information on the behavior of your workload in conditions of stress, for instance, when infrastructure resources are scarce or cannot be instantiated. Analyzing this data thoroughly will help you identify possible bottlenecks or flaws in your design.

Remediating Single Points of Failure

An essential part of improving the reliability of your workload is to eliminate single points of failure. Unless your workload has a monolithic structure, you are dealing with a distributed architecture. When you consider an end-to-end transaction sent across that distributed architecture, the data flow hops from one component to the next and then goes back to the client that submitted the request. What happens if one of these components fails? Will you be able to provide an alternate path for the current transaction and any others coming after that?

If you're not convinced this is an issue or not 100% sure that your workload would handle such a failure appropriately, I recommend you read the *Challenges with Distributed Systems* paper from the Amazon Builders' Library, which is referenced in the *Further Reading* section of this chapter.

Here you will briefly re-examine the most important aspects when architecting for reliability; for more details, I encourage you to read the *Designing for Failure* section of *Chapter 5, Meeting Reliability Requirements*. The following are some important considerations when dealing with component failure. First of all, did you carefully consider the type of distributed system you were building in your initial design? Does your workload require rapid, synchronous responses within seconds? Or are responses allowed within a broader time window, for instance, a few minutes or more? Or does your workload mostly work as an offline system through batch processing? These various system types impose very different constraints on the architecture and call for possibly very different designs.

Second, did you implement loose coupling, when it made sense, between the various components of your system? Tight coupling, unless strictly necessary, is a recipe for failure since the failure of a component directly impacts any component tightly coupled with it. Loose coupling brings more isolation between components. As such, it increases the overall system resilience, since a failed component can be replaced without impacting any other components relying on it. It also brings more agility in managing the software life cycle since component development and deployment can take place independently of other components as long as the contract (API) between them is respected. Multiple AWS services let you implement loose coupling, such as **Amazon SQS** to buffer requests, **Amazon SNS** to send notifications, **AWS Step Functions** to orchestrate workflows, and **Amazon Kinesis** to manage data streams. It is of utmost importance to select the right design pattern before picking up the technology bricks to implement it.

Third, mimic the AWS infrastructure, with redundancy at every layer, to avoid single points of failure. Make sure to distribute your workload resources across multiple AZs, as a minimum. Given the resiliency of AWS infrastructure in general, distributing your resources across multiple AZs automatically ensures strong protection against power outages or disasters such as fires, lightning strikes, floods, or earthquakes. When using AWS services, make sure to know which ones offer you direct protection against an AZ failure—for instance, **Amazon S3** or **Amazon DynamoDB** have built-in multi-AZ redundancy—and which ones need you to deploy the necessary resources across multiple AZs—for instance, **Amazon EC2**—to benefit from redundancy.

Fourth, leverage scaling mechanisms to avoid single points of failure.

Enabling Data Replication, Self-Healing, and Elastic Features and Services

To avoid losing important data, it is recommended that you replicate it across multiple AZs, as a minimum. In some cases, it might even be necessary to replicate it across multiple Regions. The way this replication takes place depends on several factors, including your requirements and the AWS services used to handle the data.

First things first, you must precisely determine your needs in terms of data replication as a part of your solution design to address your business objectives in terms of reliability. Consider the following questions: can you afford to lose some data? And how much data can you afford to lose? In case of failure, is it acceptable for your workload to run with limited functionalities (degraded mode)? How fast do you need to have it back fully operational? Is it acceptable for your business to operate in a different Region or is it prohibited (for example, by some regulatory constraints)? Do you have a business need to be operational in multiple Regions?

Determining the answers to the preceding questions will help you define your needs in terms of data replication. Two critical metrics are the **recovery point objective (RPO)** and the **recovery time objective (RTO)**. RPO refers to the maximum acceptable amount of data loss measured in time. Basically, how much data can you afford to lose? RTO is the maximum acceptable length of time that a service, application, or function can be unavailable after a disaster occurs, or downtown. Both metrics are vital for developing effective disaster recovery strategies and will be key parts of risk assessment and business impact analysis.

You have probably already gone through these considerations for the initial solution design but it might be a good time to re-visit them and your requirements with what you've learned so far.

You are now aware of the various aspects involved in assessing your architecture and their significance. Another crucial element of ensuring reliability is staying agile in your cloud operations. Regular reviews and updates to your workload monitoring strategies are crucial to ensure they remain effective and relevant. This is covered in the next section.

Adapting to Change

It is recommended to frequently review how workload monitoring is implemented so you can update it if needed, based on significant events and changes. Among the things to review, look at your metrics, KPIs, and SLAs. Verify that these metrics are still meaningful to your current workload since business priorities change over time.

Additionally, auditing your monitoring setup is another safeguard to make sure that you know whether an application is meeting its reliability objectives. Establishing regular operational performance reviews and conducting knowledge-sharing sessions is a great way to enhance your organization's ability to achieve higher operational performance.

For example, AWS service teams conduct internal weekly reviews assessing various teams' operational performance. This allows them to share learnings among teams. Since there are too many reviews to go through each time, AWS has created an application that it has open sourced, called **The Wheel**. They use this application to randomly pick a team to review. See this blog for more details—`https://packt.link/l7Nf8`.

You can think of using a similar mechanism inside your organization.

Adapting to Failure

After you have designed your solution to withstand failure, you need to test that it's actually as resilient as you expect. The following subsections discuss the various ways in which you can test your solutions.

Using Playbooks to Investigate Failures

Whenever a failure occurs, you want to react consistently and remediate promptly. This is what playbooks are about. They provide you with a list of steps to be followed in order to fully identify the issue and address it effectively.

When an undocumented failure scenario occurs, focus on addressing the issue first. After the fire has been put out, come back to the playbook and update it by adding the new scenario along with the exact steps that you took to address the issue.

Performing Post-Incident Analysis

Performing post-incident analysis, also known as post-mortem analysis, is essential in the process of failure management. It helps you potentially prevent a recurrence of the failure. The failure scenario might have revealed a weakness in your architecture or certain wrongly adjusted thresholds that prevented early remediation. In any case, you can use the conclusions of the analysis to reinforce your workload resiliency.

Additionally, don't forget to identify and add any missing tests to your quality gates that would be able to recreate the failure conditions and prevent further deployment of the workload without a fix.

Testing Functional Requirements

This was mentioned already when deployment automation was covered, testing should be part of your workload deployment pipeline(s). If the deployment fails to pass the tests, the pipeline(s) should be terminated and the project team should be notified to take action.

Functional tests can typically be categorized into unit tests and integration tests to validate the workload's functional requirements.

Testing Scalability and Performance Requirements

These tests are meant to stress test your workload so that you're confident that it can meet your scalability and performance requirements.

As mentioned earlier, once you've automated your deployment pipeline(s), it becomes straightforward to roll out a new environment where you can deploy your workload for performing stress tests.

Testing Resiliency Using Chaos Engineering

Chaos engineering is a discipline that emerged at Netflix when they wished to make their platform more resilient. They had the idea of creating perturbations to test their system's ability to handle them. It led to the creation of the **Chaos Monkey** and the **Simian Army**, which now provide resiliency testing tools. These tools have now become a best practice for improving the resiliency of distributed systems.

Thus, it is recommended that you test and reinforce your workload by injecting failures into your pre-production and production environments. The idea is to conduct an experiment that involves injecting a failure into your environment. You then form a hypothesis as to how your workload should handle the failure. Then, you execute the test and measure the impact and how your workload actually dealt with it. Whatever the outcome, you've learned something. If your workload failed the test, then it gives you an opportunity to test your recovery procedures before fixing the design or the code. If it passed, you carry on with another test, maybe with a slightly larger blast radius or tackling a different portion of the workload, and again document the results and see how things go. In this manner, you can keep augmenting the blast radius until you're confident that your workload meets its reliability requirements.

Ultimately, you will have an entirely new set of repeatable tests that can inject various types of failures and make sure to reduce your overall risks.

Now, should you use this practice in production environments?

Well, it is recommended that you do it, but take some precautions. Remember, the objective is not to break things but to reduce your risks. For instance, if your workload consists of a **video-on-demand (VOD)** service, you might not want to simulate a large-scale event during prime time. You want to make sure that your experiments don't impact your users. Don't forget, if your workload made it to production, it has already passed all the functional, stress, and resiliency tests in other environments.

On AWS, you should validate that your components that deploy resources in VPCs can handle failure in one AZ, or maybe even two AZs, depending on the reliability requirements defined by your business. Thus, if you use Amazon EC2 instances or Amazon RDS databases, make sure that your workload can handle losing one (or more) EC2 instance(s) or one (or more) RDS instance(s). You can also test a failure in your dependencies. Consider the following scenarios: what if an external dependency, local or remote, cannot be reached, times out, sends back errors, or throttles your requests? What if your DNS service fails to resolve a name for one of your dependencies?

To help test situations such as the ones mentioned previously, AWS introduced a service called **AWS Fault Injection Service** (**AWS FIS**). You can use this service to gradually impair the capabilities of some of the underlying AWS services such as Amazon EC2, Amazon ECS, Amazon EKS, and Amazon RDS. Note that FIS is integrated with Amazon CloudWatch alarms; so, for instance, in production, you could trigger an alarm and interrupt a test if it becomes too disruptive.

FIS is naturally not the only tool available for chaos engineering. There are now a number of open source and commercial offerings available out there, such as AWS FIS and AWS Resilience Hub, Azure Chaos Studio, Chaos Monkey, and ChaosBlade.

Conducting Game Days Regularly

Game days are practice drills in which teams simulate issues in their systems in a safe and controlled manner to test their own readiness to address or fix those issues. Game days are very useful to test your automated procedures and playbooks and how they respond to unexpected events. The idea is to make sure that you are ready when adverse events occur in production.

The game day environment should be as similar as possible to your production environment. Game days should involve the same teams and people who are in charge of the production environment. You will inject perturbations or failures in the environment to reproduce possible situations that these teams will have to face in production. This will help you determine improvement areas, whether with the architecture, code, runbooks, or playbooks.

By repeating game days on a regular basis, you can better help your teams build "muscle memory" so they know how to respond when failures occur in production.

Summary

In this chapter, you looked at the specifics of reliability engineering within AWS, focusing on four critical areas to ensure adherence to reliability best practices: foundational checks, architectural assessment, adaptation to change, and failure response. The chapter further explored foundational elements such as AWS resource constraints and network topology, detailing service quotas and environmental constraints that could impact AWS workloads' reliability.

In *Chapter 13*, *Improving Performance*, you will look at the various aspects of improving the performance of your cloud architecture in depth.

Further Reading

- The Well-Architected Framework's reliability pillar: `https://packt.link/4JxeS`
- The Hybrid Connectivity whitepaper: `https://packt.link/fGHgY`
- Challenges with distributed systems: `https://packt.link/zRwcj`
- Implementing Microservices on AWS: `https://packt.link/TeNlP`
- Reliability, constant work, and a good cup of coffee: `https://packt.link/ESCgk`
- Timeouts, retries, and backoff with jitter: `https://packt.link/LLOrv`

13
Improving Performance

Designing performant workloads is essential to maintain your service level agreements and to ensure customer satisfaction. This chapter will discuss the various aspects you can focus on to improve the performance of your workload on AWS. These aspects assess the current performance of your workload against your business requirements, identify and examine its performance bottlenecks, and eventually, test and recommend potential remediation solutions.

Prior to reading this chapter, you are encouraged to complete *Chapter 8, Meeting Performance Objectives*. The key concepts from *Chapter 8* are important to consider when designing your workloads for performance efficiency. Therefore, this chapter will make constant references to *Chapter 8* without explicitly repeating the concepts.

Before you start examining the best practices for ensuring performance efficiency, here is a quick refresher on the five design principles of the **AWS Well-Architected Framework** performance efficiency pillar:

- Democratize advanced technologies
- Go global in minutes
- Use serverless architectures
- Experiment more often
- Consider mechanical sympathy

These principles are meant to guide your overall approach to improving performance. For more details on each of these principles, please refer to *Chapter 8, Meeting Performance Objectives*.

This chapter will begin by discussing how you can assess the performance of your current architecture against your performance objectives, as defined by your business requirements. You will learn how to identify potential performance bottlenecks that your solution architecture may present and how you can examine and address any such bottlenecks. You will then explore how your architecture can be improved to remediate those performance issues.

Reconciling Performance Metrics against Objectives

You need to have a comprehensive understanding of where your solution architecture stands with respect to your performance objectives. So, first, you need to have a clear understanding of the performance expected from your workload. Consider whether any specific non-functional requirements related to performance have been defined for your workload. Such requirements could, for instance, include the following:

- **Response time**: Your workload must respond to incoming requests within 1 second, 99% of the time

- **Throughput**: Your workload must be able to process 500 incoming requests per minute

- **Error rate**: The percentage of requests resulting in an error should not be more than 0.1% of all requests

If no such requirements have been explicitly documented for your workload, it is strongly advised that you engage with the product owner or business leader in charge of the solution to make sure that the expectations of their customers are well understood and catered to. As a solutions architect, you are expected to challenge them so that they come up with a reasonably well-thought-out set of performance objectives.

Now, assuming that you have gathered a list of the performance objectives that your workload must meet, you need to translate those objectives into measurable metrics that you can actually monitor. To achieve that, you will define a set of measurable **key performance indicators** (**KPIs**) that best describe your performance objectives. Then, you will need to collect the necessary metrics from the various components making up your workload (compute, storage, network, etc.) in order to compute each KPI. Once all your KPIs are computed, you can then do two more things to make your life easier – first, build a dashboard to visualize your KPIs status, and second, set up alarms for when these KPIs breach the thresholds corresponding to your performance objectives.

AWS provides a number of tools to assist you with collecting metrics from the various resources you deploy as part of your workload. The best AWS service to use for that purpose is **Amazon CloudWatch**. CloudWatch brings a plethora of features, from real-time metric collection to logs collection, dashboarding, alarms, and more. Monitoring consists essentially of four distinct phases – data generation, metric calculation, processing and alarming, and storage and analytics. In the generation phase, you collect data from the various components constituting your workload that are relevant to the KPIs you want to monitor.

AWS collects various types of data from the various services that you can use. Metric data is either directly collected through CloudWatch agents or resides in the logs collected by CloudWatch. In the second phase, you leverage all the relevant data from pre-defined metrics (e.g., CPU or disk usage) or the collected logs to define and calculate your workload-specific metrics or KPIs. In the third phase, you essentially define alarms in order to be notified when some of the monitored metrics breach a given threshold. This is typically done using **Amazon Simple Notification Service (SNS)**.

On its own, CloudWatch also uses statistics and machine learning to determine what your workload baseline looks like; it then reports any behavioral anomalies over time when they occur. In the storage and analytics phase, you decide what to do with the logs collected by CloudWatch. By default, these logs are stored indefinitely, but you probably won't need to do that. Storing logs in CloudWatch bears a cost, and you only need to keep logs for as long as they're needed to meet your business or operational imperatives; you may have some operational requirements to keep the logs for several weeks or months (for instance, for analysis or investigative purposes), but you may also have regulatory constraints that force you to keep some or all logs for several years. You can alter the retention period of your logs in CloudWatch; this can be adjusted for each individual log group, depending on your own requirements, from 1 day to 10 years, unless you want to keep the default indefinite retention period.

You can also export your logs to **Amazon S3** to better cope with mid- to long-term storage, applying a storage life cycle policy on the S3 bucket where they are stored. This allows you granular control and freedom to move your logs through the various storage tiers offered by S3 (Standard- or One Zone-Infrequent Access, Glacier Instant or Flexible Retrieval, and Glacier Deep Archive). These tiers also let you optimize your log storage costs and access data in the logs, using any analytical tool that can query the data in S3 – for instance, **Amazon Athena**. Additionally, when your logs are still in CloudWatch, you can leverage **Amazon CloudWatch Logs Insights** to run analytics on your log data. CloudWatch Logs Insights lets query your log data interactively and provides simple graphical visualization to help you understand your query results. CloudWatch also provides a log event subscription mechanism to let you process your log data in real time, also allowing you to stream it to many other destinations such as log management solutions (e.g., Logstash), search engines (e.g., OpenSearch), or any other destination. For this real-time log processing mechanism, CloudWatch relies on **Amazon Kinesis Firehose**, **Amazon Kinesis Data Streams**, and **AWS Lambda**.

We recommend checking some of the resources in the *Further Reading* section at the end of this chapter if you would like to dive deeper into CloudWatch.

After you have dealt with all the aforementioned aspects, you are all set for the next step, which is to identify areas of improvement within your workload.

Identifying and Examining Performance Bottlenecks

Now that your KPIs are in place, you're ready for the next step, which is to identify performance bottlenecks and determine what could be improved in your solution architecture. These improvements are something that, as a solutions architect, you should always relentlessly look for. Even in situations where you already meet your performance objectives, there might still be some margin for improvement. You may, for instance, discover that you could potentially still meet your performance objectives using fewer resources, thereby reducing your costs and carbon footprint at the same time.

If you have set up alarms and have been notified of out-of-range KPI value(s), you know exactly which KPI(s) to use to start your investigation. If no alarm was set up or triggered but you still want to scrutinize your KPIs, start by looking at your most important KPI first, then proceed with all your KPIs in decreasing order of importance to identify any potential performance issues. If you spot anything suspicious (for instance, KPI values that are out of the expected range or close to it, or KPI values following an unexpected trend over time), then conduct an in-depth investigation of the relevant KPI(s) to identify the root cause.

Suppose your workload is an e-commerce application and one of the business KPIs you defined is the conversion rate, which describes the rate at which visitors make actual purchases. Simply put, the conversion rate equals the total number of sales divided by the total number of visits to your e-commerce application. Now, suppose you notice that the conversion rate has been consistently dropping for the past several days or weeks, despite ongoing marketing campaigns. You see no reason from a business perspective to justify this drop, especially considering that your company has hit the market with some marketing campaigns recently. So, you start investigating potential technical reasons. Is your application performing as expected? In particular, are the various screens or pages of your web application loading fast enough for each visitor?

Excessive latency on the page screen or upon loading is a common reason for visitors to give up. So, you examine the stats for each page in CloudWatch metrics, only to realize that the screen or page showing the detailed view of your catalog items has been taking more and more time to load across the past few days or weeks. You also notice, through CloudWatch, that the number of visitors has increased steadily, which is in line with your expectations given the ongoing marketing activities. Accordingly, you can leverage CloudWatch to investigate further to understand what exactly is causing that extra response latency on that particular screen or page. Has that particular component hit a limitation of resources such as CPU or memory, is processing being queued due to the increased number of incoming visitors, or has the catalog database become a bottleneck? All these are potentially valid reasons, but you won't know for sure until you look at the numbers.

CloudWatch provides you with many instruments to effectively pinpoint the root cause of a performance issue, such as the one described in the previous example. Apart from logs and metrics, alarms, logs insights, and dashboards, CloudWatch has also added a couple of key features to the family in the last few years, namely **CloudWatch Real User Monitoring** (**CloudWatch RUM**), **CloudWatch Synthetics**, **CloudWatch Evidently**, and **CloudWatch ServiceLens**.

CloudWatch RUM is a capability that lets you identify and troubleshoot issues related to performance from a client-side perspective. CloudWatch RUM's purpose is to assist application developers in reducing the **mean time to resolution** (**MTTR**) for client-side performance issues. You basically insert a JavaScript code snippet provided by CloudWatch RUM to instrument your web application, instructing RUM to collect data on a percentage of a user's requests. RUM will then work its magic, collecting data, anomalies, errors, and stack traces on your behalf from the end user perspective. You can also combine data provided by RUM from the client-side frontend of your application with the data collected by **AWS X-Ray** from the server-side part of your application. This will then provide you with an end-to-end view of the performance of your application, based on real end-user interactions with your application.

CloudWatch Synthetics is another capability that lets you create canaries (i.e., scripts that run on a schedule) to test your HTTP/S endpoints and APIs. It supports canary scripts written using Node.js or Python at the time of writing. CloudWatch Synthetics will then execute these scripts on a schedule specified by you to check the availability of your endpoints and/or APIs, assessing their latency. It will then provide you with the results, together with screenshots of your application UI. CloudWatch Synthetics also integrates with X-Ray to get an end-to-end view of requests made to your application, identifying any performance bottlenecks encountered by the canaries.

CloudWatch Evidently is a capability that enables application developers to introduce experiments and feature management in their application code. It enables essentially two use cases – implementing shadow launches (aka feature flags) and performing A/B tests. Shadow launches entail rolling out a new application feature for a percentage of your end users, measuring its impact before you roll it out to your entire user base. Evidently also lets you test multiple variations of a feature, or A/B testing, so that you can understand which variation performs best.

Finally, CloudWatch ServiceLens is a capability of CloudWatch that provides you with an overview of your application end to end. It collects data from CloudWatch and a bunch of other services, such as X-Ray, and plots that information as service maps. A service map displays your application components as nodes and attaches information about traffic, latency, and errors to each node and the connection between the nodes. It also lets you investigate further once you've identified a problematic node by showing any related detailed information it has collected, such as metrics or logs.

Coming back to the previous example with the e-commerce application, you would typically use any combination of the preceding capabilities of CloudWatch to clearly identify the culprit. In this case, it may point, without any doubt, to the database holding the catalog items, providing enough details for you to realize that the database is properly sized for writes but clearly undersized for reads.

As we mentioned in *Chapter 8*, *Meeting Performance Objectives*, you should make the most of these CloudWatch capabilities, focusing on end-user digital experience management in order to clearly identify and isolate performance bottlenecks.

Recommending and Testing Potential Remediation Solutions

Now that you have clearly identified the potential bottlenecks in your workload, it's time to devise recommendations to get rid of them. As a solutions architect, you have to devise a rationale for each recommendation that you make. The rationale will typically include a description of the issue, a set of options that you have considered to solve it, and finally, the recommended solution and the reason why you recommend it. Luckily, in most cases, you don't have to reinvent the wheel, as chances are that the community of solutions architects has already faced similar issues and devised design patterns, and anti-patterns, that you can directly utilize. AWS has collated this collective intelligence, including many best practices, in its Well-Architected Framework for you to leverage as much as you can. AWS also regularly publishes and updates reference architectures and prescriptive guidance to address determined use cases and well-known issues.

Consider the previous e-commerce application example, where you identified that the database was the actual bottleneck slowing down the catalog screen or page load. More specifically, you pinpointed that database reads didn't follow the pace of the incoming traffic. There are potentially multiple options to address the issue; each of them presents pros and cons. It is for you as the solutions architect to evaluate them, build a rationale for your choice, and then, eventually, test your recommendation. But what if you lack the data to make a clear decision between two (or more) options? Fortunately, the cloud makes experimentation easy, so instead of spending countless hours or days discussing which option is the best based on thin air, simply try them out and then back up your choice with actual data from those tests. In this case, you would, as a minimum, consider the following options:

- Increasing the size of the resources supporting the database (add more CPU, more RAM, etc.)
- Splitting the load between reads and writes by adding a read replica to your database
- Adding a cache mechanism to offload reads of the most popular items from the database
- Splitting the database, with a separate catalog database optimized for reads

This is not an exhaustive list of options but presents the major options that come to mind, based on some of the most common design patterns for two-tier or three-tier applications. The first option consists of strengthening the database infrastructure to release the pressure on the infrastructure. It is far from ideal for multiple reasons (it's not addressing the read issue specifically, it's not cost-optimized for sure, and not optimized for CO_2 emissions either), but it can work as a quick and temporary fix with nearly zero risks (there is no code change and no application re-architecting), provided that you take care of the potential downtime during the database infrastructure upgrade.

Database technical metrics (such as CPU usage, I/O per second, etc.) collected by CloudWatch provide you with some extra information to estimate what the required infrastructure bump could be. Again, that's alright for a quick fix, but as an architect, you can certainly do better and keep looking for the best option to address the issue in the longer term.

For the remaining three options, you would typically list the pros and cons of each. For instance, the second option consists of adding a read replica, which is relatively easy to do, depending on the database technology being used. For instance, if you've used the AWS relational database service **Amazon RDS** when adding a read replica, then the service creates a read-only database instance from a snapshot of the source database. It then leverages the database engine's native replication capabilities to asynchronously copy the changes made to the source database over time to the newly created read replica, in order to keep it in sync with the source database. Note that creating the snapshot from the source database can cause a short I/O suspension if your source database is deployed in a single **Availibility Zone (AZ)**, but this might not even be an issue, depending on your application's **service level agreements (SLAs)**. This second option is already more targeted at the issue at stake compared to the first one; more infrastructure resources will be used (another database instance) but specifically to boost the database read performance. Using a read replica may induce some application code changes, unless you already have a proxy in front of the database, which could be used to intelligently route reads to the read replica and writes to the primary database.

The third option contemplates the addition of a data cache to store the most frequently accessed items and reduce the load due to reads on the database. It will naturally consume additional resources, not all the reads will hit the cache (the exact proportion of reads still hitting the database will vary depending on the cache size), and this will have an impact on your application code, so you have to leverage the cache using one of the caching design patterns (write-through or lazy loading).

The last option takes a step back to re-architect the solution and consider partitioning the database, potentially using a different technology for the database dedicated to the items catalog versus the technology already in use for the existing database. Assume that you've been using a relational database so far to persist your application data. Then, it is worth considering using a NoSQL or an in-memory database specifically to store your items catalog. The data consumption patterns, the type of data, and the flexibility and scalability requirements related to your items catalog might indeed be a better fit for other technologies, so it's worth a shot.

As you just examined, each option has its pros and cons, and it's your task as a solutions architect to make an informed decision based on a clear rationale.

In general, there are many different things that you should consider when attempting to improve the performance of existing solutions on AWS. The initial question for you, as a solutions architect, is, "*Is the current architecture the optimal architecture for my performance requirements?*" If you cannot answer, or if you're not convinced that the answer is a plain "*yes*," then you should look around for existing reference architectures from either AWS, AWS Partners, or the IT architect community of practice at large. The following subsections discuss in detail some other approaches that you can adopt when evaluating solutions to improve performance.

Leveraging High-Performing System Architectures

As has been mentioned a number of times in this book already, there is no reason to try and reinvent the wheel. AWS and cloud computing have been around for over 15 years. The collective wisdom and knowledge of its community of practice have formulated prescriptive guidance in terms of design patterns and best practices. You will find much insight already in the Well-Architected Framework, which is great as a starting point to then discover other resources. For instance, you will find a lot of extremely useful material through publicly available online resources, such as the AWS blogs and white papers, the *Amazon Builders' Library* website, and the *AWS Prescriptive Guidance* website.

When looking for improvements, consider every component in your solution, and unless there was a clear rationale to select the technology that has been used to implement them, review possible alternatives among AWS portfolio of services, for there might be a better match for your performance requirements. In some cases, you might, for instance, find opportunities to eliminate inefficient operational tasks by replacing an existing component with a managed service from AWS. In some other situations, you may conclude that a component would be better served by another or a new kind of technology. Consider, for instance, in the case of a database, evaluating purpose-built database engines (e.g., a NoSQL, a graph, a time series, or a document database) and reviewing the corresponding AWS managed services (**Amazon DynamoDB**, **Amazon Neptune**, **Amazon Timestream**, and **Amazon DocumentDB**) in relation to your performance requirements, determining whether one is a better fit compared to your existing database engine.

There are several things to keep in mind when architecting for performance efficiency; here are a couple of simple tips that will help you improve your workload infrastructure's performance posture:

- Use auto scaling to automatically add or remove compute resources based on demand, in order to ensure that your application has the required resources to meet demand without over-provisioning.

> **Note**
> The preceding point does not just refer to EC2 Auto Scaling or AWS Auto Scaling; several AWS services provide auto-scaling capabilities that you might want to enable and configure to meet your needs.

- Use instance fleets to easily create and manage a mixed fleet of EC2 instances. This can help you optimize your workloads using a mix of instance types and purchase options (e.g., On-Demand, Spot, Reserved Instances, and Savings Plans).

- Use placement groups to improve the network performance of your instances. Placement groups allow you to group EC2 instances within a single AZ, enabling them to communicate with each other with very low latency.

Using Global Service Offerings

Some AWS services are there to assist you if you need to scale globally to reach performance efficiency. The following is a list of some tips for regarding what services to use:

- Use **Amazon CloudFront**, an AWS **content delivery network** (**CDN**), to deliver your static and dynamic content (e.g., HTML, CSS, JavaScript, and images) with low latency and high transfer speeds

- Use edge computing services such as **AWS Lambda@Edge** to run code globally in response to specific CloudFront events (e.g., a viewer request, response, or exception)

- Use **AWS Global Accelerator** to improve the performance of your applications by routing traffic through the AWS global network to the optimal AWS Region

Since good intentions and rule of thumb are not sufficient to support architectural decisions and technical recommendations, you want to confirm your choices by measuring their actual impact on your workload performance. So, after you have put together a set of recommendations, including a clear rationale to support each of them, it is time to test them. For that purpose, simply use the same tooling that you used to measure the current performance of your workload, and then perform some load tests to benchmark your proposed revised architecture against the current one. Please refer to the *Reconciling Performance Metrics against Objectives* section where this was discussed in more detail – in particular, the discussion of the CloudWatch family of capabilities to conduct real user testing, process custom metrics, report on your KPIs, and so on.

Summary

This chapter began with a discussion of what it means to reconcile performance metrics against performance objectives. You then reviewed the elements involved in identifying and examining performance bottlenecks. The chapter then concluded with a review of the recommendations you might provide, also focusing on the importance of testing the potential remediation solutions.

Chapter 14, Improving Security, will discuss how you can improve security in your architecture.

Further Reading

- The Well-Architected Framework Performance Efficiency Pillar: `https://packt.link/JLmg8`

- The AWS Architecture Center: `https://aws.amazon.com/architecture/`

- The Amazon Builders' Library: `https://aws.amazon.com/builders-library/`

- AWS prescriptive guidance: `https://aws.amazon.com/prescriptive-guidance/`

- Getting started guide with Amazon CloudWatch: `https://packt.link/i9Y3T`

- An observability journey with Amazon monitoring tools: `https://packt.link/27pM4`

14
Improving Security

Designing secure workloads is essential to protect your data and systems and to be able to respond in a timely way, and successfully, to security threats. As part of the continuous improvements you want to bring to an existing solution on AWS, improving security should be at the top of your priorities list. Whether the concerned application is a public-facing application or for internal use only, it is paramount to ensure that you have a trustworthy foundation in place on AWS for your application to run in a safe environment.

In w, *Determining Security Requirements and Controls*, you examined how to leverage IAM and identity federation in your solution to provide granular access control. You also reviewed the best practices to protect both your infrastructure resources—using tools such as **AWS Web Application Firewall** (**AWS WAF**), **AWS Shield**, and **AWS Firewall Manager**—and your data using encryption at rest with **AWS Key Management Service** (**AWS KMS**) and enforcing encryption in transit. Finally, the chapter covered incident detection and response for worst-case scenarios, leveraging tools such as **AWS CloudTrail**, **AWS Config**, **Amazon GuardDuty**, **AWS Security Hub**, and **Amazon EventBridge**. If you don't recall these topics well or feel unsure about some of these aspects, it's worth revisiting *Chapter 5, Determining Security Requirements and Controls,* before reading any further.

Building on *Chapter 5*'s foundations, this chapter will cover how to enhance the security of your existing AWS solutions and be better prepared to handle potential security threats. This can be done in two steps. First, you assess the existing environment where your solution is deployed for vulnerabilities and make recommendations on how to address the most common security issues you may come across. Second, you deal with not only detecting vulnerabilities but also automating responses to such issues so that you can feel more confident about your solution's security posture.

Evaluating the Environment for Security Vulnerabilities

Before you can make any adjustments to your solution or its AWS environment, you need to assess them against a set of security best practices. AWS provides a comprehensive list of security best practices, which can be found in the Security pillar of the **AWS Well-Architected Framework**. We've used this framework extensively across this book, and we will rely on its wisdom in this chapter since it represents the central reference for AWS best practices. Now, apart from this framework's security pillar, there is another essential resource at your disposal, which is the **AWS Security Reference Architecture (SRA)**. The SRA consists of a set of guidelines that helps you implement AWS security best practices leveraging the rich portfolio of AWS services. In its current form, it actually takes you through a concrete example showing how to set up your AWS environment in order to securely run a three-tier web application. It is a must-read document for anyone willing to ensure they clearly understand the various AWS concepts and services involved to follow AWS state-of-the-art recommendations in terms of security. The SRA is expected to evolve following not only the continuous evolution of AWS services and their features but also the readers' feedback, so don't miss out on an opportunity to get your voice heard if you have suggestions for AWS to improve it further. For these two key resources and more, please consult the *Further Reading* section at the end of this chapter for links to these resources.

> **Note**
>
> To submit feedback about an AWS console or specific AWS service, open the service's console, and then choose `Feedback`. See this link for more details: `https://packt.link/Ryi1V`.

Getting Started with the Evaluation

So, how should you approach the assessment of your workload for security vulnerabilities?

Well, the easiest way to get started with limited effort is most certainly by using **AWS Trusted Advisor**. Trusted Advisor is a service, available on every AWS account at no extra charge, that makes recommendations based on a number of checks it performs on the resources in the AWS account at stake. You also have the ability to get an organizational view of Trusted Advisor findings, which may be convenient when you want a consolidated view of resources within a set of accounts, within some **Organization Units (OUs)**, or even across your entire **AWS Organization**. The checks made by Trusted Advisor depend on the level of your AWS Support subscription. Core security checks are available from the lowest support level (Basic), while additional checks regarding areas such as cost optimization, performance efficiency, or resilience become available with higher support levels (Business or above). So, regardless of your support level, check Trusted Advisor to start with as it may highlight some potential security issues regarding the AWS resources used by your application. If, for some reason, your application spans more than one AWS account, then either check Trusted Advisor in each account or leverage the organizational view.

The checks done by Trusted Advisor are by no means exhaustive but cover diverse essential security aspects such as **Amazon S3** bucket permissions that may be too permissive, **AWS Lambda** functions using deprecated runtimes, and poorly configured security groups related to **Elastic Load Balancing** (**ELB**) resources. For the complete list of all the security checks performed by Trusted Advisor, please refer to the AWS documentation.

When you first run Trusted Advisor, you will be shown a dashboard summarizing the results from its checks, as illustrated in *Figure 14.1*.

Figure 14.1: Summary of Trusted Advisor checks

From this initial verification, you can then dive into more details to understand the recommended actions and the resources concerned for each of them, as illustrated in *Figure 14.2*.

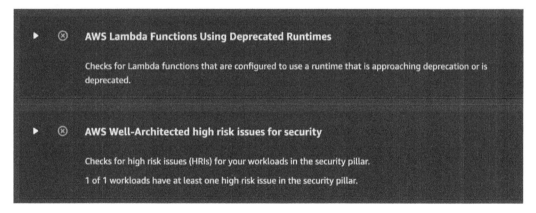

Figure 14.2: Sample list of checks from Trusted Advisor

As shown in *Figure 14.2*, Trusted Advisor lists the checks that identified security issues and, as you can see, it also collects some of its data from the Well-Architected reviews previously performed by the Well-Architected tool.

For all the checks it performs, Trusted Advisor lists the resources affected by the issues, provides evidence for the actual issue, and makes a recommendation to guide you in addressing the problem. For example, in the case of the deprecated Lambda runtimes issue above, you can review the check's details to find out more (*Figure 14.3*). To review the details of a specific check, you can click on the name or status of the check.

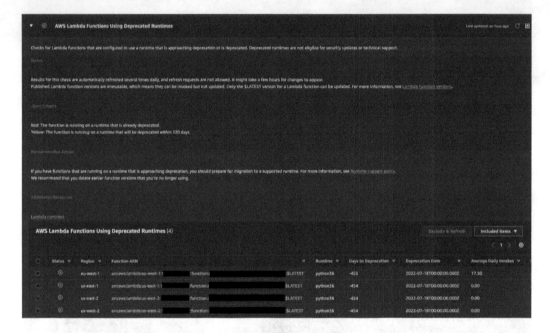

Figure 14.3: Details of a sample check from Trusted Advisor

Now, if you want to have a comprehensive understanding of which workload is affected by the high-risk issue identified in the security pillar, you can look at the included details, as shown in *Figure 14.4*.

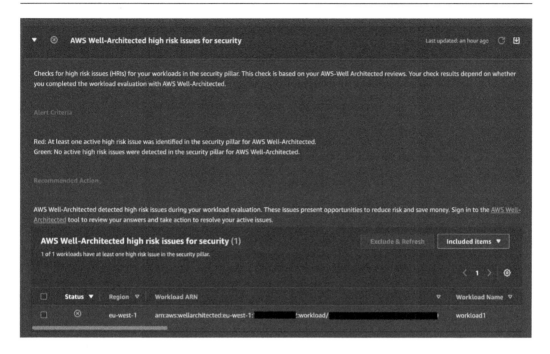

Figure 14.4: Details of a TrustedAdvisor check

The details have all the necessary information regarding the security issue for the given workload, `workload1`, in the Well-Architected tool.

This takes us naturally to the next crucial instrument that you should leverage as a starting point to identify security issues—the AWS Well-Architected Framework.

Following your initial verification using Trusted Advisor, you can now assess your workload against AWS best practices in a more systemic manner. For this, leverage the Well-Architected Framework and focus specifically on the Security pillar. Either conduct a Well-Architected review of your workload, if you have never done so before, or check whether the previous review is up to date with the current state of your application and update it accordingly. Then, look at the outcome of the review for the Security pillar. Are there any **high-risk issues** (**HRIs**) identified for your workload? Chances are that there will be some and these HRIs should be your initial focus. For each issue, the Well-Architected Framework will suggest different possible resolution paths; simply look at the recommendations made in the review results and check each one's details.

The Well-Architected Framework Security pillar currently raises the following questions:

- SEC 1. How do you securely operate your workload?

- SEC 2. How do you manage identities for people and machines?

- SEC 3. How do you manage permissions for people and machines?

- SEC 4. How do you detect and investigate security events?

- SEC 5. How do you protect your network resources?

- SEC 6. How do you protect your compute resources?

- SEC 7. How do you classify your data?

- SEC 8. How do you protect your data at rest?

- SEC 9. How do you protect your data in transit?

- SEC 10. How do you anticipate, respond to, and recover from incidents?

- SEC 11. How do you incorporate and validate the security properties of applications throughout the design, development, and deployment lifecycle?

It is possible that the list of questions may have been updated since the time of writing this book, but, in any case, these questions will give you an idea of the extensive corpus of security best practices recommended by AWS for any workload you deploy on the AWS cloud. These best practices cover multiple aspects of security on AWS, such as account management, workload operations, event detection, infrastructure protection, data protection, incident response, and application security.

In the next few sections, you will take a closer look at some of the most common security issues that need to be addressed to reinforce a typical workload from a security perspective in AWS.

Auditing an Environment for Least Privilege Access

One of the first security aspects to consider for your workload is credentials and their associated permissions that are necessary both for the workload's components to fulfill their tasks and for the end users to be able to interact with them.

In particular, you want to make sure that the permissions associated with the various applicative components or permissions allotted to end users follow the least privilege principle. Here is a refresher on this principle (for more details, see *Chapter 5, Determining Security Requirements and Controls*):

> **Least privilege principle**
> The least privilege principle recommends that you scope down the permissions that you assign to identities (users or roles) to limit their perimeter of action to their exact needs and nothing more.

If you were to remember only one access control security best practice, this should be it. Although the principle is easy to remember and understand, its implementation requires extra care. Why so? Well, essentially, IAM is a very powerful instrument that lets you control access to your AWS resources either at a coarse grain or very granularly. Therefore, it is important to make sure to do things right when controlling access to avoid opening doors for malicious actors.

What can you do, then, to make sure you do things right? Well, to start with, refrain from giving administrator access away to users and roles—chances are they don't need admin privileges to perform their tasks. Access might be given temporarily to run tests for a short period of time. Some privileges that are too permissive might get left behind. So, it's always better to do things right from the very beginning.

Limit every policy that you attach to users and roles to the essence of the latter's needs. This holds true whether the policy applies to humans or machines. Although, applying the principle of least privilege will always be easier in the case of machines as their behavior is typically more predictable. The key to determining the right level of permissions is to have a clear picture of the actions that need to be performed by the principal (user or role) and the required scope (the perimeter in which these permissions will apply). For example, if a component in your workload needs to use S3, it might be tempting, at first, for the sake of simplicity, to assign the following policy:

```
{
    "Version": "2012-10-17",
    "Statement": [
        {
            "Sid": "VisualEditor0",
            "Effect": "Allow",
            "Action": "s3:*",
            "Resource": "*"
        }
    ]
}
```

However, it is evidently overly permissive. Does the principal need to read and/or write data from/to every single object on S3 or do they only need to do so within a specific bucket or a set of buckets? The former is highly unlikely. After a careful analysis of the principal's needs, you will be able to define a reduced scope of permissions and determine what a more appropriate policy could be. For instance, it might look something like the following:

```
{
  "Version": "2012-10-17",
  "Statement": [
    {
      "Sid": "VisualEditor0",
      "Effect": "Allow",
      "Action": [
        "s3:GetObject",
        "s3:ListBucket"
      ],
      "Resource": [
        "arn:aws:s3:::mybucket",
        "arn:aws:s3:::mybucket/*"
      ]
    }
  ]
}
```

As you can tell, there's a significant difference in terms of permissions between the above two policies. The first policy gives full access to any S3 bucket on your accounts to perform any possible action on any S3 object therein. This might be feasible for performing some tests but it is far too broad for a workload in a production environment. For such an environment, the second policy is much more suitable as it restricts access to listing and reading objects within a specific S3 bucket. It allows the principal to perform s3:GetObject and s3:ListBucket actions, but only within the mybucket bucket. This is a much safer approach than granting blanket permissions to all S3 objects.

In any case, it's important to realize that defining least privilege access permissions is not something you would necessarily do right from day one and then never touch ever again. Rather, it's a continuous improvement process in which you streamline permissions management. AWS provides some tools to assist you in the creation of such least privilege policies. First of all, AWS IAM offers a visual policy editor, so you don't need to write your IAM policies directly in JSON, should you prefer not to. That said, it still requires you to have a very clear view of the scope of those permissions beforehand. If you don't have a clear view, chances are that you will generate overly permissive IAM policies to avoid permission-denied errors. Don't be afraid of starting from a very restricted set of permissions and expanding them later as needed. Also, don't hesitate to leverage AWS-managed IAM policies, which provide a good starting point to scope down the permissions of your workload.

Consider the previous example again where you need to provide your workload with read-only permissions to S3. Instead of starting from scratch, you could, for instance, create a custom policy using the contents of the **AmazonS3ReadOnlyAccess** managed policy as a starting point and then simply set your specific bucket(s) as the target resource. Alternatively, you could directly use the same managed policy and then combine it with a resource policy on your S3 bucket(s) or with a permissions boundary attached to the IAM entity (user or role) to reduce its scope to your specific S3 bucket(s). There's no single approach to put the least privilege principle into practice, and the right approach for your own case will depend on your own priorities, such as operational simplicity or reusability, as well as your environmental constraints, for instance, IAM policy quotas (the number of policies attached to an entity, the number of characters in a policy).

So far, you examined a very simple example where you just needed to define read-only permissions to an S3 bucket. However, real-life workloads will likely require you to create more complex permissions. You may not necessarily have a clear view of the exact permissions you should give to your principals (users or roles). Therefore, to help you determine what permissions your workload actually needs to perform its job properly, AWS offers the **IAM Access Analyzer** service. IAM Access Analyzer can generate fine-grained policies by looking at your CloudTrail logs. Through this analysis, it will determine the range of actions that your workload typically performs on your AWS resources and generate action-level policies for you. It can also determine, from among the identified action-level permissions, policies that support resource-level permissions and provide you with a template where you can just specify your resources' ARNs to set fine-grained resource-level permissions. For instance, you can start running your workload in a test environment with purposely overly permissive IAM policies assigned to it, and then, after some time (make sure to leave enough time for IAM Access Analyzer to capture all the actions performed by your workload), check the findings of IAM Access Analyzer to determine the current most adapted fine-grained permissions that can be assigned to your workload. In this manner, IAM Access Analyzer gives you an alternative way of scoping down your permissions.

Evaluating a Strategy for the Secure Management of Secrets and Credentials

Another essential element to consider when improving your security posture is protecting the access credentials, sometimes also called secrets, that your workload needs to access external services such as databases or third-party APIs. AWS provides several options to provide, protect, and store (when needed) these access credentials. However, as you might have already concluded, the most secure credential is the one you don't have to store. It sounds obvious but it is a key principle, as we are always unfortunately reminded by access credential leakage incidents. As already stressed in *Chapter 5, Determining Security Requirements and Controls*, it is a recommended best practice to use temporary credentials, whenever possible, when authenticating with external services. The benefit of temporary credentials is that they only live for a limited period of time, that is, they stop working eventually, thus reducing the risk of unauthorized access. You also don't need to store them permanently, which implies that there's a lower risk of them being leaked or stolen.

It is a valid principle for any type of access credential, not just for AWS access keys, as they can be specific to tasks adhering to the principle of least privilege. First, consider AWS IAM access keys. Suppose your workload needs to interact with an AWS service when it doesn't already have permission to do so. In this case, all it needs is the ability to assume a valid IAM role to access that service. Therefore, there's absolutely no reason why you would not use temporary credentials in this case. When your workload assumes another IAM role to access an AWS service, IAM manages the temporary credentials behind the scenes on your behalf. You don't need to store any credentials at all but only specify which role you want to assume to call the specific AWS service and IAM does the work for you. It's a mechanism that is used, for instance, by AWS services themselves to access other AWS services that they need to do their job. In fact, it is this same mechanism that also works when you need to access AWS resources located in an AWS account that is not yours. For instance, if your workload needs to access an S3 bucket owned by another AWS account belonging to your organization, it can simply assume a role to do so, provided it is allowed access to the bucket by its owner. You do not need to store permanent AWS credentials (access key and secret access key) to do that.

However, what if your workload consists of various components, some of which actually run outside of AWS?

This might be the case, for instance, for a mobile application. In this situation, you can leverage **Amazon Cognito** identity pools to obtain temporary AWS credentials to access the AWS services or resources you need. Cognito identity pools act as an identity federation mechanism to exchange the credentials that you provide, either from AWS or third-party identity providers (including social identity providers such as Google, Apple, Amazon, etc.), in the form of an OIDC token or a SAML assertion with temporary AWS credentials.

Or, consider a situation wherein a part of your workload still resides in an on-premises environment outside of AWS. In this case, you can leverage **AWS IAM Roles Anywhere (IAMRA)** to provide the calling party with the necessary access to AWS resources. In a nutshell, IAMRA lets you exchange a digital certificate (X.509) with temporary AWS credentials. This secures machine-to-machine access from an external source, typically an on-premises environment, to your AWS resources without resorting to using permanent credentials.

Now, what about other sensitive data, such as database credentials and API keys?

These secrets can be managed using **AWS Secrets Manager**. Secrets Manager safely stores these secrets, protects their retrieval, and can also facilitate their rotation. When storing a secret, you essentially provide a KMS encryption key, specify the type of secret you're storing (DB credentials, API key, other), the contents of the secret (e.g. username and password), and some optional settings such as resource permissions, a rotation schedule, and instructions for rotation if needed (in a Lambda function). Note that if you are already using **AWS Systems Manager (SSM) Parameter Store** in your workload, your applications can keep using the same APIs and retrieve the secrets you stored in Secrets Manager through Parameter Store. You could also directly store your secrets in Parameter Store but Secrets Manager brings in some specific features that can prove handy for the lifecycle management of your secrets, in particular their automatic rotation.

Reviewing Implemented Solutions to Ensure Security at Every Layer

To ensure security at every layer, it's crucial to regularly review the solutions that have already been implemented through a comprehensive approach that addresses various aspects of AWS services and configurations. A proactive approach to infrastructure protection is a fundamental step in building a secure and resilient system. The following subsections cover the crucial aspects and various tools that you can utilize to make your existing solutions more secure.

Improving Infrastructure Protection

Before you dive into infrastructure protection, recall a key AWS principle, the shared responsibility model. Security is indeed a shared responsibility between AWS and the customer. Essentially, AWS is responsible for the security *of* the cloud while you, as a customer, are responsible for security *in* the cloud. On the one hand, AWS secures the infrastructure supporting the services they provide, that is, the facilities (data centers) and the hardware running in those facilities, such as for compute, storage, or networking. On the other hand, the customer is in charge of securing the resources they use on AWS. What this actually entails from your perspective as an AWS customer may vary a lot depending on the AWS services utilized by your workloads.

From an infrastructure perspective, you are responsible for securing access to the AWS resources deployed in your AWS environment. The notion of a zero-trust security model has gained a lot of traction as of late. As was discussed in *Chapter 5, Determining Security Requirements and Controls*, it's particularly well-suited to securing a cloud environment. Putting in place a zero-trust model helps you prevent any unauthorized access. It also allows you to go past the traditional on-premises security model, which considers systems within the known and controlled perimeter as trustworthy. In a zero-trust approach, no system is trusted by default, and every access request needs to meet certain criteria to be authorized. If you are shifting to the cloud after having worked in an on-premises environment culture, you must take care not to reproduce the security setup of on-premises, which is mostly based on perimeter control. While AWS does provide you with the tools to implement a similar security setup, it also provides tools to do better. These tools are discussed in detail below.

AWS Shield

You can secure the perimeter of your AWS environment using **AWS Shield Standard**. This service comes free of charge and is activated by default for every AWS account, protecting your AWS resources from infrastructure attacks that are common at layers 3 (network layer, typically for the IP protocol) and 4 (transport layer, e.g. for TCP or UDP protocols) of the OSI model, such as SYN/UDP floods, reflection attacks, and so on. If your workloads deployed on AWS are likely to become particularly exposed to such external attacks, you can opt for more sophisticated protection with **AWS Shield Advanced**. Shield Advanced offers extra protection on layers 3 and 4 and also at layer 7 (application layer, e.g., HTTP/S protocols) of the OSI model.

Shield Advanced lets you be more specific about the protection of your exposed AWS resources. You can, for instance, protect your publicly accessible web applications or APIs using the integration of Shield Advanced with AWS services such as **Amazon CloudFront**, AWS WAF, **Amazon Application Load Balancer**, **Amazon Network Load Balancer**, **Amazon Elastic Cloud Compute** (**Amazon EC2**), and **Amazon Route 53**. Shield Advanced will also automatically deploy **Network Access Control Lists** (**Network ACLs** or **NACLs**) that you defined to protect resources such as EC2 instances or Elastic IP addresses, at the border of the AWS network to protect these resources against large DDoS Attacks (an order of magnitude bigger than what you could handle at the VPC level). It can also manage WAF Web ACLs on your behalf to take measures automatically against detected DDoS events at layer 7. Shield Advanced also gives you access to the AWS **Shield Response Team** (**SRT**), which can either proactively reach out to you in case of a suspected DDoS event or assist you if you're already affected by such an event. Finally, Shield Advanced provides financial protection against any extra costs incurred due to the scaling of the AWS resources under its protection. In case there is abnormal scaling and exceptional costs on AWS resources that were protected by Shield Advanced, you will be entitled to AWS credits to compensate for the generated extra costs.

AWS Firewall Manager

If you operate in an AWS environment with multiple AWS accounts and resources that require protection, **AWS Firewall Manager** can make your life easier by providing a central location where you can set up protection. You can then roll out this protection across multiple accounts and resources in your AWS environment. Firewall Manager integrates with Shield Advanced, WAF, **AWS Network Firewall**, and **Route 53 Resolver DNS Firewall**. The idea is to leverage Firewall Manager to apply the same protection baseline to your entire organization or whenever you deploy a new application in your AWS environment, making sure it complies with your security rules. This method is much easier operationally and less error-prone than repeating the configuration again and again from account to account.

Route 53

Route 53, AWS's DNS service, also provides extra security measures to protect against attacks on the DNS protocol. First, you can leverage Route 53 Resolver DNS Firewall to filter outbound DNS requests from your own VPCs. Such requests go through Resolver to resolve domain names. If one of your workloads has been compromised by an attacker, they may want to exfiltrate data from your AWS environment by conducting a DNS lookup to a domain they control. DNS Firewall lets you monitor and control the domains that can be queried from your VPCs, so you can, for instance, allow access to only the domains you explicitly trust (allow-listing) or block queries to well-known untrustworthy domains and let all other queries through. DNS Firewall manages the lists of known bad domains, keeping them up to date, to make your life easier.

Second, you can enable DNSSEC validation on Route 53 Resolver in your VPCs. This will instruct Resolver to validate the cryptographic signature of the response you get upon a DNS lookup, thereby ensuring that the response was not tampered with. Note that Route 53 Resolver does not, at this stage, return the DNSSEC response, so if you require a custom validation of that response, you would need to rely on a different mechanism for DNS resolution.

Access Control Lists and Security Groups

Check whether your AWS resources deployed in VPCs are properly protected against unwanted access too. Even though AWS protects the AWS perimeter against malicious attacks such as DDoS attacks, it remains your responsibility to protect the resources deployed in your VPCs, such as EC2 instances, **Amazon Relational Database Service** (**RDS**) databases, and VPC endpoints, against illegitimate traffic. One option to block any illegitimate traffic coming to your resources is through NACLs, which you are likely familiar with if you have had some experience with on-premises architecture. NACLs consist of rules that you define to allow or deny traffic in or out of the subnets within your VPCs. Their role is to protect the network perimeter to which they are assigned, typically a subnet. One important thing to remember is that NACLs are stateless, which means you must allow or block traffic both ways. For instance, if you allow HTTPS traffic in from port 443 within a given subnet, you must not forget to allow the same type of traffic out as well. Otherwise, any responses to incoming HTTPS requests would be blocked.

If you want more granular control over how traffic flows in your resources, you can define **security groups** (**SGs**) with rules that filter traffic to individual resources, such as EC2 instances. As opposed to NACLs, security groups are stateful so that if you let specific traffic in, the response traffic is also automatically allowed out. This is beneficial from an operational standpoint since you don't have to remember to allow or block the symmetric traffic each time you define a security group rule. Another main difference between SGs and NACLs is that SGs protect resources themselves and not through a perimeter. Think of them as virtual firewalls in front of your resources.

AWS Nitro

Another important service system that has totally revamped AWS virtualization technology is **AWS Nitro**. Nitro is the underlying system powering virtual machines, or EC2 instances, on AWS. It consists of a combination of software and hardware to optimize the performance of virtual machines and reinforce the security of EC2. A set of Nitro cards handles I/O interfaces such as networking and storage. Freeing the hypervisor layer from those tasks lets AWS make the Nitro hypervisor lightweight compared to traditional hypervisors. Consequently, AWS can deliver almost all the resources of a physical server to the virtual machines running on top of it, to the point that there is no observable difference anymore between a bare-metal instance and a virtual machine powered by the Nitro system, in terms of the resources available. The Nitro system also offers extra protections to further secure your AWS infrastructure. It has indeed been purposely designed without any shell or interactive access mode, making sure there is no possible operator access by design. That feature alone reinforces the security of your workloads by ensuring that no AWS operator (employee or subcontractor) can actually remotely log into an EC2 Nitro host to access its memory or storage resources.

But what if you want to go a step further and prevent any unauthorized software from running on your EC2 instance? That's where **AWS Nitro Enclaves** come into play. Nitro Enclaves are isolated compute environments meant to reduce the attack surface for highly sensitive workloads (for instance, to manipulate healthcare **PII** data, that is, **Personal Identifiable Information** such as names, addresses, and other sensitive types of data). They provide a hardened environment, providing cryptographic attestations for the software that you intend to deploy on them, prohibiting any unauthorized code from running. Nitro Enclaves have no interactive access (no SSH), no persistent storage, and no networking access to external resources. The only open communication channel is a local channel between the Nitro enclave and the EC2 instance to which it is attached.

Improving Data Protection

Improving data protection in AWS involves focusing on data classification, encryption, and regulatory compliance. Classifying data helps apply appropriate security measures based on its sensitivity and regulatory requirements. Encryption, both at rest and in transit, is crucial to prevent unauthorized access to data. Further, adherence to regulations ensures that your data is legally compliant and aligns with best practices. Remember, data protection is an ongoing process that requires regular updates, and staying informed about the latest practices is vital for such continuous efforts. The following subsections discuss some areas of improvement that can help you better protect your data.

Data Retention, Data Sensitivity, Data Privacy, and Data Regulatory Requirements

One of the first areas of focus when improving your security posture with respect to data protection is to examine the data that you are already processing or planning to process on your AWS environment through the prism of your data classification scheme. Many enterprises already have existing frameworks to classify their data according to their needs. If you don't have a framework in place, it might be time to consider having some data classification in place. A basic data classification scheme needs to account for the type of data, its sensitivity, and the regulatory requirements for that data that your organization has to comply with. Many regulations already provide some basis for establishing data categories. For instance, **GDPR** – the European Union **General Data Protection Regulation** – considers the following categories of data:

- **Sensitive**: for data with the highest degree of sensitivity that would cause considerable damage if exposed

- **Confidential**: for data that could cause some damage if exposed

- **Private**: for data that might not cause damage to the organization but still needs to be kept private

- **Proprietary**: for data that might be disclosed outside of the organization on a need-to-know basis

- **Public**: for the least sensitive type of data

This above is just an example. Your own organization's data classification scheme is likely to consider data category definitions from multiple regulations with which it needs to be compliant. For more details on this topic, please refer to the data classification whitepaper in the *Further Reading* section at the end of this chapter.

Once you have determined the data classification levels that apply to your workload, you first need to check the existing recommendations in your organization or best practices in your industry to protect the corresponding data. Then, and only then, can you look into potential measures to protect the relevant data.

There are multiple steps you can take to protect your data, one of the most basic steps being encrypting your data at rest and in transit, which is covered in the next two subsections. This is the least you can do to protect sensitive data. Additionally, your specific use case may require that you take special care of PII data and track and monitor its use across your workload for security and auditing purposes. Amazon Macie can help you with when the sensitive data potentially at risk is stored on Amazon S3. Macie uses pattern matching and artificial intelligence techniques to uncover PII data in your S3 buckets. Macie continuously monitors your S3 buckets for PII data and reports its findings in a dashboard in the AWS console, sending them to Amazon EventBridge, as well as to AWS Security Hub if you ask it to. Macie also lets you consolidate its findings from multiple AWS accounts in a central account for better visibility across your AWS environment. Macie supports multiple file formats, such as plain text and PDF as well as Parquet or Avro files (and more), stored on various S3 storage types for analysis. However, do check Macie's documentation to validate that your S3 objects can be properly analyzed. In some cases, you might be required not just to detect and know where sensitive data is located and used in your workload but also to mask it or desensitize it.

Encryption at Rest

Encryption at rest and in transit was already introduced in *Chapter 5, Determining Security Requirements and Controls*. If you're unsure about the details, please read the section about protecting your data in that chapter again.

Here is a quick refresher: AWS KMS is the central service that allows you to manage your encryption keys and provides encryption capabilities for the services and applications that manipulate data on AWS. KMS manages the cryptographic material and lets you conduct all the necessary operations (such as key lifecycle management, key generation, encryption, and decryption) in **Hardware Security Modules** (**HSMs**). HSMs can be of two types, shared or dedicated. Shared HSMs consist of a fleet of HSMs managed by the KMS service and shared among multiple customers. Dedicated HSMs consist of your own fleet of HSMs, which you can elect either to run on AWS (**Cloud HSM**) or host externally (**External Key Store**, aka **XKS**) on-premises or at one of the KMS partners providing the option. The HSMs provided by AWS, whether shared or dedicated, are tamper-proof security hardware validated under the US NIST **Federal Information Processing Standard** (**FIPS**) *140-2* program, which is in charge of assessing cryptographic modules. The HSMs are currently all validated at *FIPS 140-2* level 3.

Now, deciding which flavor of HSM to use is one of the first steps when using KMS, and this decision has to be made no later than when you need to create your first key to encrypt your data. Remember, you are by no means restricted to using a single flavor of HSM in your AWS environment to store your key material. Deciding which type of HSM to use depends, on the one hand, on the degree of control you want to have and, on the other hand, on the specific requirements of your AWS workloads. There might be some workloads for which you need to store the cryptographic material solely outside of the AWS perimeter for various possible reasons (auditing, legal jurisdiction, etc.). This is also referred to as **Hold Your Own Key** (**HYOK**) in IT literature and it is a need that XKS addresses in the first place. However, you must be aware of the higher risks accrued, especially in terms of availability and latency. Since the HSMs are now totally outside of the AWS perimeter and control, your operational responsibility increases as you have to manage their security, availability, and performance.

That's why, if you absolutely need to rely on XKS, its use should be restricted to workloads for which you really have no other option. For all the other workloads for which you assess that the increased operational, performance, and reliability risks far exceed the benefits XKS would bring, you must decide between a shared or a dedicated HSM fleet hosted on AWS. Again, opting for CloudHSM—a custom dedicated keystore on AWS—will depend on your appetite for control, your constraints in terms of compliance, and potentially other aspects such as increased flexibility for auditability, for instance. Further, the choice again can be made on a workload basis, depending, for instance, on the sensitivity of the data it has to handle or on the local regulation under which it will operate. When using CloudHSM, you must be aware of the operational impact it will have on your organization, since it then becomes your responsibility to manage and, in particular, maintain the scalability and reliability of your HSM fleet. If, on the other hand, you opt to use KMS backed by its default shared HSM fleet, KMS will transparently handle all operational aspects on your behalf. That's the beauty of a fully managed service, the compromise being that you have less control over how things are done.

Now, after having carefully weighed your options and having decided which keystore to use for your workload, you are ready to start using KMS. Whichever keystore you opt for, KMS gives you the ability to import your own cryptographic material to seed your encryption operations. This is an option often referred to as **Bring Your Own Key** (**BYOK**) in IT literature. This option lets you manage and generate your cryptographic material (keys) in your own external keystore, either on-premises or hosted by a third party. You can then import the material in KMS for use as a basis to generate the keys that will be used to encrypt your data. It is a valuable option that you may want to put in place as an extra protection layer for your data if you believe that AWS might lose your cryptographic material, which would then render your data unusable unless you have a backup of those keys. KMS has not lost any cryptographic material of its customers since its inception but should that be a requirement from your compliance department, that's something that can be easily fixed. Again, be aware that this increases your operational burden since you now have to take care of protecting your cryptographic material outside of AWS. So, only make this decision if you are ready and able to manage those keys securely outside of AWS.

Encryption in Transit

While it's paramount to encrypt any sensitive data that you store in the cloud, it's also important to encrypt data when it moves around. Encryption in transit is set by default whenever you access AWS services' APIs—traffic is encrypted using TLS. AWS services' endpoints have been using TLS for years and older versions are being deprecated over time. TLS 1.2 has now become the minimum TLS version required to connect to AWS services, while TLS 1.3 support should be completely rolled out across all AWS services' API endpoints by the end of 2023. In any case, when you use the AWS SDK or the AWS CLI, you already benefit from TLS 1.3 without having to do anything, given that AWS services will anyhow negotiate the highest TLS version supported by the requester.

As a rule of thumb, always use the highest TLS version supported by your workload, and a minimum of TLS 1.2 or possibly TLS 1.3, as the latter brings notable improvements both in terms of performance (single round trip for the initial handshake) and security (only supporting more secure ciphers that ensure perfect forward secrecy). In fact, AWS services, such as CloudFront, Application Load Balancer, API Gateway, and AppSync, that you can use in front of your actual workload to secure incoming traffic can all handle TLS 1.3. So, the bottom line is that there's no reason to leave your workload exposed to security threats that could be easily prevented by upgrading your TLS version.

Additionally, AWS has also been working with various organizations to prepare for a post-quantum world to make sure the cryptographic algorithms in use can resist attacks against a (as yet hypothetical) large-scale quantum computer in the years and decades to come. One of the areas most vulnerable to attack, when it comes to communications over the internet, is the **key exchange mechanism (KEM)** used by TLS to determine a key that is then used to protect the data transmitted over the TLS session. AWS's implementation of TLS, called s2n-tls, gradually incorporated post-quantum KEMs as NIST issued its various drafts of the new standard for post-quantum cryptography standardization. Critical AWS services such as KMS, Secrets Manager, and ACM all support post-quantum key exchange in TLS, so you can already leverage the post-quantum KEMs implemented, for example, in s2n-tls to connect to those services—something to keep in mind if post-quantum safety of your most critical data is essential to you.

Reviewing Comprehensive Traceability of Users and Services

Apart from all the measures mentioned in the previous section and other measures that you can take to secure your AWS environment and resources, it is also paramount to keep track of all the activities that take place within your AWS environment. This is where AWS CloudTrail plays a key role. CloudTrail provides traceability of all the actions (API calls) taken in your AWS environment. It essentially logs all the events related to your resources in a log stored on S3. It also natively integrates with AWS Organizations, which allows you to manage a central trail of all events across your entire organization. Following the best practices from the Security Reference Architecture (see the *Further Reading* section of this chapter), you are encouraged to keep a central trail of all the events in your organization in a separate Log Archive account that can be used for security or auditing purposes. Additional logs should also be stored centrally for safekeeping or auditing, such as VPC flow logs, Route 53 DNS logs, and S3 access logs.

Improving Incident Detection

You have already seen how incident detection is important for increasing your security posture. However, in order to remain agile and continuously adapt to new threats in terms of both actors and methods, you should be constantly looking at ways to optimize your incident detection methodology. In the following subsections, you will go through some ways of doing this.

Security Logging (Segregated Bucket, Dedicated Account)—Config, CloudTrail

Improving incident detection starts with a review of your workloads for reliability and operational excellence. Leverage the AWS Well-Architected Framework to assess your current posture and benefit from its recommendations regarding corrective actions. One of the things you need to do, unless you've already done it, is to define metrics and alarms to ensure you get clear visibility of the application and infrastructure layers of your workloads. This makes it easy to find and prioritize issues during an incident. Early incident detection is key to making sure you have control over your workload's behavior after you deploy it into production. If you have subscribed to AWS Enterprise Support, you can use **AWS Incident Detection and Response**, which essentially consists of a team of experts that will take care of incident response on your behalf. Otherwise, there's no AWS feature or service that does it for you and you have to define the right metrics and alarms to take good care of your workload once it's in production. Suppose your workload can be accessed, through the internet, by end users located within specific countries. Then, you might want to be notified about any repeated failed login attempts or any successful connections from unexpected geographical areas. Similarly, if your workload filters access to data based on the end user's role, you may want to be alerted about unusual data access patterns. Either of the cases might be a sign of a security incident.

Now, how can you do that? First of all, you want to activate logging and log collection at various levels: it's essential to log your workload activity and generate custom metrics from that activity whenever it makes sense, to be able to monitor the state of any specific processing inside your workload. Especially for security purposes, it's essential that you activate logging on your AWS account, AWS CloudTrail at least, that captures all activities within a given AWS account, as well as other essential logs such as VPC Flow Logs or Route 53 DNS logs. CloudTrail trails enable the delivery of events as log files to an Amazon S3 bucket that you specify. When you create a CloudTrail trail, you associate it with an S3 bucket of your choice, and from then on, CloudTrail records all activities on your account and delivers them to log files in that S3 bucket. Each entry in your CloudTrail log files represents a request that has been made by an entity on your account to perform a specific action and contains information about when the action took place and the context of the request (such as the requester identity, the request parameters, the response elements, and so on). So, in the case of any malicious use of your account, you can detect and track the malicious actions.

Preserving Logs

To preserve the essential logs, it's highly recommended that you set a mechanism for all logs from various key AWS services (such as AWS CloudTrail, AWS Config, Amazon VPC, and Amazon CloudWatch) to converge to a central dedicated account where they can be archived and used both for incident detection as well as forensic investigation whenever needed. If you don't use, or don't plan to use, Control Tower across your AWS environment, it is still strongly recommended that you centralize all your logs for the aforementioned reasons. Further, the logs should be stored in an S3 bucket in an AWS account dedicated to log archival, with access restrictions in place to prevent data tampering. There should also be a process, one that is again already put in place automatically by Control Tower, to ensure that every new account in your AWS environment is systematically baselined with the log collection mechanism before any workload is allowed to land in it. That baselining should be part of your account factory process, wherein you enforce a number of security rules and controls for the various kinds of AWS accounts that you will use (from sandbox to production accounts).

Now that you have the log collection mechanism in place that centralizes all the trails from the various accounts of your AWS environment, how can you utilize all that information? The information is substantial so you cannot manually review it every now and then. You need an automated process in place to detect security issues, report them, alert the right people and teams, and potentially immediately take corrective actions. The first issue to tackle, then, is security issue detection.

Central Security Hubs

Multiple AWS services come into play to help you with detection, but AWS Security Hub is certainly key. First of all, it can implement controls in your AWS environment for security issues related to a set of standards including, but not limited to, the following:

- **AWS Foundational Security Best Practices (FSBP)**
- **Center for Internet Security (CIS) AWS Foundations Benchmark**
- **National Institute for Standards and Technology (NIST)** security and privacy controls
- **Payment Card Industry Security Standard (PCI-DSS)** best practices

Security Hub also offers you the possibility to deviate from its standard behavior by delegating the management of the security controls it implements to another AWS service through service-managed standards. For instance, Control Tower gives you the ability to define the very controls that you want Security Hub to implement on your behalf. In any case, whether you prefer to use Security Hub's built-in standards or rely on service-managed standards, it's up to you to decide which sets of standards you want Security Hub to leverage when it performs its duty. Further, as an architect, you are expected to make that choice based on your business context and security requirements. For more details on which controls each of these standards deal with, please refer to Security Hub documentation.

Additionally, Security Hub can accumulate findings from multiple AWS or third-party sources. It gives you the ability to centralize security findings under a single umbrella so that you don't miss out on anything. It can collect findings from AWS sources such as AWS Config, Amazon GuardDuty, AWS IAM Access Analyzer, Amazon Macie, AWS Firewall Manager, and others. It also supports numerous third-party products as security findings sources, including solutions from vendors such as CheckPoint, CrowdStrike, IBM, Snyk, SumoLogic, and many others.

> **Note**
>
> The list of sources capable of sending events to Security Hub has been growing over the years, so if you're interested in leveraging such vendor solutions, make sure to check out Security Hub's documentation for the currently supported integrations. Additionally, you can also build your own custom integrations, should you need to send security events to Security Hub from sources for which there is no support yet.

Now, you are aware of the importance of centralizing all logs. Doing so feeds Security Hub with vital information that it can use to identify potential security issues among all the activity taking place across your entire AWS environment.

Finally, Security Hub can also send the security findings it collects to multiple AWS services and third-party solutions. However, this will be covered in more detail in the next section, about improving how you respond to security events.

Single Security Accounts

Since Security Hub integrates with AWS Organizations, another important consideration is to centralize your Security Hub setup in one AWS account, typically the one dedicated to the security team, that has the necessary tooling and configurations for that team to take actions and investigate security issues. Having such a central Security Hub setup allows you to have a single control pane where you can define the security standards that you want to be applied to all your AWS accounts across your cloud environment and across AWS regions. Then, if needed, you can centrally define different policies for the OUs and member accounts in your environment. You can also delegate that configuration to the OUs and member accounts for cases where you would have them manage their own security standards, for instance, in a decentralized governance approach or for isolated accounts such as sandbox accounts where no sensitive data or workload is exposed. In any case, configuring Security Hub centrally protects you against a drift of your security standards across your AWS accounts, unless you let it happen on purpose.

So, now that you have centralized your logging capabilities and improved your capability to detect security events, you will likely start to see quite a few such events pop up on your radar in the beginning. It might be because you simply did not capture them previously or because you still need to adjust your detection threshold. For the latter, you may, at least initially, see findings that you consider false positives, things that Security Hub may consider a security risk when they actually don't represent a major issue for you. Security Hub processes all incoming events and matches them against the set of controls that you have defined. It might be, though, that the set of controls that you have picked initially is more restrictive than you actually intended. For instance, you may have enabled more standards than you actually needed to or enabled more controls within these standards than you needed to. It is not necessarily a bad thing (it's better to have more security controls than not enough), but the risk of having too many security findings that include many false positives is that you might get distracted and miss some important findings. You can naturally adjust your security standards as you go by enabling or disabling some of the security standards, enabling or disabling controls across the standards you enabled, and customizing parameters for the enabled controls.

Improving the Response to Security Events

The previous section focused on improving security event detection. You are now ready to examine the measures to take to improve your response to security events.

Prioritizing Automated Responses to the Detection of Vulnerabilities

Based on the configuration policies that you've defined, Security Hub correlates the findings it collected from the various sources that you've integrated with it, such as AWS Config, Amazon GuardDuty, Amazon Macie, AWS Inspector, and others. Each finding has attributes documenting the context in which they were created and the issue they identified, such as severity level (from low to critical), workflow status (the status of your investigation regarding this finding), record state (is it active or has it been archived?), region, account, product (that reported the finding), resource (affected by the finding), and other aspects depending on the type of finding. You can then leverage these attributes by applying filters to extract only those findings bound to a specific account or a given region, or of a specific severity level, and so on, for further analysis or processing.

For instance, you can use these attributes to define automation rules to update findings matching a given combination of attribute values. Security Hub provides pre-built rules for the most common scenarios; for instance, it will automatically elevate the severity level of findings detected in your production accounts or affecting your critical resources. Additionally, you can define custom rules to automatically update findings based on your own criteria. This allows you to refine the categorization of findings according to your own preferences and eventually take further actions.

Security Hub sends all new and updated findings to Amazon EventBridge as events, which gives you the opportunity to route these events to a specific destination for further analysis or processing, executing, for instance, an AWS Lambda function or an **SSM Automation document**. You can also define custom actions in Security Hub that can be later applied to selected findings and attach these custom actions to EventBridge rules to perform specific actions on a set of findings. In order to trigger these rules, you need to explicitly select one or more findings in Security Hub and apply the custom action of your choice. This will send an event corresponding to that custom action to EventBridge for each of the selected findings and then automatically trigger any rule attached to that particular event type. These features allow you to build your own playbooks that you can then leverage to automate remediation for specific security vulnerabilities identified in your AWS environment.

Further, you don't have to start from scratch since AWS maintains a library of playbooks called the Automated Security Response on AWS. These playbooks come with prebuilt remediations for the security standards supported by Security Hub, such as the AWS FSBP, CIS AWS Foundations Benchmark, and PCI-DSS. You can select the playbooks that are of interest to you and only deploy those to start with. For more details on this library, please refer to the *Further Reading* section of this chapter.

Remediating Non-Compliant Resources

Apart from Security Hub, AWS Config also provides a mechanism to directly remediate non-compliant resources. AWS Config, which provides findings to Security Hub, is in charge of monitoring the configuration of the resources in your AWS environment. It records the configuration changes of your AWS resources, such as EC2 instances or VPCs, keeping track of the history of changes. You can think of AWS Config as a managed **Configuration Management Database** (**CMDB**) service on AWS that tracks configuration changes of resources identified as **Configuration Items** (**CIs**). When you set up AWS Config, you can specify the resource types you want it to track and ask to be notified, through **Amazon Simple Notification Service** (**SNS**), of any configuration changes. You can also define rules reflecting your ideal configuration settings that Config then uses to evaluate whether your resources are compliant or not with these rules. For instance, if you want to detect RDS database clusters that do not encrypt data at rest, you could enable a Config rule to notify you whenever it happens. For a complete list of the resource types supported by AWS Config, please refer to the AWS documentation: `https://packt.link/5LRkV`.

By default, AWS Config provides built-in rules managed by AWS, in charge of evaluating your resources against the most common best practices, such as the example that was just used about RDS databases and encryption at rest. Config gives you the ability to configure every rule as you deploy them with the settings that make up your standards. Then, if built-in rules are not enough to cover all your needs, Config lets you define custom rules by either writing Lambda functions or using **AWS CloudFormation Guard**, a **domain-specific language** (**DSL**). Guard is an open-source declarative language that lets you define policies as code. You can use DSL to validate YAML or JSON configuration files, such as CloudFormation templates, Terraform, or even Kubernetes configuration files. Additionally, you can also use it to evaluate the conformity of CMDB resources such as CIs managed by AWS Config.

Config not only identifies non-compliant resources based on the rules you deploy but also lets you take corrective actions. Remediation actions in Config are defined using the SSM Automation document, which lists the actions to be performed manually or automatically on resources that a Config rule flags as non-compliant. Similar to rules, Config provides built-in remediation actions that you can leverage out of the box to apply remediation to non-compliant resources.

Config also offers hundreds of built-in rules to cover the various resource types it supports. If you start crafting your own custom rules, you may end up with tens or hundreds of them, which is great and gives you a lot of fine-grained control over the compliance of your AWS resources. However, from an operational perspective, it lacks ease of use. This is where Config **Conformance Packs** can help. Conformance packs are collections of Config rules with remediation actions packaged together, making it easier for you to manage resources or operations at scale and roll out across your AWS environment. AWS offers prebuilt conformance packs to kickstart your experience with Config. You can choose from a selection of operational and security best practices packaged as a conformance pack, covering standards such as the AWS Well-Architected Framework, the **Cybersecurity Maturity Model Certification** (**CMMC**) levels 1 to 5, CIS AWS Foundations Benchmark, and many more. You will also find Config conformance packs focusing on best practices for a given AWS service, such as RDS, **Amazon Sagemaker**, **Amazon Redshift**, and many more.

The next section focuses on a crucial aspect of maintaining any IT infrastructure—designing and implementing a patch and update process. This process is essential for ensuring the security, performance, and reliability of your AWS resources. The section discusses how you can formulate an effective patch management strategy, implement it using AWS services, and ensure that your applications and systems are always up to date.

Designing and Implementing a Patch and Update Process

Putting together a patching process is also essential to keep your operating systems up to date with the latest security patches, and it's not something you typically put in place from the very beginning when you start using the cloud. In the cloud, it becomes natural to use immutable environments by defining your infrastructure as code and re-creating your entire workload stack from scratch for every deployment. In this case, mutable environments are the exception more than the rule, and one tends to forget about patching. Further, it is also very natural to use managed services, which often take care of the patching on your behalf, especially when such services are serverless and you have little to no control over the supporting infrastructure.

However, the fact is that patching is as important for immutable environments as it is for mutable environments, and it is also super important for managed services for which you sometimes have to, as a minimum, decide when to apply minor updates. For immutable environments, patching simply happens differently, as it has to be done on the virtual machine images (**Amazon Machine Images** or **AMIs** in AWS terminology) or on the container images used to run your workload. However, patching still needs to be done, although it may seem less obvious at first. It is something that can become part of your build phase, your CI/CD process, and offline, so to speak.

For mutable environments, however, it is something that you need to factor in as part of your maintenance process once your workloads are already online. The process typically consists of deploying a patch during a maintenance window of your workload. You can leverage SSM Patch Manager to simplify the maintenance task for you by defining patch policies to apply your patch baselines globally across your entire AWS estate.

Remember that patching is part of your security improvement process, even if you use only immutable environments or serverless technology. Although, for the latter, the job is AWS' responsibility instead of yours, you still need to factor it into your maintenance process and take care of defining maintenance windows for AWS to do the patching safely.

Designing and Implementing a Backup Process

Now, to discuss the next extremely important topic: establishing a backup process. In principle, making backups of your important data is something that you have been doing from the very beginning, before deploying a workload in production for the first time, that is, when you can't afford to lose data. However, it is also possible that you pushed your workload into production with a basic backup scheme in place without considering the more granular aspects.

There is no point in backing up your data simply for the sake of doing so. Every backup is meant to serve some recovery objectives. Backup is an essential part of any recovery strategy that will be activated in case of a major failure affecting your workload. So, the first thing is to clarify your recovery objectives, such as **Recovery Time Objective** (**RTO**) and **Recovery Point Objective** (**RPO**). The RTO describes the acceptable amount of time that the recovery of your workload can take. The RPO defines the acceptable amount of data that might be lost once your workload is restored upon failure. Both objectives can be expressed in minutes, hours, or days depending on the criticality of your workload and of the data it manipulates.

> **Note**
> RTO and RPO can be calculated through Resilience Hub. This service evaluates the infrastructure and reports whether it meets RTO and RPO.

This was already covered in the chapters of this book dedicated to improving resiliency, but here backup is considered from a security perspective. You need to make sure that your backup strategy resists the eventuality of one or more security issues. Think of cases such as ransomware, where an attacker gets access to or, worse, takes control of one of your AWS accounts, even if partially. What if your backups were stored in the same account? That's not a situation you want to look forward to.

Therefore, it's essential that you review your existing backup strategy, or that you define one, in the light of the potential security issues you may face, to make sure that you can indeed meet the recovery objectives that were set for your workload. You will have to consider a more or less granular backup strategy, depending on the criticality of your workload, involving things such as continuous backups with **Point-in-Time Recovery** (**PITR**), in particular for databases, object or file-level recovery, storage volume-level recovery (typically using crash-consistent snapshots), or instance-level recovery. You will also want to look into making multiple copies of your backups and, in particular, protecting your backup vaults with extra care. All these elements can be put in place with AWS Backup, from a central pane of glass, for your workload and beyond for your AWS environment across multiple accounts and multiple regions.

Finally, don't forget to draw lessons from previous incidents to keep making improvements to your workload architecture and incident response plans. This is part of a continuous improvement cycle that you should put in place to iteratively improve the security of your workloads.

Summary

This chapter has built on *Chapter 5*'s foundations, covering how to improve security for a solution already in place on AWS. You examined how you can monitor the existing environment for vulnerabilities and how you can address the most common security issues. Then, you explored remediation techniques in terms of how to improve the detection of vulnerabilities and automate responses to such issues to improve your solution's security posture.

In the next chapter, you will look at the various aspects of improving your deployment.

Further Reading

- The AWS Security Reference Architecture (SRA): `https://packt.link/7OO3x`
- The Well-Architected Framework Security Pillar: `https://packt.link/KJopk`
- The AWS Security documentation: `https://packt.link/6zS9w`
- The AWS Architecture Center best practices for security: `https://packt.link/DI7EM`
- The AWS data classification whitepaper: `https://packt.link/eGpAe`
- The Automated Security Response on AWS: `https://packt.link/0gfYG`

15

Improving Deployment

This chapter covers the deployment strategies and capabilities of AWS that can help you improve deployment for an existing solution. Being able to keep deploying workloads efficiently over time is key to successfully managing your solution stack and meeting your business requirements.

In this chapter, you will first review your existing deployment strategy and then learn how to leverage a couple of additional AWS services to make improvements to it.

Reviewing Your Deployment Strategy

The idea is to review your existing deployment strategy and assess your established practices to deploy workloads, including rolling out new releases over time.

The major deployment strategies and most relevant AWS services for supporting those strategies were introduced in *Chapter 9, Establishing a Deployment Strategy*. If you don't recall much of the concepts discussed or haven't read *Chapter 9* yet, please take a moment to refer to that chapter before reading this one. As a quick refresher, the most popular deployment strategies you may come across are discussed briefly here.

DevOps has long been the de facto standard approach to software engineering across the industry. Naturally, the degree of adoption and form or flavor of DevOps—GitOps—will vary from one organization to the next, but all organizations typically adopt some form of DevOps in their software engineering practice. As you have learned already, DevOps is a set of engineering practices whose objective is to enhance organizations' capabilities to deliver better quality software at a higher pace. Some of the DevOps practices are already well known to us now, such as **continuous integration (CI)**, **continuous delivery (CD)**, and **infrastructure as code (IaC)**. CI entails improving software quality and delivering software faster, while CD is about delivering new software releases into production faster and safer. Both concepts are often combined into a single term, CI/CD, which encompasses the management of the entire software lifecycle, from inception to production. Further, because software cannot run on thin air, we also need to consider the infrastructure necessary to run it. DevOps prescribes treating infrastructure the same way you treat application code, and that's where IaC comes in.

At this stage, you've already built and deployed some workloads to the cloud and want to keep improving the way you deliver software to production. You may now wish to consider the following questions: is there a clear deployment strategy in place in your organization? Has the choice been clearly made to follow a standard process? If so, have the tools of the trade been selected for the process?

Depending on the governance in place within your own organization, there might be a central delivery process with strict rules to follow, or, on the other hand, the deployment responsibility might be delegated to the individual project teams with significant freedom of choice, with all degrees of variations between the two extremes. Whichever governance model you may have, the software delivery best practices applied will largely be the same. There will naturally be variations in terms of responsibility within the organization—think of a **Responsible, Accountable, Consulted, Informed (RACI)** matrix defining roles and responsibilities—but there will also be some invariants.

You also want to consider whether you are already using separate environments for the different stages of a software release. At a minimum, you should have two environments to isolate your production environment from any other non-production environments. This clear separation between production and non-production workloads is one of the few non-negotiable best practices. There is no reason not to have this clear separation in the cloud, not that there is any good reason to do any differently on-premises either, but in the latter case, at least you might have the excuse that the infrastructure available is clearly more limited. Usually, you have three or more such separate environments: for instance, **integration (INT)**, where you deploy your freshly built latest application release and then perform integration tests; **user acceptance testing (UAT)**, which is possibly an exact replica of your production environment where you typically perform acceptance, resilience, and performance tests; and finally your **production (PROD)** environment.

You will also have to consider whether you automatically or semi-automatically deploy your workload in one or more of the above-mentioned environments. You hopefully already have implemented a CI pipeline to automatically build releases and perform quality checks whenever new code is pushed into your source code repository. If you haven't, it's time to consider this as there won't be any CD without going through CI first. So, after the software has been built and tested during the CI stage, what happens? Is there any form of CD already in place? How do you currently package your workload to deploy a new release into the relevant environment(s)? If no part of the process is minimally automated, there is obviously room for improvement. But even if it is already automated, there might still be room to do things better.

Infrastructure Provisioning

As mentioned earlier, DevOps practices advocate treating infrastructure as code. This is paramount if you want to always be able to deploy your workload in any circumstances starting with a clean slate. In most cases, you will merely have to deploy a new release of your workload on a normal day without any issues. However, someday, you might need to achieve the same deployment after your workload has been severely incapacitated due to an issue, such as an infrastructure breakdown, a network outage, or a security incident. In such a situation, where you might want to react very quickly under possibly a lot of stress, you don't want to start building and packaging software or creating infrastructure components manually. It will be more feasible to launch, as easily as possible, a clean environment, preferably automatically at the push of a button, where you can then (re-) deploy the latest stable release of your workload and start afresh. This is where IaC really stands out.

If all the infrastructure components that your workload ever needs to run smoothly can be automatically created, the only worry you have left, so to speak, is to back up any relevant data that may be needed to start the workload in the proper state. If you can achieve such automation, you'll be able to face most situations you'll ever come across. The rest, especially protecting your workload against data loss or maintaining business continuity, naturally depends on your **service level agreements** (**SLAs**), and you may want to check out *Chapter 6, Meeting Reliability Requirements*, and *Chapter 12, Improving Reliability*, of this book for a closer look at how best to tackle reliability aspects.

Now, that said, does your current deployment strategy involve leveraging an IaC approach? Assuming that you see the value of adopting an IaC approach, the first step often takes the most effort, but then you are ready to automatically deploy your infrastructure resources. Is it going to be enough though to manage the lifecycle of these resources? Well, not quite. As a matter of fact, since your infrastructure is now code, you can manage it exactly how you manage code, leveraging all the tooling you already use to manage your software lifecycle, including integrated development environments, source code repositories, testing tools, and so on.

But what do you do at every release? Will you reuse any pre-existing infrastructure resources? Or will you seize the opportunity to consider your infrastructure as immutable and deploy every new release of your workload on fresh infrastructure resources? The latter is a good approach to limit infrastructure configuration drift and can also help you easily stay up to date with the latest OS version, patch version, or virtual image. You won't completely get rid of the necessity of patching environments but you will certainly have to do it less often. Similarly, this approach won't suppress the need to manage the configuration of your resources but it will make it easier to avoid drift over time. So, essentially, taking an immutable infrastructure approach is a great way to make your life easier, and that's easily done in the cloud. It is a change of mindset compared to the traditional on-premises infrastructure management where servers are typically long-lived resources as you need to amortize their costs over multiple years. In the cloud, you don't have this problem—it's the cloud provider's problem, not yours, and there's absolutely no reason to make it yours again.

Here, you might be wondering, can you easily work with immutable infrastructure though? Especially over the lifetime of your workload, you may initially find it challenging to deploy new infrastructure resources again and again at every release. Essentially, your infrastructure is now code, so as you do for every release for the rest of your workload software, you just need to deploy new code. In that sense, working with an immutable infrastructure can be easy enough. Further, you can reuse your existing CI pipelines to build and package infrastructure resources and you're ready to deploy a brand new environment where your next workload release will land. If your workload ever needs to maintain state between releases, you'll simply have to factor that into your deployment process.

However, you may say, what about the possible configuration drift of these infrastructure resources? Well, this risk doesn't entirely go away, but because your infrastructure resources are short-lived, the risk of having configuration drift over time is considerably reduced. For every new release of your workload, your infrastructure is in the ideal configuration state that you've specified. Nevertheless, you still have to ensure that no drift occurs between releases. Therefore, configuration management is still crucial. However, because your infrastructure is now code, there are fewer reasons than before to go and make changes to the configuration of resources manually. If a configuration change is required, in most cases, your infrastructure code should reflect that change, which should trigger a new release to handle it once all the testing and validation has been done. This approach will help you avoid any drift away from your ideal configuration state. There might be some edge cases, though, where configuration drift is not necessarily a bad thing. For instance, consider a hidden feature that you activate at runtime, often called a feature flag. Activating the feature will likely result in a configuration change, even possibly a change in the configuration of your infrastructure; for instance, perhaps new resources will be created or the state of existing resources will change. In this case, you will not do a new release, as it would defeat the purpose of feature flags in the first place. However, you may still want to track any related configuration drift in your resources; it will just not trigger any corrective action.

Now that you have gone through the components involved in the provisioning of your infrastructure, it is time to review the way you are currently deploying your application software.

Application Deployment

The two most important questions to consider are as follows: is there any automation in place? Is software building already fully automated? There's no point in trying to automate deployment any further if the build phase is not fully automated yet. Therefore, building your software automatically is the first thing to do. What you want is a mechanism in place that builds your software continuously, not on demand. This requires putting together a workflow in charge of building your application software that gets triggered every time new code is pushed to your source code repository. Any error in the building process interrupts the process and the development team is notified of the error for fixing. If the build is successful, then you go through the next stage of the building process, which typically consists of a series of unit and other validation tests and some quality checks. An error in this test phase also interrupts the whole workflow and the development team is notified of the error so they can for adjustments.

Does this reflect your current building process or is your process different? Are you building software without performing quality checks? If that's the case, this is a clear opportunity to improve your software delivery process right away. If your process has the prerequisites mentioned here already, then you're ready to look into potential improvements in the rest of your CI/CD process.

The questions to consider are the following: what happens in your current CI/CD process after the initial test phase completes successfully? Is a deployment stage coming next? And, if so, do you create infrastructure resources on the fly or do you rely on some pre-existing infrastructure?

Having gone through the discussion so far, you are ready to examine what you should expect from the CI/CD process. After the building stage and the initial test phase are completed successfully, you are ready to deploy your workload in the first non-production environment in your environment chain (for instance, INT -> UAT -> PROD). This means that you first need to provision the environment, that is, the infrastructure resources, before deploying the application software. If you have taken an IaC approach, that's easily done, as you just have to deploy the IaC templates that will automatically create your infrastructure. Additional tests, such as integration functional tests, would be carried out. If anything breaks during either infrastructure provisioning, application software deployment, or integration testing, then again the whole workflow is interrupted and the development team is notified to identify and fix any outstanding issues. If, on the other hand, the rollout in the first environment and the tests performed were both successful, then the workflow moves on and the workload deployment in the next non-production environment is triggered, for instance, UAT. That deployment again first requires the infrastructure resources to be provisioned before the actual application software can be deployed and the next flight of tests can be carried out. Again, in the case of any issue, just like the previous phase, the process is interrupted, the development team is notified, and so on and so forth until you reach the last non-production environment.

At this stage just before the PROD environment, many organizations opt to go through a manual validation step where a person in charge, for instance, the product owner or the technical lead, makes the decision to go ahead and deploy the workload in the production environment, which could be executed right away or scheduled at a more convenient time. Less frequently, organizations automate deployment all the way down to the production environment, putting in place a continuous deployment process, where they decide to systematically release every new feature as it comes. There's no obligation to adopt this approach and not every organization is comfortable with it, especially large organizations that have long built a habit of strictly controlled IT management processes with manual approval at multiple stages. In any case, implementing continuous deployment requires very high confidence in your CI/CD process. However, if a robust CI/CD process is baked into your organization's DNA, then the process is as easy as any other.

Now suppose you have your whole CI/CD process clearly figured out and you're ready to go; there's still one important decision to make when it comes to releasing your workload into PROD. We've discussed immutable infrastructure in the previous section, so you're aware of a first decision point, that is, whether to reuse/update the existing infrastructure or create a new one. But what about the application software? How are you currently deploying a new release? You will want to consider the following: do you have the new software release replace any pre-existing release all at once and in-place? Or do you progressively roll out the new release, updating your workload in batches (aka blue/green deployment)? Or, while you progressively roll out your new release, do you keep any pre-existing release up and running until you have fully validated that the new release works as expected (canary deployment or linear deployment)?

Your organization's software development culture is bound to influence your choice; this culture will vary from one organization to the next and even within the same organization from one project to the next. In any case, answering the preceding questions is the first step to determining your deployment strategy. For instance, you may opt to use immutable infrastructure with an IaC approach and then enforce canary deployment for every new software release in production. In fact, this is something to decide upfront and is vital for your production environment. You should test whatever you end up choosing within your deployment process. You don't want to discover, on the very day of an important milestone release, that your blue/green deployment to production is actually not functioning. To avoid this, you would have to use the exact same rollout mechanism to deploy in UAT to be sure to detect any issues in the rollout process.

Reviewing Deployment Services

In the previous section, you reviewed your current deployment strategy and possibly identified opportunities to bring some improvements. Now it's time to look into the tooling and AWS services that can help implement your strategy.

Evaluate Appropriate Tooling to Enable Infrastructure as Code

As discussed earlier in the previous section, DevOps and AWS best practices do recommend adopting an IaC approach. This was covered in *Chapter 9*, Establishing a Deployment Strategy, already but, to summarize, AWS lets you choose your preferred IaC technology between **AWS CloudFormation**, the **AWS Cloud Development Kit (CDK)**, and **Terraform by HashiCorp**. Both CloudFormation and Terraform are low-level language representations of cloud resources, while the CDK lets you define and provision AWS resources with the programming language you're most familiar with such as TypeScript, JavaScript, Python, Java, C#, and Go. Behind the scenes, the CDK actually synthesizes CloudFormation templates that will be used to provision the relevant resources. Note that there also exists an implementation of the **CDK for Terraform (CDKTF)**, which can instead synthesize Terraform templates. The CDKTF is not as mature as the CDK at the time of writing, so please refer to the CDKTF project page before widely adopting it. In any case, both the CDK and the CDKTF provide higher-level abstractions to define your AWS resources, which bring the following benefits:

- They let you use the programming language that you are more familiar with, instead of using **Yet Another Markup Language** (**YAML**), **JavaScript Object Notation** (**JSON**), or **HashiCorp Configuration Language** (**HCL**). This is especially useful for developers, although possibly a little less so for other professionals such as system engineers.

- They also let you create your own higher-level abstractions to help you manage complexity. This is particularly handy if you would like to create a library of reusable infrastructure patterns fitting your own standards, composed of multiple AWS resources.

Ultimately, it's better to carefully weigh the pros and cons of going with one or another IaC technology. This is especially important because you're likely to make a single choice for the entire organization, at least initially, to make it easier to manage and spread IaC standards, libraries, and best practices.

There are some advantages to choosing CloudFormation or Terraform. For one, CloudFormation is the native AWS IaC representation. It also does not require anyone to be a programmer to write their own CloudFormation templates. This is a natural choice if it's important for you to stick as close as possible to the source (AWS) and if you also don't want to force everyone to use a programming language to define infrastructure. Think, for instance, of your typical system engineers team: they may have some familiarity with markup languages and JSON and might also be comfortable writing shell scripts, but they might not prefer writing code either in TypeScript or Python or anything else. On the other hand, Terraform can be used as an IaC tool to deploy cloud resources across various cloud providers. This is a legitimate choice if you plan to use or even already use multiple cloud providers in your organization. Terraform allows you to standardize IaC practices and share best practices on a broader scale.

However, the alternatives having certain advantages doesn't mean you shouldn't adopt the CDK or CDKTF. The CDK and CDKTF will be beneficial especially if you adopt a true DevOps practice where developers and operations team actually collaborate within the same project or product team and can help each other. Therefore, developers, who are certainly more comfortable writing code in any of the aforementioned languages supported by the CDK, can contribute to writing much of the code covering infrastructure resources. To dive deeper into IaC aspects, please refer to *Chapter 9*, Establishing a Deployment Strategy, of this book or the *Further Reading* section for additional guidance.

Test Automated Deployment and Rollback Strategies

You may have already gained experience in your current delivery process with a deployment technique that was previously mentioned, such as in-place, blue/green, or canary deployments. You may have also made up your mind regarding which one you like the most. However, your selection doesn't need to be set in stone. Your organization may enforce some rollout standards and leave you no choice. However, if you can opt for the most appropriate technique for your workload, there are a number of aspects to consider before making a decision.

As discussed in *Chapter 9, Establishing a Deployment Strategy,*, some AWS services leave you full freedom of choice, while some others are rather opinionated when it comes to deployment. The following presents a quick recap of the services discussed.

Consider **AWS Elastic Beanstalk**, which is an application management platform targeting web applications. It is a great option for newcomers because it abstracts away a lot of the underlying complexity of setting up a complete stack for a web application. When it comes to deploying a new application release, Beanstalk offers you multiple options, from all-at-once in-place deployment and rolling an update in-place to blue/green deployment either in-place or on totally new infrastructure resources, or a combination of both.

AWS App Runner also aims to simplify web applications and API management on AWS and goes a step further than Elastic Beanstalk by offering a software deployment environment that is fully managed by AWS. However, it lets you choose from a smaller set of options to customize your application environment. In terms of deployment, you can opt to let App Runner manage your CI/CD pipeline on your behalf or you can manage it yourself. The choices are much more limited compared to the multiple deployment options you have with Elastic Beanstalk.

If you prefer to fully control the deployment process, you may well opt to use **AWS CodeDeploy**. CodeDeploy lets you fully automate the deployment of your applications, whether you want to deploy them on **Amazon Elastic Compute Cloud** (**EC2**) instances, on-premises servers, containers, or **AWS Lambda** functions. It supports a variety of source code repositories, such as **AWS CodeCommit**, GitHub, and Bitbucket. CodeDeploy is often used in combination with the other members of the AWS Code family of services (**AWS CodeCommit**, **AWS CodeBuild**, and **AWS CodePipeline**) to form an end-to-end CI/CD solution. Additionally, it natively integrates with other AWS services such as **AWS Auto Scaling** and **Elastic Load Balancing** (**ELB**) and with third-party configuration management tools (for instance, Ansible, Chef, Puppet, and Salt) and CI tools (for instance, Circle CI and Travis CI). CodeDeploy allows you to deploy application software either via in-place deployments or blue/green deployment (including canary), depending on the target infrastructure platform. In-place deployment is only supported if you deploy to EC2 instances or on-premises servers. Blue/green deployment is available for all the targets supported by CodeDeploy: EC2 instances or on-premises servers, **Amazon Elastic Container Service** (**ECS**), and **AWS Lambda**.

Another important option that can also be used for application deployment is CloudFormation (or CDK). It is primarily intended to deploy infrastructure resources but can indeed be used to define resources such as Lambda functions, Beanstalk applications, or containerized applications running on ECS.

> **Note**
> Please refer to *Chapter 9, Establishing a Deployment Strategy,* for more details on the numerous, most commonly used AWS services available for deploying workloads.

The previous section briefly touched upon the notion of configuration changes that you want to manage and monitor. It was also mentioned that, in some cases, those changes are not necessarily a bad thing, a fact that was highlighted by discussing feature flags. Feature flags can be seen as a form of deployment, given that you actually roll out a new feature. The software code behind a feature is actually already deployed but not activated until you make a specific configuration shift and switch the related feature flag. This is something that you can implement in a number of ways, but AWS provides a feature called **AppConfig** within **AWS Systems Manager** for this purpose.

With respect to configuration changes, you also need to keep an eye on the configuration of your AWS resources besides feature flags. You want to monitor their state over time between releases to make sure that any undesired configuration drift away from their initial ideal state (right after creation) doesn't go unnoticed. For this purpose, it is strongly recommended that you activate **AWS Config** across your AWS production environment. Config records and tracks the evolution of your AWS resource configuration over time, which you can use to take actions such as countermeasures in the case of unexpected changes.

Regardless of the services or tools you eventually line up to deploy your application software, the most important consideration is that they're able to support your deployment strategy, whether it involves an in-place update mechanism or a blue/green canary deployment with specific amounts of traffic shifts over a given period of time till the new release. You also want to specifically emphasize testing multiple deployment failure scenarios to make sure that you can do a clean rollback when your deployment process fails. There's indeed no point in having to clean things manually after your workload has been left hanging in an unknown state following a release deployment that failed in the middle of it.

Therefore, thoroughly testing your strategy is crucial. You can even test multiple strategies, not just for rollback but for the different variants of the entire deployment strategy, to find the one that brings you the most benefits. In the end, your strategy should help you meet your delivery objectives, whether the emphasis is on producing better quality software, delivering at a higher pace or for a lower cost, or a combination of these and other factors.

Summary

In this chapter, you learned how to examine your existing deployment strategy and explored the possible improvement angles to consider, leveraging various services AWS offers.

In *Chapter 16, Exploring Opportunities for Cost Optimization*, we will delve deeper into the world of AWS cost optimization. Building upon the foundation laid so far, you will explore advanced strategies and techniques for cost optimization. This includes a detailed analysis of various AWS services and their cost implications, as well as an in-depth look at how to effectively manage and control AWS costs.

Further Reading

- The Well-Architected Framework operational excellence pillar: `https://packt.link/zCNWV`

- Overview of deployment options on AWS: `https://packt.link/kwNZB`

- Introduction to DevOps on AWS: `https://packt.link/xL7VW`

- AWS DevOps blog: `https://packt.link/eTs4k`

16
Exploring Opportunities for Cost Optimization

In the previous chapter, you learned about the essential cost optimization principles to keep in mind when establishing governance and cost control across your enterprise organization. We also touched upon tagging services and tools, along with some best practices. Finally, you learned how to establish alerts, notifications, and reports for tracking resource utilization as well as forecasting.

In this last yet important chapter of this section, we shall explore some organizational best practices for creating a workload review process, followed by a quick review of the importance of decommissioning resources in AWS as well.

The following main topics will be covered in this chapter:

- Developing a workload review process
- Decommissioning resources

Developing a Workload Review Process

As was discussed earlier, cost optimization is not a one-off process. It is a continuous exercise that involves regular reviews of existing architectures, coupled with evaluating opportunities for cost reductions by either using newer, more improved managed services or decommissioning resources that are no longer required. In this section of the chapter, you will explore some key steps that a solutions architect must keep in mind when designing and developing workload review processes.

When it comes to running workloads on the cloud, many organizations that start out with a traditional lift-and-shift approach often encounter the daunting challenge of reducing the costs of running the workloads as well as the operational costs. Migrating your workloads to the cloud may have saved you a lot as you would no longer need to make the capex investments that you would make in your traditional datacenters. However, these cost savings are short- to medium-term at best. The real question you should be asking yourself here is how you can plan for a long-term, cost-optimized workload hosting strategy, and the answer to this starts with a very simple exercise: reviewing the workloads.

Reviewing your workloads not only ensures that you are aware of the resources that are running in your cloud environment but also provides opportunities to improve an application's overall architecture. The process of reviewing workloads can be split into various sub-categories, such as the following:

- **Reviewing infrastructure opportunities**: Once an application or workload is migrated to the cloud using a lift-and-shift strategy, the underlying compute is often a good starting point to look for costs as well as performance optimization. How? Well, take for example any EC2 instance family. You will notice that AWS frequently launches newer instance generations marked by numerical values. Examples include C4 instance types, followed by C5, and now we even have C6 and C7 types available. Each of these instance types provides an improved layer of performance and also operates at a reduced hourly cost as compared to the previous generation. You could also review infrastructure optimization opportunities by moving away from the standard x86-based architecture to an ARM-based one, such as the AWS Graviton instance type, in order to reduce costs. Storage options could also be evaluated similarly in order to reduce costs. For instance, you could move away from warm storage options to more infrequently used storage tiers if your application does not have a very high throughput or frequent read/write requirements.

- **Reviewing application improvement opportunities**: Since lift-and-shift exercises often mean there are little to no application changes made, you could potentially review your application and look out for any modernization opportunities. These could include opportunities to improve the application's runtime or platform by switching to the latest version or potentially even decoupling the application and moving away from monoliths into microservices.

- **Reviewing operational opportunities**: Although EC2 instances provide a far better operational experience in terms of agility and flexibility compared to standard on-premises infrastructure, they are still resources that need to be managed, and any form of management results in operational costs. The AWS cloud provides numerous automation and managed services, such as AWS Systems Manager and AWS OpsWork, that allow sysops administrators to create simple runbooks and automate most repetitive tasks, such as patching operating systems. However, customers can realize far better cost savings if they move their applications to higher-level managed services such as Fargate or even Lambda functions. Remember that the more advanced your managed service, the more refactoring or rearchitecting of your application you may need.

- **Reviewing licensing opportunities**: Most traditional workloads rely on commercially available off-the-shelf software to run and operate. However, the cloud provides numerous opportunities to evaluate such workloads and either choose to move to a more cloud-native service or even an open source variant. Take for example an Oracle database running on an EC2 instance and managed by an organization's ops and DBA teams. This Oracle database can be ported to either Amazon RDS for Oracle (thus reducing operational overheads of managing the database as well as rightsizing the database) or even to Amazon Aurora for PostgreSQL (an open source database with up to five times the performance of a standard PostgreSQL database instance). Again, in each of these cases, evaluating the application and selecting the right option for your workloads is key to optimizing resources as well as costs.

You have just explored how workload reviews can be divided into various categories, but how do you actually implement this? The following are a few basic steps to keep in mind when implementing a workload review:

- **Understand requirements and plan accordingly**: Each application workload is unique and subject to its own complexity. Some workloads may take months to years to review depending on their legacy nature, whereas some workload reviews may be as short as a few days. What is important in each of these cases is having a clear understanding of each application's requirements and planning for that application's optimization by setting up short-term (days to weeks), medium-term (months), and long-term goals (years).

- **Defining a review frequency**: Once the review goals are set up, the next step is to identify and define how frequently the workloads may be reviewed. Review frequencies can be set either reactively—for instance, an organizational requirement to move away from commercial databases that results in a frequent review of all application workloads—or on a proactive basis, based on each application line owner's requirements.

- **Setting up a review advisory board**: With the requirements laid out and planned for and a review frequency set, the next most important step in setting up a workload review process is the creation of an organization-wide review advisory board. The goal of this board is to act as cloud ambassadors, foster optimization activities across each of the organization's departments on a proactive basis, and provide the right set of guidance, best practices, and knowledge for application modernization and optimization. This board usually comprises cloud or solutions architects, operational leads, and key business stakeholders from across the organization.

- **Create tracking and monitoring mechanisms**: In order to ensure that workloads are optimally reviewed, it becomes ever more important to track resources across a cloud environment, tag them appropriately, and monitor their usage and consumption patterns frequently. This, again, can be achieved by setting up cost allocation tags (as mentioned in *Chapter 4, Ensuring Cost Optimization*), leveraging automation using services such as AWS Organizations and AWS Control Tower to enforce tagging policies in each AWS account, and monitoring and reporting on usage consumptions using AWS Cost and Usage Reporting and Amazon CloudWatch billing and utilization alerts.

In the next section of this chapter, you will examine some best practices and implementation steps required to optimize costs by decommissioning resources.

Decommissioning Resources

Chapter 15, Improving Deployment, discussed at length the importance of tagging and monitoring resources in order to better understand resources and their usage patterns. This not only helps track resources but also helps identify the orphaned ones that, in most cases, are decommissioned. In this last section of the chapter, you will explore a few key best practices for decommissioning resources on AWS as well as how to implement them. The following are the key considerations:

- **Tracking and monitoring resources**: In order to effectively track resources, tagging can be set up automatically using automation or AWS Managed Services and performed manually by developers and operational teams on a case-by-case basis, for example, by using cost allocation tags (*Chapter 4, Ensuring Cost Optimization*). Tags can be set up based on an application's modernization goals, and an automated workload review and decommission process can also be initiated accordingly, for example, using a workload created with an automated tag called "created on" with a value of the date on which it was first created on AWS. Administrators can run simple reports and filter resources based on such tags to identify when these applications are either set for a review (short-/medium-term goal) or decommissioning (long-term goal). Administrators can also leverage AWS monitoring services such as Amazon CloudWatch to effectively trigger alerts and notifications based on an application's overall resource consumption patterns. This too could be taken into account to decide whether to review the application and perform a few optimizations on it or decommission it instead.

- **Designing a decommissioning process**: With tracking and monitoring enabled, you can additionally automate the decommissioning of your AWS resources as well based on automation with services such as Amazon CloudWatch and AWS Auto Scaling. The best part of leveraging a service such as Auto Scaling is that the service can be used to gracefully terminate unwanted resources from your fleet of EC2 instances without impacting the overall performance or availability of the workloads.

With that, we come to the end of yet another chapter, but before we wind up, here is a quick summary of the aspects discussed so far.

Summary

Throughout the previous few chapters, we have discussed various aspects of cost optimization, including AWS services, design best practices, and implementation scenarios. So far, you have reviewed various AWS pricing models and how to monitor and report on the costs and usage for your very own AWS accounts. You also learned the importance of right-sizing resources and went through some key best practices and recommendations as well. You reviewed the key cost optimization principles, followed by a review of essential tagging best practices and strategies. Later, you explored how to optimize costs by setting up simple alarms, alerts, notifications, and reports using various AWS services. Finally, in this chapter, you explored the best practices for creating a workload review process and examined the importance of decommissioning resources in AWS.

In *Chapter 17, Selecting Existing Workloads and Processes to Migrate*, you will move to the part of this book that discusses the various aspects of migrating your workloads, including migration preparedness, application discovery, and the evaluation of application portfolios and strategies for choosing and prioritizing workloads for migration.

Further Reading

- AWS Cost Optimization Pillar: `https://packt.link/KdS7s`

17

Selecting Existing Workloads and Processes to Migrate

Migrating enterprise workloads to the cloud remains one of the most common business cases for most organizations today.

This chapter will discuss some of the common reasons why customers choose to migrate their workloads to the cloud and provide an overview of some of the essential steps required to execute a successful migration. You will also be exploring some commonly adopted migration strategies, along with their pros and cons, and examining the AWS methodology for migrating workloads using the AWS **Cloud Adoption Framework** (**CAF**) and AWS **Migration Acceleration Program** (**MAP**) programs. Finally, the chapter will conclude with some key best practices to keep in mind before performing any large-scale migrations to the cloud.

The following main topics will be covered in this chapter:

- The need to migrate workloads to the cloud
- Steps for a successful cloud migration
- Migration strategies and their associated pros and cons
- Best practices to keep in mind for a successful cloud migration

The Need to Migrate Workloads to the Cloud

Before you can learn about the steps required to undertake a successful cloud migration, you first need to have a clear understanding of some of the common factors that drive a move to the cloud:

- **Cost reduction**: Perhaps the most common driver for many large organizations, cost reduction can be viewed from two perspectives: first, reduction in operational costs from operating legacy and aging data centers or a reduction of infrastructure costs incurred due to large investments made in acquiring data center equipment such as servers, storage, and network; second, specialized or proprietary software that involves complex license agreements and upfront investments. The cloud can prove to be a major factor in reducing overall costs incurred by organizations as, unlike traditional infrastructure that you need to procure in advance and also operate and maintain, the virtualized infrastructure provided by the cloud can be obtained in an on-demand form, scaled as required, and you pay only for the resources consumed and for the duration they were running.

- **Agility and staff productivity**: Organizations also look to the cloud to obtain agility. Unlike traditional data centers, where it could take anywhere between a month or a year to obtain new infrastructure, with the cloud this is made possible almost instantaneously with just a few clicks. This particular way of obtaining managed infrastructure allows organizations to focus more on improving their applications and actually bringing about innovation rather than focusing on procuring, operating, and managing infrastructure components.

- **Improved security and resiliency**: A common misconception around cloud adoption and migration is the lack of visibility and security. Many organizations look to the cloud as it provides a multitude of managed services as well as security measures to ensure that data is protected and made available at all times. Features such as the encryption of data at rest and in transit, along with the introduction of auditable policy-driven controls, ensure that data is always secured against most forms of threats, including DDoS and cyberattacks.

- **Hardware/software end of life**: Many organizations also migrate their workloads to the cloud due to aging data center hardware or end-of-life support for their software license agreements.

- **Data center consolidation**: Managing and operating data centers comes at a significant cost to organizations. Besides the obvious costs of running and managing data center equipment and ensuring its availability, there are also the costs of people, power, specialized software, backup, and disaster recovery solutions that need to be taken into account to build a business case for migrating to the cloud.

- **Digital transformation**: A traditional data center may be able to provide infrastructure-as-a-service to some extent, with the help of automation and other specialized software; however, it's still very complex and difficult for a data center to provide a platform or software as a service. By leveraging the cloud's agility and broader use of its managed services, organizations can look to adopt newer technologies such as containerization, serverless, IoT, and AI/ ML much more quickly than they would in a traditional environment.

- **Going global quickly**: With the cloud's ability to provide customers with a wide array of digital services spread across its global regions, organizations can look to quickly deploy and scale out their applications and workloads across multiple geographies without having to spend time and effort setting up their own points of presence as well as making heavy upfront investments for the same.

These are again just some of the common business drivers that lead organizations to migrate their workloads to the cloud. However, any and all migration journeys start with a helpful understanding of some important and well-defined steps. The next section presents some of these steps that you, as a solutions architect, should keep in mind when working toward a successful cloud migration.

Steps for a Successful Cloud Migration

Before actually starting a migration journey, there are a number of prerequisites that any and all organizations need to have established for a smooth adoption and transition to the cloud. These are as follows:

- Assess

- Mobilize

- Migrate and modernize

The Assess Phase

The *Assess* phase forms the foundation of all migration projects and is an essential step in understanding, planning, and building a business case for change within an organization. As the name suggests, this phase is used to assess an organization's readiness for cloud adoption and plan for subsequent steps required in the *Mobilize* and *Migrate and Modernize* phases.

AWS provides a comprehensive set of best practices and experiences that an organization can leverage in the assessment phase to identify, prioritize, align, and launch a successful cloud transformation and adoption journey. This framework is known as the **Cloud Adoption Framework (CAF)** and is essentially built on the basis of six perspectives: **Business**, **People**, **Governance**, **Platform**, **Security**, and **Operations**, as shown in *Figure 17.1*:

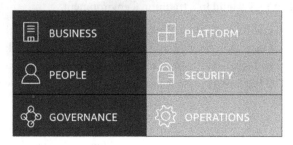

Figure 17.1: The Cloud Adoption Framework's perspectives

These perspectives are explained in detail here:

- **Business**: Helps ensure that the organization's investments in its cloud and digital transformation enablement are accelerated to meet the required business outcomes.

- **People**: Like any other significant or large-scale project, a successful implementation and outcome is not only based on the technology or process that is selected but also, more importantly, on the people who work on it. This perspective acts as a bridge between the technology and business perspectives, helping organizations scale more rapidly by introducing a culture of change and innovation specifically targeted toward senior leadership as well as project stakeholders.

- **Governance**: With any large-scale transformation or migration project, there are bound to be a variety of potential risks that need to be understood and mitigated. This perspective helps with standardizing as well as orchestrating an organization's cloud initiatives while minimizing the overall risks and challenges.

- **Platform**: A cloud transformation and migration journey is only as good as its underlying technology platform. This particular perspective helps organizations build a strong foundational platform for the cloud that provides mechanisms to not only lift and shift workloads from the on-premise data center to AWS but also look at modernizing workloads with the help of cloud-native services.

- **Security**: The security perspective helps organizations create a strong security foundation to ensure the confidentiality, integrity, and availability of their data and workloads on the cloud.

- **Operations**: Finally, with the migrations completed, there is a need to operate and manage the new workloads as well as ensure that their operational effectiveness meets the business requirements. The operations perspective of the CAF helps organizations design, develop, and run incident and problem management, change management, application, and infrastructure observability, among other operational aspects, effectively on the cloud.

As a part of the outcome from the *Assess* phase, a **Migration Readiness Assessment (MRA)** is used together with the six CAF perspectives to analyze the completeness of the organization's transformation initiative as well as where it is currently as a part of the cloud onboarding strategy. The MRA also helps organizations identify their strengths and weaknesses in terms of cloud readiness and, last but not least, provide a plan of action to plug in the gaps identified during the assessment phase. *Figure 17.2* shows a typical MRA, showing the status of various crucial components of a migration readiness plan. The assessment status of each is marked by a symbol and a color, with the standard traffic light system. Green means that additional work is required. Both yellow and red mean that these components need attention, with red being the most urgent.

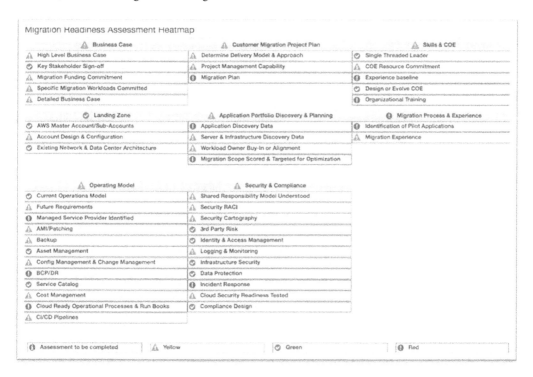

Figure 17.2: A sample migration readiness assessment output in the form of a heatmap

Now that you have a thorough understanding of your organization's cloud readiness assessment, the next step in a successful cloud migration program is to initiate the *Mobilize* phase. This is discussed in the next section.

The Mobilize Phase

This phase involves organizing and preparing the resources required to migrate workloads to the cloud, and it is conducted by categorizing individual tasks into one or more workstreams. These workstreams help create the right impetus as well as a cultural mindset for an organization to migrate workloads to the cloud. This phase helps to lay a robust foundation for migration using relevant tools, processes, and services. Some of the workstreams are as follows:

- **Detailed business case**: Before starting any large migration project, it is important to perform a **total cost of ownership** (**TCO**) exercise to understand the organization's current IT spend, what their spend on the cloud would look like in projected one-year, three-year, or five-year periods, and how much the migration initiative would cost on its own. These details help senior stakeholders understand, plan, and align on short-term and long-term goals such as identifying a pilot application for the migration, identifying migration waves, their priorities, and so on.

- **Application discovery**: Most organizations have an existing **configuration management database** (**CMDB**) that historically contains a list of all the organization's hardware and software assets, including applications, databases, integrations, and dependencies. The application discovery workstream helps collect application-specific data points and map them against the 7 Rs of migration strategies (discussed in detail in the *The 7 Rs of Migration Strategies* section of this chapter). As an outcome of this workstream, the organizational teams are provided in-depth insights into their application's current as well as future state, along with detailed prioritization and categorization, and plans to undertake each of the migration waves.

- **Landing zone**: As the name suggests, a landing zone is essentially a platform that provides structure, governance, and security for any new workload deployment to the cloud. A landing zone follows AWS-recommended best practices such as designing and implementing a multi-account strategy that provides the organization with essential building blocks such as networking capabilities with the on-premises data center, centralized logging and audits, a baseline of security services that are automatically instantiated based on the guidelines set down by the security teams, and so on. AWS provides a managed service called **AWS Control Tower** that many organizations leverage to build bespoke landing zones as part of their cloud onboarding activities. Control Tower essentially provides an easy mechanism for creating multiple AWS accounts using Account Factory, which is powered behind the scenes by two other AWS services—AWS Organizations and AWS Service Catalog.

- **Operations**: The operations workstream helps organizations understand their current IT operational model and plan and create a forward-looking cloud operations unit that can support and scale as more workloads are migrated to the cloud. This could also involve training and certifying the current operational teams with cloud knowledge, creating cloud-specific automation and runbooks, and even standardizing the tools and processes that work across the on-premises and new cloud environments.

The last, and perhaps the longest phase in terms of duration to complete, is the *Migrate and Modernize* phase. The following section discusses the workings of this phase in depth.

The Migrate and Modernize Phase

This phase essentially builds on the outputs of the *Mobilize* phase and ensures that the right set of tools, people, and processes are in place for conducting a successful cloud migration. This phase also helps accelerate the migration of identified workloads to the cloud using the concept of Cloud Migration Factory—a structured approach to cloud migration where each step involved in migration is standardized for efficiency and the mitigation of errors. The main purpose of Cloud Migration Factory is to assimilate the application findings gathered from the earlier phases along with AWS best practices and migration patterns and create a scalable and repeatable mechanism for accelerating migrations to the cloud. The factory can include tools, scripts, or a combination of one or more AWS services, such as AWS **Database Migration Service (DMS)** or AWS **Application Migration Service (AMS)**.

Following the three phases is key to any successful migration to the cloud. With this in mind, you are now ready to have an in-depth look at some of the key strategies to keep in mind when working toward a large-scale migration project.

Migration Strategies

So far in this chapter, we have discussed the need for migrating workloads to the cloud, as well as some of the essential steps to perform when journeying toward a successful migration project. In this section, you will explore some important migration strategies and learn about some best practices to keep in mind when it comes to selecting the right strategies for your workloads.

The 7 Rs of Migration Strategies

The 7 Rs of migration strategies are as follows:

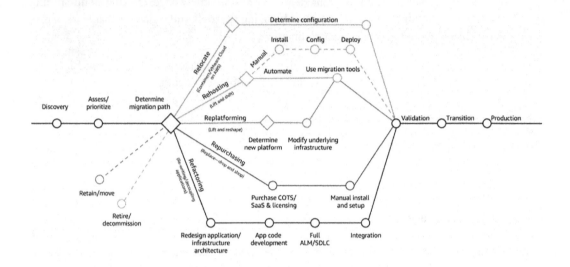

Figure 17.3: Application migration process

The 7 Rs are discussed in detail here:

- **Retire**: Often, an organization will find a few applications during the *Assess* and *Mobilize* phases that are not used frequently or are simply too old to be migrated to the cloud. The fact that a majority of such applications are built using legacy or aging hardware means they are often very difficult and expensive to develop, maintain, and upgrade. Applications that are no longer needed within the organization are perfect candidates for decommissioning or retirement.

- **Retain**: Organizations sometimes choose to retain some of their legacy applications in their on-premises data centers, either due to strict compliance or security reasons or due to the time and effort required in migrating or modernizing the application to the cloud.

- **Relocate**: Relocation is sometimes confused with the more commonly employed rehost migration strategy. Relocation primarily involves the migration of workloads using the hypervisor as a common medium. For example, an on-premises data center with VMware ESXis as the choice for hypervisors can migrate **virtual machines** (**VMs**) running on it to AWS using VMware Cloud on AWS, which is a managed service that essentially provides VMware ESXi hypervisors along with other essential VMware services such as NSX in the form of a managed service. Since the underlying hypervisor platform is the same, migration in these cases can often be done quickly with a low to minimalistic amount of effort.

These types of relocation strategies are excellent for organizations that wish to migrate their workloads to the cloud quickly either due to an upcoming data center lease expiry or a data center hardware refresh cycle. The obvious downside to this strategy is that there is little to no scope for modernizing the workloads as the migration is still focused on a standard lift-and-shift mechanism.

- **Rehost**: By far the most commonly employed strategy by many large organizations, rehosting allows application workloads to be migrated to the cloud without making any significant changes to the applications, however, still using the cloud provider's native compute capabilities instead of relying on the underlying hypervisor (relocate) technology. In this case, workloads running in an on-premises data center can be migrated to AWS services such as Amazon EC2, which offers runtime capabilities similar to that of a VM but without the added costs of running or managing a hypervisor.

  Besides the ease of lift and shift, rehosting also provides organizations an easier pathway to modernize their workloads once they are running in the AWS cloud. This is also one of the reasons many organizations choose to leverage this strategy for migrating their applications to the cloud.

- **Repurchase**: Repurchasing involves the process of dropping traditionally licensed software from an in-house developed solution for a more scalable SaaS-based application. There are many reasons for an organization to select this particular strategy, including a lack of developers or skills to manage the custom in-house application or expensive traditional licensing models, which often require large upfront payments as well as long-term commitments.

- **Replatform**: Replatforming allows organizations to attain cloud benefits such as scale and agility by adopting certain parts of the application with cloud-native services. Replatforming comes in two forms. It could either be replatforming an application, say, from Windows to Linux in order to save costs while achieving better scaling and performance. Or, it can be replatforming from Amazon EC2 to another higher form of a managed service that provides additional features on top of your standard EC2 instances, such as replatforming an SQL server that ran traditionally on an EC2 instance to an Amazon RDS managed SQL server. Depending on the pathway chosen for a particular application, replatforming can require small to significant efforts; however, the long-term benefits greatly outweigh them.

- **Refactor**: The last of the 7 Rs, refactoring an application requires the highest amount of effort as, in this case, an application is more often than not completely rearchitected using cloud-native services to get best-of-breed performance, agility, and cost savings. Refactoring can also be split into two types: either a complete rearchitecting of the application to work with cloud-native services or refactoring an application to leverage an open-source variant of a commercial product to reduce overall costs, for instance, moving away from an Oracle database to an Amazon Aurora PostgreSQL compatible database.

The consideration that you are now left with is which strategy is suitable for your organization. Well, the answer to this is very much what all solutions architects will tell you – *it depends!* It really does depend on a number of factors, such as the long-term vision or strategy of your organization, whether your organization is cloud-mature and ready for a migration based on the CAF and MRA assessments, the timelines for migration, the complexity or legacy nature of applications, the team structure, culture, and willingness to adapt to changes and newer technologies, and so on.

Organizations can always choose a mix of strategies depending on their applications' assessments as well as business requirements. Strategies such as rehosting and relocating can prove to be a faster way to cloud adoption but bring in the least amount of cloud-native benefits. In both cases, the workloads are just lifted and shifted to AWS. On the flip side, strategies such as replatforming and refactoring can bring in significant cloud benefits but usually take the longest time to adopt and build upon, as demonstrated in *Figure 17.4*:

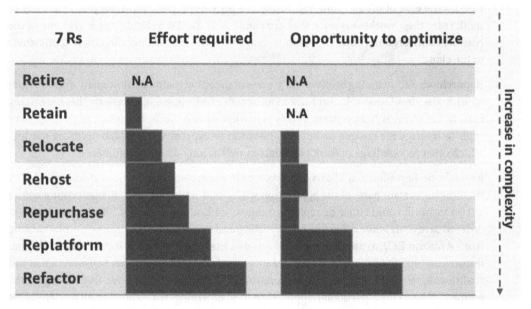

Figure 17.4: Cloud migration strategies and their complexities compared

Now that you have explored all the strategies, you are ready to examine some essential best practices to keep in mind when working on a cloud migration program.

Best Practices to Keep in Mind for a Successful Cloud Migration

This chapter has so far discussed the need for organizations to migrate workloads to the cloud and some of the steps and strategies to consider when deciding how to actually perform the migration. This section presents some essential high-level best practices that all solutions architects should be aware of when it comes to successfully implementing a migration to the cloud. These practices are explained here:

- **Obtaining senior leadership support and alignment**: Migrating workloads to the cloud is much more than just technological choices. A successful migration, as well as a modernization program, hinges on a variety of other factors, including getting the organization's senior leadership to buy in. This could be achieved by demonstrating the cloud's true capabilities as well as delivering a comprehensive cloud adoption strategy, business case, and migration plan.

- **Creating top-down goals**: Once the senior leadership buy-in and sponsorship have been obtained, identifying and implementing top-down goals that are quantifiable and measurable throughout the organization is critical in ensuring a faster cloud adoption and migration program. These goals are often aggressive and force an organization to move at a faster pace than it would organically.

- **Training teams**: With the senior leadership support and top-down goals set, the next important step in any migration and cloud adoption journey is training your staff and ensuring that the right cultural mindset is set in place before the cloud adoption journey actually begins. Developers and operational teams can be upskilled with newer processes, tools, and technologies, and enterprise architects and business owners can be trained to identify the right strategies and patterns for migrating workloads to the cloud.

- **Identify the right migration strategy**: Crafting a detailed strategy that identifies the right set of tools, processes, and practices for migrating your application is an essential and crucial step for any organization looking to move its applications to the cloud. The strategy choice will depend on numerous factors, such as the complexity of the overall application, its usage patterns, licensing or hardware constraints, security, and governance requirements.

- **Start small, scale big**: Once the migration strategy is identified, the next step is to start developing and rolling out small-scale pilots or **proof of concepts (PoCs)** that demonstrate and validate the strategical outcomes as per your organization's requirements. This is done by identifying and targeting the low-hanging fruits first or, in this case, applications that are easy to migrate as they have little to no complexity or constraints. The important aspect to keep in mind here is to perform these pilots quickly and through a repeatable mechanism.

- **Set up a Cloud CoE:** As the organization slowly but steadily starts with its cloud adoption and migration journey, there is now a need to create and standardize cloud architectural patterns, operational practices, and security measures across all the individual teams.

  This can be achieved by forming a core team of individuals from across the organization containing cloud-trained and certified developers, architects, engineers, business stakeholders, operational leads, and so on in a **Cloud Center of Excellence (CCoE)** group. The function of this group is to act as the organization's cloud advocates, provide guidance, eliminate barriers to migration, and support overall cloud adoption.

As you are aware, there is no single remedy to solve all problems, and the same applies to complex and large migration projects as well. Any migration project comes with its own set of challenges and blockers, in the form of people, processes, or technologies. However, with the right set of recommendations and best practices, such projects can be achieved with real business benefits as well.

Summary

This chapter started with a discussion on some of the common drivers that motivate an organization to migrate its workloads to the cloud. This was followed by an exploration of the essential steps involved in carrying out migrations in the form of the *Assess*, *Mobilize*, and *Migrate* and *Modernize* phases. You then dove a bit deeper into some commonly used migration strategies along with the key considerations to keep in mind when selecting the right strategy for your workloads. Finally, the chapter concluded with some essential best practices to keep in mind when working toward any successful large-scale migration projects.

In *Chapter 18*, *Selecting Migration Tools and Services*, you will explore some of the essential AWS services and tools that you can leverage to jumpstart a migration project.

Further Reading

- Evaluating migration readiness prescriptive guidance: `https://packt.link/89prr`
- Mobilizing an organization in order to accelerate large-scale migrations: `https://packt.link/Ra6ts`
- Automating large migrations using Cloud Migration Factory: `https://packt.link/VyEWa`

18

Selecting Migration Tools and Services

In the previous chapter, you learned about some essential steps and strategies that all cloud solutions architects should keep in mind in order to plan and execute a successful application migration from on-premises data centers to the cloud.

In this chapter, we'll take that knowledge and dive a bit deeper into the various tools, services, and strategies that AWS provides to facilitate such migration endeavors using the *Assess*, *Mobilize*, *Migrate*, and *Modernize* phases. Additionally, you will also explore mechanisms and operational best practices using which you can migrate the essential components of any application workload to the AWS cloud, such as raw data, the application's backend systems, such as databases, and finally, the frontend servers that host the application and web tiers, respectively.

The following main topics will be covered in this chapter:

- Selecting an appropriate database transfer mechanism
- Selecting an appropriate data transfer service
- Selecting an appropriate server migration mechanism
- Applying the appropriate security methods to the migration tools

Selecting an Appropriate Database Transfer Mechanism

Migrating database workloads to the cloud can be a daunting challenge even for the most seasoned database administrator or cloud architect. Once an organization builds a strong business case for migration, the next steps involved are assessing the workloads, identifying the interdependencies, selecting the right migration strategy, and finally undertaking the migration itself using one or more specialized sets of tools or services.

In this section of the chapter, we shall explore each of the Assess, Plan, Migrate, and Operate phases that are required to undertake a database migration from on-premises data centers to the AWS cloud.

The Assess Phase

As was covered in the previous chapter, the Assess phase forms the backbone for any and all migration as well as modernization activities that an organization undertakes. With database migrations, the Assess phase helps the organization analyze database workloads and their various interdependencies, schemas, or custom code, and subsequently plan for the right migration strategy to accompany it. To assist with the process, AWS provides a simplified workload qualification tool called the AWS **Workload Qualification Framework** (**WQF**). WQF is specifically designed to evaluate a database workload and its interdependencies and helps categorize the workload into one of the five categories shown in *Table 18.1*. These five categories later form the basis for which the right database migration strategy can be planned.

The five categories are as follows:

Categories	Workloads	Effort Required
1	ODBC/JDBC workloads	< 50 manual changes
2	Light, proprietary feature workloads	< 200 manual changes
3	Heavy, proprietary feature workloads	> 200 manual changes
4	Engine-specific workloads	No changes possible
5	COTs/non-profitable workloads	No changes possible

Table 18.1: Workload categorization and effort calibration

The following points present details about each category:

- **Category 1**: This category is assigned to databases that operate with standard, open source drivers such as ODBC or JDBC instead of proprietary ones provided by commercial-grade databases. The databases that fall under this category are likely to have little to no complex integrations or specialized code as in stored procedures and so on, and require the least amount of effort in migration or refactoring as well.

- **Category 2**: Category 2 databases often leverage a few proprietary features and add-ons that a commercial database such as Oracle or SQL Server may provide; however, the effort required to refactor the database code is still minimal and modest.

- **Category 3**: Database workloads that are heavily tied to commercial databases and leverage all of their proprietary features are classified as Category 3 workloads. Such workloads are generally complex to migrate and modernize, requiring significant effort both in terms of time and costs.

- **Category 4**: Category 4 database workloads leverage a very specific set of services or features provided by a commercial/proprietary database vendor and only provide integration capabilities with other products or services provided by the same database vendor. Such workloads can include, but are not limited to, Oracle Forms, Oracle Reports, Oracle **Application Development Framework** (**ADF**), and Oracle **Application Express** (**APEX**).

- **Category 5**: The last category of database workloads is very specifically catered toward **common, off-the-shelf** (**COTs**) products that provide no alternatives or mechanisms to migrate from a commercial database to a cloud-native option.

With the Assess phase discussed comprehensively, you are now ready to examine the Plan phase as well.

The Plan Phase

With the Assess phase completed and a WQF categorization done as well, the next step in any database migration is identifying and planning for the right migration strategy. Seven key migration strategies or pathways were discussed in *Chapter 17*, ranging from *Rehost* to *Relocate*, to more complex ones that involve changing the underlying platform or architecture such as *Replatform* and *Refactor*, respectively. This particular section of the chapter will detail each of these strategies and present some best practices to keep in mind when selecting the right strategy for your database migration.

At a high level, there are two main options available for a database to be migrated to the cloud—a homogeneous migration, referring to a like-for-like migration, or a heterogeneous migration, where a different database target is selected from the source. Selecting either of these pathways depends on numerous factors, such as how much time and effort the organization is willing to invest in the migration and the organization's long-term goals, business requirements, and technical constraints.

Table 18.2 provides a high-level view of the common database migration strategies alongside their corresponding options and use cases:

Migration Strategy	Migration Option/Type	Use Case	Example
Rehost	Homogeneous	Moving the database with zero to minimal changes	Migrating an on-premises Oracle database to Amazon EC2
Relocate	Homogeneous	Moving a database with zero to minimal changes, but using the hypervisor as the common migration medium	Migrating an on-premises Oracle database from a VMware Virtual Machine (VM) to a VMware Cloud on AWS VM

Migration Strategy	Migration Option/Type	Use Case	Example
Replatform	Homogeneous	Migrating a database to a managed service on the cloud	Migrating an on-premises Oracle database to Amazon RDS for Oracle
Repurchase	Heterogeneous	Migrating the data from the database to a new SaaS application	Migrating data from an on-premises Oracle database to Snowflake
Refactor	Heterogeneous	Migrating the data from the database to a cloud-native, purpose-built database	Migrating data from an on-premises Oracle database to Amazon Aurora Serverless with PostgreSQL compatibility

Table 18.2: Common database migration strategies

It is evident from the preceding table that with each migration use case/strategy that is selected, there is an underlying effort in terms of time, complexity, and costs that organizations need to consider. Strategies such as Rehost and Relocate provide a much faster pathway for migration as in these cases, there is little to no change in the underlying database, its code, operating systems, and method of operations. On the other hand, Replatform, Repurchase, and Refactor strategies each bring an added layer of complexity in terms of the effort required to change the database from one platform to either a SaaS product or a cloud-native database.

Figure 18.1 depicts the correlation between each of the migration strategies and the outcomes of the WQF categorization:

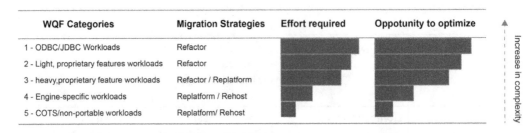

Figure 18.1: Migration and optimization effort calibration

Now that you've gone through the factors involved in the Plan phase, it's time to swiftly proceed to the crucial Migrate phase. The next section discusses the Migrate phase in detail.

The Migrate Phase

We will now move on to the actual migration phase, where the source database is migrated to a target database (homogeneous or heterogeneous) using one or more tools and AWS services. This section will talk about two such powerful AWS tools, the first being the AWS **Schema Conversion Tool** (**SCT**).

Converting the Schema

Working with a simplified Rehost or Relocate database migration initiative—for example, moving an Oracle database from an on-premises data center to an Amazon EC2 instance—requires little to no change in the database's underlying schema. However, if the database were to undergo a Refactor or Replatform exercise wherein the source and the target databases are two different entities (heterogeneous migration), a specialized tooling or service is required to assess, plan, and migrate the schema from the source to the new destination. This is where AWS SCT comes into play.

AWS SCT is a Java-based, UI-driven tool that automates the conversion of a source database schema along with its objects, such as database views, stored procedures, and functions, into a format that is compatible with the selected target database. This is made possible by SCT's unique ability to analyze a source database schema and generate a modernization/conversion report. The report clearly highlights the source database objects that can be automatically changed by the tool itself, along with the objects and code that may require manual intervention. Additionally, SCT is also capable of scanning application code for embedded SQL statements and helps convert those to relevant objects as part of a database schema conversion project.

Some of the database schema conversions supported by SCT are mentioned in *Table 18.3* (source: `https://packt.link/egmFl`):

Source Database	Target Database on Amazon RDS
Oracle Database	Amazon Aurora MySQL-Compatible Edition (Aurora MySQL), Amazon Aurora PostgreSQL-Compatible Edition (Aurora PostgreSQL), MariaDB 10.5, MySQL, PostgreSQL
Microsoft Azure SQL Database	Aurora MySQL, Aurora PostgreSQL, MySQL, PostgreSQL
Microsoft SQL Server	Amazon Redshift, Aurora MySQL, Aurora PostgreSQL, Babelfish for Aurora PostgreSQL (only for assessment reports), MariaDB, Microsoft SQL Server, MySQL, PostgreSQL
Oracle data warehouse, Teradata, IBM Netezza, Greenplum, HPE Vertica, Snowflake, Azure Synapse Analytics	Amazon Redshift
MySQL	Aurora PostgreSQL, MySQL, PostgreSQL

Source Database	Target Database on Amazon RDS
PostgreSQL	Aurora MySQL, Aurora PostgreSQL, MySQL, PostgreSQL
IBM Db2 LUW	Aurora MySQL, Aurora PostgreSQL, MariaDB, MySQL, PostgreSQL
IBM Db2 for z/OS	Aurora MySQL, Aurora PostgreSQL, MySQL, PostgreSQL
Apache Cassandra	Amazon DynamoDB
SAP ASE	Aurora MySQL, Aurora PostgreSQL, MariaDB, MySQL, PostgreSQL

Table 18.3: Database schema conversions supported by SCT

Getting started with SCT is relatively straightforward. You start by downloading and installing the SCT binary locally on your workstation. Once installed, you create a database migration project using the tool. The project helps establish a connection with the source as well as the target database using established credentials. Then, the project lets you select one or more schemas from the source database for conversion. During this phase, SCT examines in detail all the objects in the schema of the source database and provides you with an assessment report that highlights the list of objects that it can convert automatically and the ones that would need manual intervention. An example is shown in the screenshot in *Figure 18.2*:

Figure 18.2: Schema conversion using SCT

When the report is generated, you can choose to apply the schema changes automatically or apply them manually too. The red warning labels indicate the issues and changes that are required to be addressed to match the schema with the target database. Once the changes are done, you can apply the new schema to your target database using SCT itself.

The next section discusses the second tool that is crucial during the Migrate phase—**Database Migration Service (DMS)**.

Migrating the Database

With the schemas' compatibility and configurations out of the way, it is finally time to start the actual migration of the data from the source database to the target database in AWS. For this purpose, you leverage AWS DMS. DMS is yet another managed service that allows organizations to migrate their on-premise databases, supported data warehouses, and other types of datastores to the AWS cloud with minimum downtime. DMS offers added migration capabilities that allow you to either migrate all your data in one go or perform synchronous replications between the source and target databases over a sustained period of time. Together with SCT, DMS allows users to perform both homogeneous as well as heterogeneous database migrations with relative ease, offering built-in features such as managed replication software and infrastructure, automated failovers, and encryption of data at rest and in transit.

How does DMS work? DMS essentially provides a completely managed data replication software packed with an EC2 instance in the form of a replication server. The job of this replication server is to essentially connect the source and target datastores, load the data from the source into its memory (in some cases, it can also write larger transactions to disk), and finally write that data to the target datastore as well. The connections between the source and target datastores are maintained by a source and target endpoint and each migration activity is represented by a task, as depicted in *Figure 18.3*:

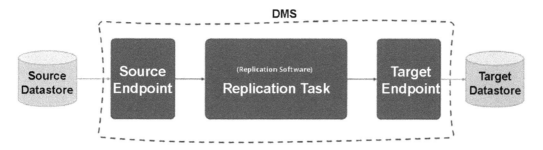

Figure 18.3: High-level overview of DMS

Each replication task runs inside a managed instance, and similar to your standard EC2 instances, the replication instance type can be selected based on your migration requirements. A single replication instance itself can handle one or more replication tasks. However, if your migration involves a significantly large database containing multiple sets of tables, it is best to opt for a high-memory replication instance type to compensate for the large number of concurrent replication tasks. You could also deploy the replication instance in a multi-AZ deployment model to provide an additional layer of availability for your migration tasks. In this case, DMS ensures that a secondary standby replica instance is provisioned and running in a different AZ from the primary one. The following steps teach you how to create a replication instance:

1. Sign in to the AWS Management Console and launch the DMS console by visiting https://packt.link/qwTTC.

2. Once DMS is launched, select the Replication instances option from the navigation pane on the left and then select the Create replication instance option.

3. This will bring up the Create replication instance page; here, you configure details for your replication instance such as the name for the instance, a description, the instance class based on the migration requirements, the allocated storage for the instance, as well as some other networking configuration details such as the VPC in which the replication instance will be launched and its corresponding subnet group. *Figure 18.4* shows the Create replication instance screen:

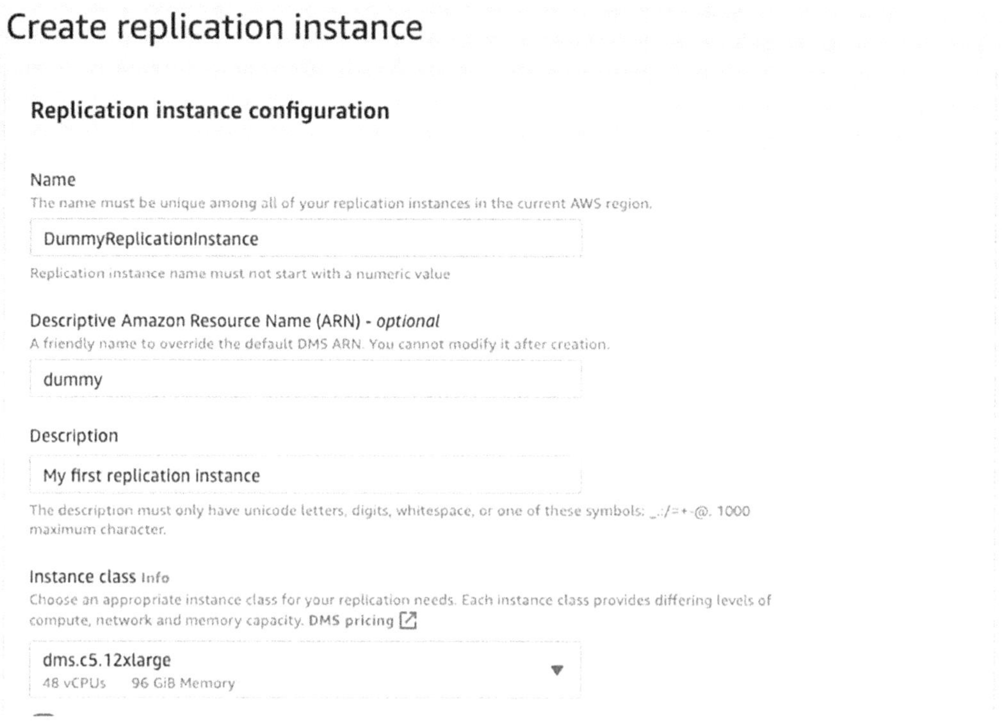

Create replication instance

Replication instance configuration

Name
The name must be unique among all of your replication instances in the current AWS region.

DummyReplicationInstance

Replication instance name must not start with a numeric value

Descriptive Amazon Resource Name (ARN) - *optional*
A friendly name to override the default DMS ARN. You cannot modify it after creation.

dummy

Description

My first replication instance

The description must only have unicode letters, digits, whitespace, or one of these symbols: _.:/=+-@. 1000 maximum character.

Instance class Info
Choose an appropriate instance class for your replication needs. Each instance class provides differing levels of compute, network and memory capacity. **DMS pricing** ☑

dms.c5.12xlarge   ▼
48 vCPUs   96 GiB Memory

Figure 18.4: Creating a replication instance using the DMS management console

4. You can additionally also choose to provide additional networking and security parameters such as the VPC security group to use and the AWS KMS key for encrypting the data stored at rest on the instance itself. Once done with your selections, click on the Create option to spin up your first DMS replication instance! It's that simple!

With the instance up and running, you can now go ahead and configure the two endpoints that will be used for connecting the source and target datastores. Follow these steps to learn how to do that:

1. From the same DMS console, select the `Endpoints` option from the navigation pane on the left. Then, click on `Create Endpoint`.

2. Here, on the `Create Endpoint` page, start by selecting the `Source` option as the first endpoint. Provide a suitable endpoint identifier and select the appropriate source engine of the source database, as shown in *Figure 18.5*:

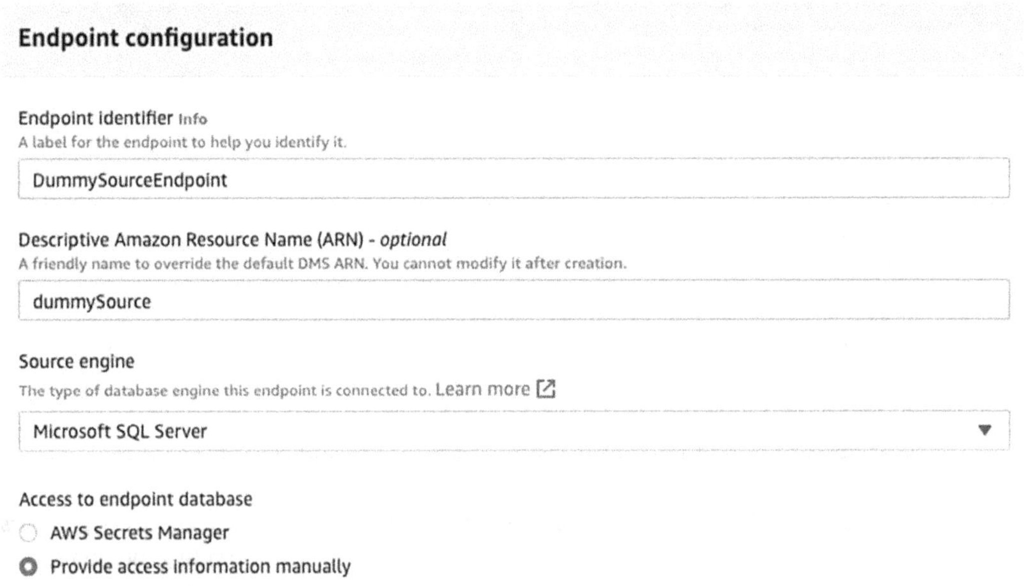

Endpoint configuration

Endpoint identifier Info
A label for the endpoint to help you identify it.

> DummySourceEndpoint

Descriptive Amazon Resource Name (ARN) - *optional*
A friendly name to override the default DMS ARN. You cannot modify it after creation.

> dummySource

Source engine
The type of database engine this endpoint is connected to. Learn more ☑

> Microsoft SQL Server   ▼

Access to endpoint database

⭘ AWS Secrets Manager

🔘 Provide access information manually

Figure 18.5: Endpoint creation

3. Next, provide the source database credentials using either `AWS Secrets Manager` or even the `Provide access information manually` option. If you select the manual option, you will be prompted to enter the source database's server name or its IP address, the port to connect to, whether to leverage `SSL Mode` to secure the connection, and finally the database credentials in the form of the username, password, and database name to connect to.

4. With all the necessary details furnished, choose to test the connectivity (optional but recommended) with the newly created replication instance. This will ensure that the source database connects with the instance successfully.

5. Once the test executes successfully, select the `Create endpoint` option to complete the creation of the source endpoint. Follow the exact same steps using the same `Create endpoint` page and create a target endpoint for the migration task as well.

With the replication instance and the two endpoints created, the final step of the process is to create the migration task. As the name suggests, the purpose of the task is to start the data migration between the source and target endpoints using the replication instance created a while back. Follow these steps to create the migration task:

1. From the DMS console, select the `Database migration tasks` option from the navigation pane, followed by the `Create task` option.

2. In the subsequent `Create database migration task` page, provide a suitable name for the task identifier and select the newly created replication instance, the source database endpoint, and target database endpoint, respectively. Once done, select the migration type that best suits your database migration workload. The types are described as follows:

   - `Migrate existing data`: Selecting this option ensures that the full load of the source database is copied to the target database in a single go. Do note that by selecting this option, the performance of your source database will be impacted and in some cases may require a planned downtime as well. For a one-off database migration, this option works best.

   - `Migrate existing data and replicate ongoing changes`: As the name suggests, this option performs a full data load while capturing changes on the source. Once the entire database load is completed, the replication instance kickstarts change data capture on the target database. This ensures that any new data written to the source database is captured to the target database as well.

   - `Replicate data changes only`: This option is useful to ensure that both the source and target databases are always in sync with each other. It maintains this sync by copying only the changed data over to the target database while the migration task is in effect.

3. DMS additionally provides options that you can leverage to configure your target database, such as preparing the target database's tables. To do so, select either of the following three options:

   - `Do nothing`: If you choose this, DMS will create new tables at the target database if they do not exist. However, if the tables already exist at the target, then they remain unaffected.

   - `Drop tables on target`: If this option is selected, DMS will drop the tables present in the target database and attempt to create new tables in their place.

   - `Truncate`: If you select this option, DMS will simply remove the data from the tables at the target database, while leaving the tables and their metadata intact.

4. With all the prerequisite steps completed, choose to enable the remaining settings if necessary, such as `Enable CloudWatch logs`, or choose the migration startup time to be automatic after this task is created or manual at a later given date and time.

5. Select `Create task` to complete the process, and there you have it, a simple database migration workflow setup using AWS DMS!

In the next section, you will explore some other essential AWS services that you can leverage to initiate data migration from on-premises data centers to the AWS cloud.

Selecting an Appropriate Data Transfer Service

AWS DMS along with AWS SCT facilitates a rapid and secure database migration to the cloud. However, there are pockets or silos of data marts within an organization that aren't stored on traditional database servers, including data such as email archives, images and videos, and flat files. This section will cover some data transfer options (online and offline) offered by AWS for migrating large sets of data from an organization's on-premises data center to the AWS cloud.

Online Data Transfer

AWS provides a wide assortment of services that organizations can leverage to migrate data from on-premises data centers to the AWS cloud via the internet (online mode). Some of these services are described here.

AWS Storage Gateway

AWS Storage Gateway provides a hybrid connectivity and storage delivery medium between an organization's legacy on-premises hardware and the storage services on the AWS cloud. AWS Storage Gateway can be deployed as a standalone VM or even as a preconfigured hardware appliance in an on-premises data center. Once deployed, you can integrate Storage Gateway with almost any on-premises storage solution as long as it supports one of these standard storage protocols: SMB, iSCSI/VTL, and NFS. Storage Gateway also provides an in-built, fully managed caching solution that enables applications to have access to the most recently written or read data. This feature comes in really handy when you wish to reduce the overall read/write latency of applications storing data on AWS. Storage Gateway comes in four different types, each briefly explained as follows:

- **Amazon S3 File Gateway**: S3 File Gateway allows you to store any file data as objects on Amazon S3 with a one-to-one mapping. This type of gateway comes very much in handy for migrating backups and archival data from on-premises to the AWS cloud as well as for advanced workloads such as machine learning and data lakes. S3 File Gateway supports both NFS and SMB-based access to data stored on Amazon S3.

- **Amazon FSx File Gateway**: This particular file gateway provides native access to Amazon FSx using SMB-based directory shares from your on-premises networks, thus allowing applications and users to store and share files across a wider organization.

- **Tape Gateway**: As the name suggests, Tape Gateway essentially replaces the physical tape infrastructure by leveraging one or more virtual tape libraries backed by Amazon S3 archiving tiers to store data over long periods of time. Tape Gateway supports almost all leading backup applications using iSCSI-VTL and provides caching functionalities too for low-latency workloads and applications.

- **Volume Gateway**: Last but not least, Volume Gateway is used to provide block storage volumes using iSCSI to applications running in your on-premises data center. This particular gateway is also backed by Amazon S3. It offers the following two modes of operation, which you can select based on your requirements:

  - **Cached Volume Gateway**: Frequently accessed data is cached locally for reduced latency workloads, while the actual data remains stored in Amazon S3.

  - **Stored Volume Gateway**: The entire dataset is stored locally and made available for use while it is asynchronously backed up to Amazon S3.

The obvious question now is when do you use AWS Storage Gateway? Well, if you are looking to retain your on-premises storage hardware or already have an existing legacy backup appliance/solution for your on-premises environment, then AWS Storage Gateway may be the right option for you. You can use AWS Storage Gateway to establish hybrid connectivity between your on-premises workloads and leverage the AWS cloud storage services to back up your data to Amazon S3. Alternatively, in some cases, you can use it to extend your on-premises storage capacity for users and applications by leveraging services such as Amazon FSx, for example.

AWS DataSync

AWS DataSync is a versatile online data transfer service that helps you migrate data between on-premises systems and AWS, transfer data across AWS storage services, as well as transfer data between AWS and other locations. The service can copy data to and from NFS, SMB file shares, **Hadoop Distributed File System (HDFS)**, Amazon S3 buckets, and Amazon FSX, including options for Lustre, OpenZFS, and NetApp ONTAP storage systems too. This versatility makes AWS DataSync not only ideal for data migration use cases but also popular for archiving cold data stored in on-premises systems to Amazon S3 Glacier and migrating data from hybrid systems to AWS storage services for just-in-time analytics, rendering, mining, and so on.

How does it all work? To put it simply, it depends on the use case. To transfer data from an on-premises system to AWS, AWS DataSync requires an agent to be installed in the on-premises storage system, as depicted in *Figure 18.6* (VMware ESXi, Microsoft Hyper-V, KVM hypervisors, as well as VMs/ EC2 instances are supported). On the other hand, transferring data between AWS services does not require the use of agents and is handled by AWS DataSync directly.

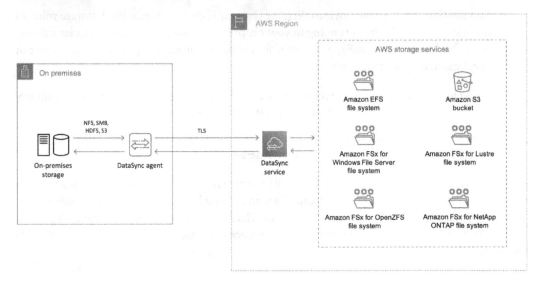

Figure 18.6: DataSync data migration from on-premises systems to AWS storage services

AWS Transfer Family

Even though we have been talking about NFS and SMB file shares as well as object storage throughout this section, there are a lot of enterprise customers out there that still rely on file transfer protocols such as **File Transfer Protocol** (**FTP**) and **Secure File Transfer Protocol** (**SFTP**) to migrate data from one system to another. Specifically for such workloads, AWS provides a fully managed service called AWS Transfer Family that allows organizations to transfer files from on-premises FTP, SFTP, and **File Transfer Protocol Secure** (**FTPS**) protocols directly into Amazon S3 and Amazon EFS storage services. The use cases for leveraging AWS Transfer Family range from data distribution to content management systems to simple transfer of data to the cloud, using either of the three supported file transfer protocols.

AWS S3 Transfer Acceleration

Although not seen as a primary service for migrating workloads to the cloud, AWS S3 Transfer Acceleration can still be leveraged to migrate large volumes of data from various locations across the world to a centralized Amazon S3 bucket effectively and quickly. How is this possible? Essentially, AWS S3 Transfer Acceleration leverages the globally available Amazon CloudFront edge locations to accelerate data migrations. When a user uploads data using AWS S3 Transfer Acceleration, the data is routed to the nearest edge location. There, upon arrival, the data is transferred to an Amazon S3 bucket using a highly optimized, internal network rather than the internet. This ensures that your bandwidth is optimally utilized and the end-user performance of uploading the data is consistent regardless of the user's geographical presence. AWS S3 Transfer Acceleration is really useful when you have large volumes of data to be uploaded to the AWS cloud from across multiple geographically diverse locations.

Amazon Kinesis Data Firehose

So far, almost all the services that we have explored are designed for migrating data in large volumes or chunks, but what about streaming data? That's precisely where Amazon Kinesis Data Firehose comes into the picture. A fully managed service, Amazon Kinesis Data Firehose allows organizations to load streams of data directly into services such as Amazon S3 and even Amazon Redshift. How does it all work? Simple! You just leverage the AWS Management Console to create a Kinesis delivery stream and data producers start streaming data to it. Kinesis Data Firehose is especially useful for use cases that require the **extraction, transformation, and loading** (ETL) of streams of data originating from multiple data sources and endpoints, such as IoT devices and social media.

Having explored all the online data transfer mechanisms, you can now examine the various ways in which data can be migrated to the AWS cloud in an offline manner.

Offline Data Transfer

To transfer data to AWS in an offline mode, AWS provides the Snow family of services that you can leverage based on the particular use case as well as on the amount of data you wish to migrate. These services can be categorized as follows.

AWS Snowcone

The smallest appliance in the AWS Snow family, AWS Snowcone is a specialized device that allows organizations to collect, store, and process data at the edge or in workloads that run in a disconnected/air-gapped environment. The device itself weighs around 2 kg and is provided with a ruggedized outer casing that can withstand up to 6Gs of force impact on it. A single Snowcone device provides up to 8 TB of HDD storage and up to 14 TB of SSD storage. The device, however, does not support clustering and as a result is more often used for standalone use cases such as industrial equipment, IoT, and drones.

AWS Snowball

Although versatile and compact, the AWS Snowcone device does not provide either large compute or storage capacities for migrating workloads larger than 14 TB in size. For workloads demanding higher volumes of data, it is advised to opt for AWS Snowball instead. The AWS Snowball service comes in the following two formats that you could choose based on your requirements:

- **Snowball**: A ruggedized device that can be used for importing as well as exporting data into and from Amazon S3.

- **Snowball Edge**: A ruggedized device that also can be used for data import and export jobs but provides additional edge computing capabilities. Snowball Edge is further available in the two following variants:

  - **Snowball Edge Compute Optimized**: Designed to provide compute power at the edge, the Edge Compute Optimized device is capable of supporting a maximum of 52 vCPUs, 200 GB of RAM, as well as an NVIDIA Tesla V100 GPU for running advanced machine learning workloads together with an approximate 40 TB of HDD storage capacity.

  - **Snowball Edge Storage Optimized**: The Storage Optimized variant also packs a 40 vCPU and 80 GB compute capacity. However, it is ideally used to migrate large datasets (up to 80 TB HDD capacity) from an on-premises environment to the AWS cloud in an offline manner.

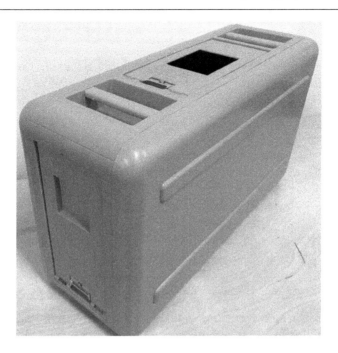

Figure 18.7: A Snowball Edge device

Both devices need to be ordered using the AWS Management Console. Once delivered (within 4-5 days), you simply connect the devices to your data center's network using either a standard RJ45 connector or optical interfaces that can support up to 100 Gbps data throughput.

So, which device should we choose when? *Table 18.4* highlights some use cases and scenarios where an organization can leverage a Snowball or Snowball Edge device:

Use Case	Snowball	Snowball Edge
Export from Amazon S3	*	*
Import into Amazon S3	*	*
Durable local storage		*
Local compute		*
Cluster support		*
NFS support using GUI		*

Table 18.4: Differences between a Snowball and Snowball Edge device

> **Note**
> Both the Snowball and Snowball Edge devices are not the most cost-effective solutions if your data volumes are less than 10 TB.

AWS Snowmobile

We covered data storage and movement from gigabytes all the way up to petabytes. But what if an organization has even higher volumes of data to migrate? In these special cases, AWS also provides a 40-foot mobile and secure shipping container offering called AWS Snowmobile that is capable of storing exabyte-scale data volumes. Each container supports up to 100 PB of storage and can be plugged directly into an organization's data center network for an optimum high-speed data migration. Once the data transfer to Snowmobile is complete, the container is shipped to the AWS Region of your choice and the data is securely uploaded to select AWS services, such as Amazon S3 or Amazon S3 Glacier, depending on your initial requirements. If your datasets are less than 10 PB in volume, it's best to leverage a cluster of Snowball or Snowball Edge devices for a more cost-effective solution.

Table 18.5 summarizes the various data transfer mechanisms that you can leverage both online and offline depending on your migration requirements:

Use case	Online	Offline
One-time transfer	AWS Transfer Family, AWS S3 Transfer Acceleration	AWS Snow family
Ongoing transfer	AWS Storage Gateway, AWS DataSync	
Continuous streaming	AWS Kinesis Data Firehose	

Table 18.5: Use cases for online and offline devices

You now have a good grasp of which storage migration services to use in order to transfer data both in an online and offline manner to AWS. The next section presents an overview of key server migration mechanisms.

Selecting an Appropriate Server Migration Mechanism

So far in this chapter, we have covered migration tools and services for both databases and data workloads. In this section of the chapter, we shall have a quick look at some of the services and tools provided for conducting server migrations to the AWS cloud.

VMware Cloud on AWS

VMware Cloud on AWS (**VMC** on AWS) is a jointly engineered solution provided by both AWS and VMware. It is specifically designed to accelerate migrations from an on-premises VMware-backed data center to the AWS cloud. Each VMC on AWS offering provides a single-tenant managed infrastructure running on i3 or i3en.metal instances, backed by VMware vSphere, vSAN, NSX, and vCenter Server. This form of migration service falls under the Relocate strategy, where VMs or servers are simply relocated from one location to the cloud using the hypervisor as the underlying medium for migration. However, apart from the accelerated migration use case, organizations also leverage VMC on AWS for a variety of other workloads, such as disaster recovery, hybrid data center capacity extension, and modernization of legacy applications using cloud-native services.

AWS Application Migration Service

AWS **Application Migration Service** (**MGN**) provides organizations with a simplified and accelerated migration experience to the AWS cloud. Unlike VMC on AWS, MGN is designed to work across your physical as well as virtualized infrastructure, supporting both VMware vSphere and Microsoft Hyper-V as the underlying hypervisors.

AWS MGN is highly recommended for any re-host migration project from an on-premises data center to the AWS cloud.

> **Note**
>
> AWS additionally provides AWS **Server Migration Service** (**SMS**) as well as CloudEndure for migrating server workloads; however, all new and future migrations are recommended to leverage AWS MGN instead.

How does it all work? Essentially, MGN requires a replication agent to be installed and configured on the source servers, following which one or more replication tasks can be configured using either the AWS Management Console or even the AWS CLI. Each replication task contains one or more replication servers that receive data from the replication agents running in the on-premise servers and writes the data to Amazon EBS volumes present in a staging area. MGN ensures that the data is synchronously copied from the source to the destination using block-level data replication technologies. Once the data is successfully copied over, you can plan and schedule a cutover where MGN will convert the source server and natively run it on Amazon EC2 instances backed by the newly created EBS volumes as well.

Figure 18.8 presents a schematic representation of the process of data migration from a company's data center or any cloud platform to the AWS cloud via AWS migration (MGN) services. The migration process is divided into two phases: ongoing, block-level data duplication and automated orchestration coupled with system transformation.

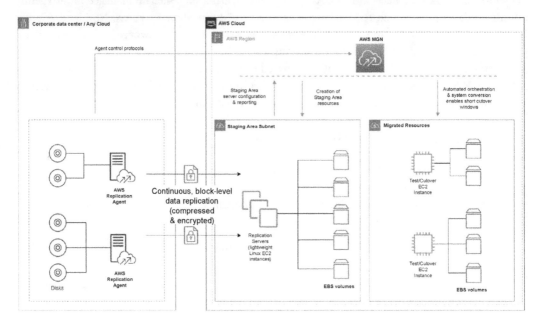

Figure 18.8: AWS Application Migration Service – under the hood

The next and last section of this chapter discusses the key security best practices and methods to keep in mind when working with any migration tool or service.

Applying the Appropriate Security Methods to the Migration Tools

Security is crucial for any successful migration journey to the cloud, and there are standard security practices that a solutions architect should always keep in mind when determining the right tool for an effective migration job.

The following are some of these key security practices to remember when working with migration tools and services:

- **Deploying services within a VPC**: To start with, always ensure that AWS resources are always created within a customer-owned and managed VPC (not a default VPC) wherever applicable. For example, services such as AWS DMS can be further made secure by launching the replication instances within. A customer-managed VPC should support the right set of network ACLs and security groups. For other services, leveraging a secure private connection using AWS PrivateLink further improves the overall security posture of your migration projects; for example, you can connect a data center with AWS MGN using AWS Direct Connect, VPC peering, or even interface VPC endpoints powered by AWS PrivateLink.

- **Managing authentication and access control**: There are numerous best practices when it comes to managing the authentication and authorization of users as well as resources on AWS, but the generic guidelines will always remain the same, such as ensuring least privileged access, leveraging IAM roles, creating and assigning identity-based policies, enforcing MFA, and controlling access to resources with tags, as demonstrated in the following example:

```
{
  "Version": "2012-10-17",
  "Statement": [
    {
      "Effect": "Allow",
        "Action": [
            "storagegateway:ListTagsForResource",
            "storagegateway:ListFileShares",
            "storagegateway:DescribeNFSFileShares"
        ],
        "Resource": "*",
        "Condition": {
            "StringEquals": {
                "aws:ResourceTag/AllowAccess": "yes"
            }
        }
    },
    {
      "Effect": "Allow",
      "Action": [
          "storagegateway:*"
      ],
      "Resource": "arn:aws:storagegateway:region:account-id:*/*"
    }
  ]
}
```

In this example, the role only allows authenticated users to perform actions on the attached storage gateway if the tag on the gateway's resource has its key set to `AllowAccess` and its corresponding value set to `yes`.

- **Ensuring auditability**: As a best practice, it is always advised to enable AWS CloudTrail within customer AWS accounts for auditability as well as tracking of API calls.

- **Encryption**: AWS migration tools and services alike support encryption of data both at rest and in transit. *Table 18.6* summarizes the various mechanisms employed by different migration services for encryption:

	AWS Service	**Encryption at Rest**	**Encryption in Transit**
Databases	AWS Database Migration Service (DMS)	The encryption of the DMS replication instance and its endpoint connection can be performed using encryption keys managed by AWS KMS. Additionally, the target database can be encrypted using AWS KMS keys as well.	AWS DMS leverages Secure Sockets Layer (SSL) to secure endpoint connections with Transport Layer Security (TLS).
	AWS Schema Conversion Tool (SCT)	AWS SCT does not store customer data within itself. It, however, uses client certificates to store your database credentials securely.	AWS SCT leverages SSL TLS to encrypt data from an on-premises connection to the AWS cloud.

	AWS Service	Encryption at Rest	Encryption in Transit
Storage	Amazon S3 File Gateway	Uses Amazon S3-Managed Encryption Keys (SSE-S3) to server-side encrypt all data stored in Amazon S3.	All Storage Gateway services leverage SSL/TLS to encrypt data between the gateway appliance and AWS storage service (FSx, S3, etc.).
	Amazon FSx File Gateway	Uses AWS (KMS) to encrypt Amazon FSx filesystems.	
	Tape Gateway	Uses AWS KMS and the Storage Gateway API to encrypt tape data stored in the cloud.	
	Volume Gateway	Uses AWS KMS and the Storage Gateway API to stored and cached volumes in the cloud.	
	AWS DataSync	Since AWS DataSync is solely managing the transfer of data from one source system to a destination, it doesn't store or encrypt data at rest.	AWS DataSync leverages various RLS ciphers to encrypt the data when it is transmitted between various services. The TLS cipher used will depend on the type of endpoint used by the DataSync agent.
	AWS S3 Transfer Acceleration	AWS S3 Transfer Acceleration supports both client-side encryption and server-side encryption for data at rest using either server-side encryption with AWS Key Management Service (SSE-KMS) or Amazon S3 bucket keys.	AWS S3 Transfer Acceleration uses SSL/TLS to protect data in transit. Customers can additionally use client-side encryption as well.
	Amazon Kinesis Data Firehose	Amazon Kinesis Data Streams automatically encrypts data using server-side encryption with AWS KMS.	Amazon Kinesis Data Streams uses SSL/TLS to protect data in transit. Customers can additionally use client-side encryption as well.
	AWS Snowcone	All data stored on AWS Snow devices is automatically encrypted with 256-bit encryption keys managed by AWS KMS.	Not applicable.
	AWS Snowball		
	AWS Snowball Edge		
	AWS Snowmobile		

	AWS Service	Encryption at Rest	Encryption in Transit
Servers	VMware Cloud on AWS	VMware vSAN encrypts all user data at rest in VMC on AWS using AWS KMS.	Data transferred between on-premises and VMC on AWS can be protected by an IPSec VPN tunnel.
	AWS Application Migration Service	Since the application data is stored on Amazon EBS volumes, use AWS KMS to encrypt the data at rest.	All communication between the AWS MGN replication instance and the MGN agents is secured using TLS 1.2.

Table 18.6: Mechanisms employed by different migration services for encryption

Summary

You have come to the end of yet another chapter in your journey toward understanding and mastering migrations from on-premises environments to the AWS cloud. Before moving on to the next chapter, let's go through a quick summary of what you have learned so far!

The chapter started by discussing various tools and services that you can leverage to effectively assess, plan, and migrate database workloads to the AWS cloud. You also explored various data migration options for both online and offline data transfers and the various services to use for your particular use case. The chapter concluded with a quick review of some server migration services provided by AWS as well, and finally listed some essential security best practices to keep in mind when working with any AWS migration tool or service.

Chapter 19, Determining a New Architecture for Existing Workloads, will continue the migration journey and discuss how you can effectively determine and design a new cloud architecture for applications on the AWS cloud along with a brief overview of how you can determine target database platforms.

Further Reading

- Migration tool comparison – AWS Prescriptive Guidance: `https://packt.link/80Jb4`
- Best practices for security, identity, and compliance: `https://packt.link/fBZHY`

19

Determining a New Architecture for Existing Workloads

In the previous chapter, you learned about the various tools and services that you as a solutions architect can leverage to plan, prepare for, and execute a successful migration.

This chapter will focus on the new cloud environment in which the migrated workloads will operate. We will cover the architectural aspects of your new cloud environment, such as selecting the right compute for your new workloads, how to transition the newly migrated workloads to managed container platforms, and how to select the right storage and database services to match the newly migrated workload's cloud environment. Throughout the chapter, we will be assuming a standard, three-tiered application as a potential migration and modernization candidate. However, most of the design considerations and services discussed in this chapter can apply to other modernization use cases as well.

The following main topics will be covered in this chapter:

- Understanding your candidate application
- Selecting the appropriate compute platform
- Selecting the appropriate storage service
- Selecting the appropriate database platform

Understanding Your Candidate Application

In order to understand the best possible choices a solutions architect can undertake for an application's migration and modernization journey, you first need to have a candidate application.

In this case, we will be leveraging a fairly straightforward three-tiered application that is built using legacy .NET code, for example, and backed by Microsoft SQL Server. The application is also connected to a shared file system that allows users to upload data into their respective directories and access them via their internal networks. The application is a monolith by design and does not have too many moving parts. However, it does have the standard networking and security features enabled in the form of a gateway router, a network firewall, and a load balancer.

Figure 19.1 depicts the various components:

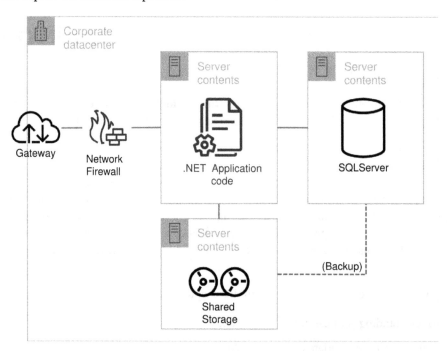

Figure 19.1: Typical legacy web application design hosted in a physical data center

The application is hosted in an aging on-premises data center that leverages VMware as the underlying virtualization technology. This is akin to what most organizations will have running on-premises as well.

Now, your role as a solutions architect is to first analyze this application, find the perfect mechanism to migrate it with the least possible disruption, and additionally suggest potential modernization pathways to the customer. So, as you have seen in the previous chapters, the first and most important step that you need to undertake here is conducting the *Assess* phase, through which the requirements and the application's complexity are analyzed using various tools and frameworks, such as the **Migration Readiness Assessment** (**MRA**) and the **Schema Conversion Tool** (**SCT**) to obtain an idea of the effort required to migrate the database schemas to the AWS cloud.

Following this, in the *Mobilize* phase, you would devise a plan for migrating the application to the AWS cloud using either the re-host pattern, where the application and its components are simply re-hosted into the likes of Amazon EC2 instances with few to no changes, or you could re-platform the application by moving it away from the legacy .NET into something that works on multiple platforms, such as .NET Core. Alternatively, you could even decouple the application and convert it from a monolith to a microservices architecture.

But with so many options available out there in the AWS cloud, which is the right service to choose for our candidate application, and what are some of the design considerations that we need to keep in mind when selecting these services? Let us answer each of these questions in the following sections!

Selecting the Appropriate Compute Platform

Considering the fact that the application is a monolith, the first few options that you have for hosting this application in terms of compute are pretty straightforward. In this section, you shall explore some of these compute services along with potential modernization options, starting with the simplest service of them all—Amazon EC2.

Amazon EC2

Amazon EC2 provides raw compute hosting for your applications and is suited for a variety of workloads, including web-based applications, batch processing, AI-ML workloads, mobile backends, and much more. Hosting a legacy .NET application on Amazon EC2 would require selecting the appropriate Windows Amazon Machine Image (AMI) and booting an EC2 instance off it. The only downside to this is that the end customer is responsible for the management and operational aspects of their EC2 instances as per the AWS shared responsibility model. Although migrating workloads to EC2 does not significantly reduce overheads in terms of operations, it still has benefits if the right set of features is incorporated into the design, for instance, leveraging autoscaling groups to spread the EC2 instances across multiple availability zones for better availability and scalability of the application.

Customers can also look at EC2 as a mechanism to modernize their legacy applications by choosing to refactor them. Take, for instance, the AWS Graviton processors. Introduced in 2019, Graviton processors are based on ARM64 CPU architecture and provide a far better price-to-performance ratio for applications and workloads that run on them. Customers who essentially have full control over their application source code can choose to refactor it and take advantage of the Graviton processors for up to 25% better performance than comparable x86-based EC2 instances. The latest Graviton3 processors even support running .NET 5 workloads and provide numerous instance types ranging from the burstable class (t4g) to general purpose (M6g), and memory (R6g) and compute-optimized (C6g) as well.

However, the biggest drawback of selecting this compute platform directly is the operational overhead. That's where the likes of Amazon Lightsail and Amazon Elastic Beanstalk come into the picture.

Amazon Lightsail and Amazon Elastic Beanstalk

Amazon Lightsail and Amazon Elastic Beanstalk are both designed to provide a more managed experience when it comes to application hosting by abstracting a lot of the management overhead from the end users.

Amazon Lightsail provides an easy-to-get-started experience for developers and small-to-medium businesses who wish to run simple workloads without having to overengineer or manage a lot of the underlying compute infrastructure. Lightsail allows users to quickly deploy instances, containers, load balancers, and databases without having to learn or understand cloud terms and terminology. The workloads themselves run on burstable compute infrastructure and hence are ideal for web-facing workloads as well as simple batch-processing jobs.

Amazon Elastic Beanstalk, on the other hand, provides a more feature-rich experience for developers by packing the most common runtimes and platforms such as Java, .NET, and Node.js all into EC2 instances and providing these EC2 instances as managed services. Elastic Beanstalk also simplifies deployments and the rollout of applications by providing developers with a single pane of glass where they can upload their .NET application code and configure the application's deployment strategy, such as rolling, immutable, or blue-green deployments. Amazon Elastic Beanstalk additionally provides support for Graviton as well as container workloads.

Amazon ECS, Amazon EKS, and AWS Fargate

Another possible alternative for application modernization is to decouple the monolith and host it as individual microservices using containers as the underlying technology. AWS provides two primary container orchestration services that can help customers run container workloads. These are as follows:

- **Amazon Elastic Container Service (ECS)**: An opinionated way of performing container orchestration, Amazon ECS allows customers to run Docker and Docker swarm-based containers at scale with easy integrations with other AWS native services as well. Amazon ECS supports the management of Windows containers. Amazon ECS also supports both Linux and Windows workloads by providing optimized **Amazon Machine Images** (**AMIs**). These AMIs are specifically designed for improved container performance and provide detailed visibility of the container as well as the EC2 instance's CPU, memory, and other vital metrics.

- **Amazon Elastic Kubernetes Service (EKS)**: As the name suggests, Amazon EKS is designed for workloads that are suited to running on Kubernetes. Unlike Amazon ECS, Amazon EKS requires knowledge and experience of working with Kubernetes, and as such can have a slight learning curve for some. On the flip side, since the platform runs vanilla Kubernetes, customers can choose to integrate their container workloads with a variety of third-party tools, open source and enterprise software, and AWS services for enhanced deployments, monitoring, security, and much more.

- **AWS Fargate**: Both Amazon ECS and Amazon EKS provide managed control planes, which means that the underlying infrastructure that powers the "brains" of the services is managed by AWS itself. However, the data plane is always the responsibility of the end customer. Customers can choose to run their container workloads on Amazon EC2 instances, but they are also responsible for patching these instances and maintaining their lifecycle. For customers looking to get rid of this overhead, AWS introduced Fargate as a serverless container compute platform that works with both Amazon ECS and Amazon EKS. With Fargate, customers do not have to worry about the host OS, its patching, scaling, availability, and so on. This greatly simplifies deployments while making the applications more secure by design as well. The important point to note here, however, is that containers that run on Fargate do not have persistent disks. This means that once a Fargate container is terminated, the underlying disks and data can potentially be lost as well unless an external storage/file system is mounted to the container. We shall discuss more of these storage options in the *Selecting the Appropriate Storage Platform* section.

Figure 19.2 presents a sample application deployed using containers on Amazon ECS with AWS Fargate as the underlying compute option:

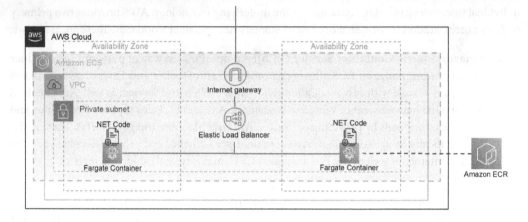

Figure 19.2: Containerized application hosting on AWS Fargate

AWS Lambda

The last of the compute services that customers can consider for their application's modernization is AWS Lambda. Yet another serverless platform, AWS Lambda operates in a similar fashion to AWS Fargate, but with one major difference. AWS Fargate is a container compute engine that simply runs the containers that are deployed on it while managing the underlying host infrastructure on the customer's behalf. AWS Lambda, on the other hand, is a serverless event-driven compute engine that relies on specific events to trigger your application code.

AWS Lambda also supports a host of frameworks and platforms, including the latest .NET 6, and provides customers with the choice of running their workloads on x86- or ARM64-based Graviton processors. AWS Lambda deploys application code using one or more small "functions," with each function performing a very specific task in the application.

Table 19.1 summarizes the various compute services as well as some of the considerations and traits to watch out for during a migration/modernization exercise:

AWS Service	Migration Strategy	Modernization Scope	Considerations
Amazon EC2	Re-Host	Low	The easiest option to migrate with few to no changes to the application code. A range of EC2 instance families to choose from specific to the workload's requirements.

AWS Service	Migration Strategy	Modernization Scope	Considerations
Amazon Lightsail	Re-Host	Low	A managed service offering little abstraction for application hosting. Ideal for small-medium businesses and simple workloads such as simple frontend systems, databases, websites, and so on.
Amazon Elastic Beanstalk	Re-Host, Re-Platform	Medium	A managed service providing customizability options. Can be used as a re-platforming mechanism with some effort in application redesign. Provides both Amazon EC2 as well as Docker support for running containerized workloads.
Amazon ECS Amazon EKS AWS Fargate	Re-Factor, Re-Platform	Medium-High	Managed services offering a full containerized orchestration and runtime platform. Require decoupling of application code and can also be used as a medium to re-platform with some efforts in application redesign.
AWS Lambda	Re-Factor, Re-Platform, Re-Architect	High	A managed service offering a fully event-driven serverless compute platform. Requires complete decoupling of the application as well as a change in the application's architecture (event-driven).

Table 19.1: Compute platform considerations

Selecting the right compute platform for your workload can also depend on other application-centric factors, such as the following:

- **Portability**: Applications may need to be ported from one cloud provider or platform to another due to a variety of reasons: compliance with certain laws, cost control, legacy backend systems, and so on. Such applications are ideally suited for either EC2 instances, in which the code is simply lifted and shifted without any significant changes to it, or containers in which the application code is decoupled and containerized using Docker.

- **Application dependencies**: Docker containers are ideal for applications that need to have all their libraries and dependencies packaged into a single, portable unit. On the other hand, AWS Lambda can leverage containers too now as an application packaging and deployment model. This makes it easier for developers to develop and release application code in a faster, repeatable way.

- **Application execution time**: When it comes to selecting the underlying compute for your workloads, it's important to understand the application's actual execution time as well. Why? Well, the reason is simple! With the cloud, you essentially pay as you go. So, if you are using an EC2 instance to run your applications, you pay for the duration for which the EC2 instance is up and running irrespective of whether the application was being used or not. This same model can also be applied to containers since they can be leveraged for long-running tasks as well. However, when it comes to Lambda functions, a single function can run for up to 15 minutes only before it is terminated. For such short-lived application code, Lambda functions are the ideal choice since you only end up paying for the duration of your function's execution time, which can be measured down to milliseconds as well.

With this, we come to the end of the discussion on compute services. The next section will take you through some of the important considerations to keep in mind when selecting the right AWS storage service.

Selecting the Appropriate Storage Platform

In this section of the chapter, you will learn about the various storage options that AWS provides for your applications, both in terms of migration and modernization. The services are discussed in detail in the following subsections.

EBS multi-attach

Amazon **Elastic Block Store** (**EBS**) is perhaps the most commonly used storage service for migrating workloads to the AWS cloud. The service provides block-level storage volumes that you can mount to EC2 instances just as you would mount a physical drive to a physical server. Once mounted, you can choose to format them as per the required file system and store application data on them as required. The best part of having an EBS volume is that it can persist independently from an EC2 instance as well, which comes in handy if your EC2 instance were to get corrupted or terminated accidentally.

AWS recently announced the availability of EBS multi-attach that allows you to attach a single provisioned IOPs EBS volume to multiple EC2 instances (max. 16 Linux instances) that are present in the same **availability zone (AZ)**. This design allows for increased durability and availability of application data that requires concurrency. The only downside to multi-attach is that it cannot span across AZs, and is still bound by the EBS design laws, which means it's not automatically scaled up as per the requirements. But fret not, that's where AWS provides two shared file storage services in the form of Amazon EFS and Amazon FSx.

Amazon EFS

Amazon Elastic File System (EFS) is a highly scalable **network file share (NFS)** as a service with which customers can attach multiple EC2 instances spread across AZs. Customers who do not wish to incur cross-AZ traffic costs can additionally choose to opt for the EFS One-Zone storage class, which, as the name suggests, creates a single mount point in the AZ that you have provided. This mechanism is very similar to the EBS multi-attach functionality; however, One-Zone storage can still scale the EFS volume dynamically based on its usage and growth.

Based on the requirements of your candidate application, EFS can provide the necessary storage for users to store their files as well as share them across the organization as required.

Amazon EFS can easily be mounted to both Amazon EC2 instances and AWS Fargate-backed containers for a shared storage experience. The only obvious downside to this is that EFS is an NFS-based file-sharing service and as such cannot be directly mounted to Windows operating systems. For that, you need to explore a different family of shared file storage in the form of Amazon FSx, discussed next.

Amazon FSx family

Amazon FSx is a highly scalable and performant file system storage service that comes in four different flavors, as explained here:

- **Amazon FSx for NetApp ONTAP**: Ideal for migrating data and workloads from an on-premise storage system, Amazon FSx for NetApp ONTAP allows you to quickly create and manage a fully functional NetApp ONTAP file system in the cloud. It supports multiple shared-file storage protocols, including NFS, **Server Message Block (SMB)**, and **Internet Small Computer Systems Interface (iSCSI)**, and can be used for a variety of use cases, including backup and DR, caching, migrations, and so on.

- **Amazon FSx for OpenZFS**: Amazon FSx for OpenZFS provides OpenZFS; an open source storage platform as a completely managed service. FSx for OpenZFS leverages standard NFS protocol for data accessibility and management and can also integrate with Linux, Windows, and Mac operating systems.

- **Amazon FSx for Windows File Server**: Amazon FSx for Windows File Server provides a highly scalable and durable shared file storage system supported by the SMB protocol. The file system can be spread across multiple AZs and provides add-on features such as integration capabilities with Microsoft **Active Directory (AD)**.

Keeping in mind the candidate application and its underlying compute options (Amazon EC2 and AWS Fargate), the containers can easily mount to an Amazon FSx for a Windows Server volume, as demonstrated in *Figure 19.3*:

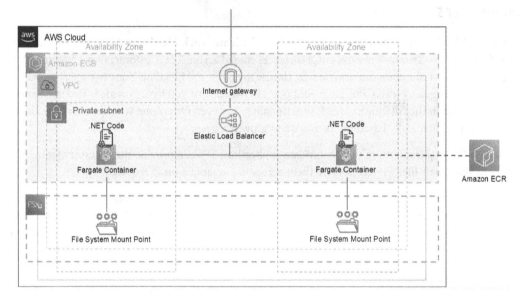

Figure 19.3: Containerized application with backend storage on Amazon FSx for Windows File Server

- **Amazon FSx for Lustre**: For workloads that require high performance in the form of high throughput, for example, parallel file systems, or accelerated machine learning, AWS provides Amazon FSx for Lustre as a fully managed, scalable, and performant Lustre file system.

Table 19.2 presents a simple comparison of the AWS services discussed in this section. It also provides a quick summary and some key considerations and traits to keep in mind when selecting a storage service for a migration/modernization exercise. For a complete breakdown of the comparison, please visit the following link: `https://packt.link/GlXn4`.

AWS Service	Regional Availability	Client Support	Considerations
Amazon EBS	Within an AZ	Windows, Linux, macOS	The easiest option to migrate with few to no changes needed to the application code. A single AZ is supported and cannot scale automatically. Ideal for EC2 instances and container workloads that need 1-2-1 mapping.
Amazon EFS	Within an AZ, Multi-AZ	Linux, macOS	A managed service offering NFS as a service. Can support both single AZ as well as multiple AZs; however, no support for Windows OS. Supported on both EC2 instances and Fargate containers and ideal for shared file system workloads.
Amazon FSx for NetApp ONTAP	Within an AZ, Multi-AZ	Windows, Linux, macOS	A managed service providing the NetApp ONTAP file system as a service. Ideally suited for accelerating migration from on-premises environments to the AWS cloud. Supported on both EC2 as well as Fargate container workloads with popular access protocols including SMB, NFS, and iSCSI.
Amazon FSx for OpenZFS	Within an AZ	Windows, Linux, macOS	A managed service offering the OpenZFS file system as a service. Suited for workloads that require high-throughput and IOPs such as data analytics, and so on. Supported only within a single AZ.
Amazon FSx for Windows File Server	Within an AZ, Multi-AZ	Windows, Linux, macOS	A managed service offering the Windows file system as a service. Supports popular access protocols such as SMB and can be scaled dynamically as per requirements.
Amazon FSx for Lustre	Within an AZ	Linux	A managed service offering the Lustre file system as a service. Ideally suited for HPC/compute-intensive workloads.

Table 19.2: Storage platform considerations

With this, we come to the end of the discussion on storage services. You are now ready to proceed to the final section of the chapter, wherein you will explore some of the key considerations for selecting the right AWS database service.

Selecting the Appropriate Database Platform

Selecting the right database platform for your workloads can be a challenging aspect at times, considering the fact that AWS provides six different database hosting options to choose from. In this section of the chapter, you will examine these six database categories and learn how to determine which database is particularly suited to handle application-specific workloads.

Relational Databases

Relational databases have been around for a long time now and most organizations rely on commercial-grade databases such as Microsoft SQL Server and Oracle to run mission-critical workloads. AWS provides Amazon **Relational Database Service** (**RDS**) as a managed service to host commercial-grade as well as open source databases at scale while providing abstraction from complex operational tasks such as managing the underlying database instance, installing and upgrading the database software, and so on.

Migrating to Amazon RDS becomes a de facto choice for customers who do not wish to rearchitect their workloads to suit more modern databases as it requires a lot of time and effort. However, Amazon RDS can still be leveraged as a mechanism to modernize a workload's database by choosing to replatform rather than rearchitect. For example, customers can choose to migrate away from SQL servers and Oracle databases and choose to run their workloads on open source variants such as MySQL or PostgreSQL (both of which are supported by Amazon RDS), or even choose to replatform to Amazon Aurora, a purpose-built database that provides the same performance as that of commercial databases but without the complexities of licenses or added costs.

In our candidate application's case, selecting a managed SQL server using Amazon RDS is just one of the options for the customer. If the customer chooses to migrate away from SQL Server into the open source world, they can still choose to do so with the help of Babelfish for Aurora PostgreSQL. This is a unique feature that allows Aurora PostgreSQL *to understand* SQL Server's wire protocol. This implies that applications originally built for SQL Server can interact with Aurora PostgreSQL with minimal changes. *Chapter 20, Determining Opportunities for Modernization and Enhancements,* will discuss Babelfish in depth.

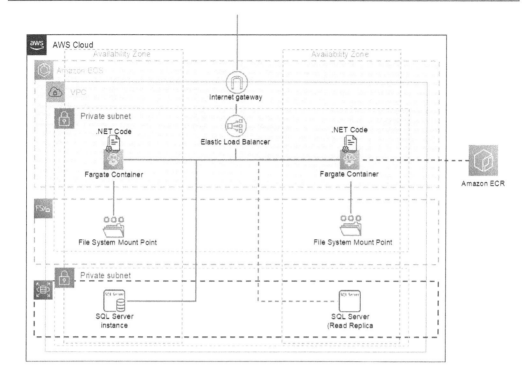

Figure 19.4: Managed SQL Server database using Amazon RDS

Document and Key-Value Databases

Although relational databases have been around for a long time, they are now slowly being replaced by more scalable and performant document and key-value databases. Key-value databases such as Amazon DynamoDB are designed to scale horizontally due to the way data is stored in them (partitioned with the help of key-value data). There is no infrastructure to manage when it comes to DynamoDB. You simply create tables, define partition keys, add data, and query it—simple, isn't it?!

On the other hand, AWS also provides a managed database that is specifically designed to store and handle JSON documents known as Amazon DocumentDB. DocumentDB is designed to provide high-throughput, low-latency data management for JSON documents and is also compatible with the industry's standardized open source document database MongoDB (v3.6 and v4.0 drivers and tools). This compatibility significantly reduces friction when modernizing a MongoDB-backed database to an AWS-managed service.

Graph Databases

Making sense of data originating from multiple sources can be a daunting task even when considering the most widely used and accepted relational database out there. Modern, web-based applications rely on a network of highly connected datasets such as social networks or knowledge graphs to identify and build new connections, run analytics, and provide better insights as well. For handling such specific workloads, AWS provides Amazon Neptune, a fully managed graph database as a service that developers can use to create highly scalable graph applications backed by open source APIs such as Gremlin, SPARQL, and openCypher.

In-Memory Databases

As the name suggests, in-memory databases provide applications with in-memory caching that can be used for a variety of real-time use cases such as gaming, streaming, social media, and much more. AWS provides a couple of options for applications that need to cache data and improve application performance:

- **Amazon ElastiCache**: A fully managed, in-memory cache as a service supported by Redis and Memcached.

- **Amazon DynamoDB Accelerator** (**DAX**): An in-memory cache specially for improving the performance of Amazon DynamoDB tables.

Search Databases

Not all databases are designed specifically to index and handle large volumes of data. Data that requires large-scale indexing, aggregation, and analytics requires a purpose-built database specifically designed to handle such complex workloads. AWS provides the Amazon OpenSearch service, which is designed to provide near-real-time analytics and search capabilities for extremely large datasets along with a built-in Kibana dashboard as well.

Table 20.3 provides you with a quick summary of the various database services available and some key considerations and traits to consider when selecting a database service for a migration/modernization exercise.

AWS Service	Database Type	Ideally Suited For	Considerations
Amazon RDS	Relational	E-commerce, web applications, CRM, ERP systems	By far the most convenient option for a database lift and shift requiring minimalistic efforts. Can be used as a modernization medium for replatform based on the workload requirements.

AWS Service	Database Type	Ideally Suited For	Considerations
Amazon DocumentDB	Document	Content management, web catalogs	Ideal for storing data in the form of JSON documents. Simplifies migrations from an on-prem MongoDB solution.
Amazon DynamoDB	Key-Value	Mobile backends, web applications, IoT	A NoSQL, highly scalable, serverless database that can support most modern application workloads. Additionally, provides in-built caching functionality in the form of Amazon DynamoDB Accelerator (DAX)
Amazon Neptune	Graph	Knowledge graphs, social media, recommendation engines	A purpose-built graph database designed for high throughput and low latency. Ideal for storing and analyzing complex data with multiple relationships among them.
Amazon ElasticCache	Cache	Gaming/chat applications, sessions store	Ideally suited for improving web and mobile application performance. Can be leveraged together with the likes of Amazon RDS as a backend system as well.
Amazon OpenSearch Service	Search	Log analysis, data mining	Purpose-built search database ideally suited for large datasets requiring indexing and search capabilities

Table 19.3: Database platform considerations

Keeping all the aforementioned services in mind, it's also important to be aware of some common characteristics when it comes to selecting a database for your workloads:

- **Understand the data**: As stated earlier in this section, selecting the right database for your workloads can be challenging, especially since there are so many options available. One of the best ways to select the right tool for the right job is by ensuring that you have a clear understanding of the data requirements. The key aspects to consider are the data model, whether the data is transactional, how frequently the data is accessed, the functional aspect of the data, and so on. Once you understand the requirements, you can map them to the right database option for your workload.

- **Metrics and monitoring**: Even after databases are migrated to the AWS cloud, you constantly need to monitor and capture performance-related metrics in order to assess the database and improve its operational functionalities. This becomes especially critical for database workloads that have been modernized, for example, replatformed from SQL Server on-premises to Amazon Aurora PostgreSQL compatible or rearchitected from Oracle to a combination of Amazon DynamoDB and Amazon ElastiCache.

- **One size doesn't fit all**: Gone are the days when every problem required a standard Relational database as a fix. With modern applications and data storage requirements, applications as simple as websites can have a combination of two or more databases of different types – SQL, NoSQL, graph, and so on, with each database playing a key role in the overall application's design, scaling independently, and providing the right set of features and services to get the job done.

Summary

With this, we come to the end of yet another chapter, but before we head on to the next, go through this quick summary of the things you have learned so far.

This chapter started off by discussing some compute options that you can leverage for your workloads; an example of a simple web application was used as a migration candidate. You learned about AWS container services along with a brief overview of AWS Lambda as well. This was followed by some key considerations to keep in mind when selecting a compute platform. Next, you examined the storage and database options provided by AWS and explored some common questions to ask yourself so that you employ the right tool for the right job.

Chapter 20, Determining Opportunities for Modernization and Enhancements, will expand upon some topics from this chapter. It will discuss aspects such as how to actually convert a monolith to microservices, migrate from a commercial database to an open source variant, and finally, cover some interesting and real-world patterns for modern workloads on AWS.

Further Reading

- Modernizing Windows workloads on AWS: `https://packt.link/hzcqN`
- Selecting the right database for the right job: `https://packt.link/Z58uv`

20
Determining Opportunities for Modernization and Enhancements

In *Chapter 19*, *Determining a New Architecture for Existing Workloads*, you explored the architectural aspects of a new cloud environment, namely understanding and selecting the right compute, storage, and database options for your new workloads, with a practical web application used as an example.

In this chapter, we take the learnings from the previous chapter and try and understand the patterns and opportunities to actually decouple an application by breaking it down into more manageable microservices. You will also learn how to host these microservices using containers or AWS serverless solutions. You will also explore similar opportunities for modernizing your sample application by selecting the right backend database, and, last but not least, learn how to tie all these loosely coupled components together using AWS integration services and patterns.

The following main topics will be covered in this chapter:

- Identifying opportunities to decouple application components
- Identifying opportunities for containers and serverless solutions
- Identifying opportunities for purpose-built databases
- Selecting the appropriate application integration service

Identifying Opportunities to Decouple Application Components

In this section of the chapter, we will also consider the same legacy .NET application monolith code used in *Chapter 19, Determining a New Architecture for Existing Workloads,* as a baseline for modernization. However, before we dive into refactoring and rearchitecting the application into microservices, let us quickly explore a few options that you as a solutions architect can leverage for rehosting and replatforming.

Opportunity 1: Rehosting

So far, we have been emphasizing that rehosting is by far the easiest way to migrate a legacy workload into the AWS cloud and is indeed the most frequently used pattern for migration as well since it involves little to no changes in the underlying application code. Often, the primary drivers for this migration can be a lack of resources and time required to modernize the application or the fact that the application is simply a **Commercial Off-The-Shelf (COTS)** variant that cannot be further decoupled into smaller, more manageable components. In this case, the end customer is left with two choices: to either go for the standard Amazon EC2 instances backed by Windows OS, which need to be patched and maintained by the customers themselves, or select Amazon Elastic Beanstalk as a hosting platform where the underlying infrastructure is managed and provided in the form of a service. *Figure 20.1* depicts the choices in flowchart format:

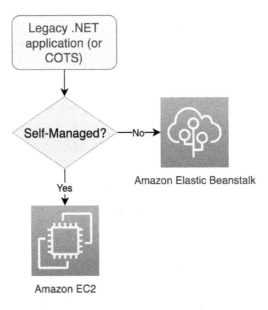

Figure 20.1: Rehosting pattern

Opportunity 2: Replatforming

Replatforming here indicates moving away from the standard EC2 instance-based approach into a more modular, containerized medium for hosting the same application. By replatforming the application to work with containers, you essentially improve the application's overall rollout strategy and security, as each of its dependencies is packed together with the application in the form of a container image. This approach is useful especially when customers are looking to keep the application code more or less intact but wish to bring out better operational effectiveness for their applications. The effort required to replatform a legacy .NET workload to containers, however, still depends on the complexities of the application's dependencies as well as on whether the existing Windows licenses are compatible with containers.

> **Note**
>
> Replatforming does not involve the migration of legacy .NET code to .NET Core. As a result, all containers for this particular pattern rely on Windows Server-based container images.

With AWS, customers can choose to run Windows-based container images on both Amazon **Elastic Container Service (ECS)** as well as **Amazon Elastic Kubernetes Service (EKS)**, as demonstrated in *Figure 20.2*.

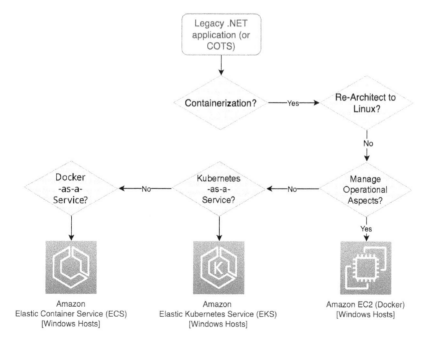

Figure 20.2: Replatform pattern

Customers can alternatively even choose to run the Windows containers on self-managed Amazon EC2 instances if they wish to have full control over the underlying infrastructure. However, this option does come with its own set of operational overheads. For the containerization of the legacy code, however, AWS provides a simple command-line tool called AWS **App2Container (A2C)** that evaluates your .NET and Java standalone code, identifies and packages the application together with its dependencies into a container image, and then optionally deploys it to a container orchestration service of your choice, be it Amazon ECS or Amazon EKS.

Opportunity 3: Rearchitecting

Although replatforming does bring about some operational and security benefits for your application now that it is running as a container, it still doesn't help solve the issues with regard to scale, performance, and costs as the underlying operating system is still Windows-based.

In this case, rearchitecting the application from legacy .NET to the more modern .NET Core might be a long-term design decision that customers can take to modernize their workloads. The obvious downside to this is that the effort required both in terms of time and costs can be significant depending on the application's code, dependencies, and so on.

Luckily, AWS provides yet another tool that can be leveraged in conjunction with A2C, known as the Porting Assistant for .NET. As the name suggests, the Porting Assistant for .NET tool runs an analysis on your legacy .NET code and provides a .NET Core compatibility assessment report. This helps reduce the manual effort in analyzing application code, and at the same time provides a detailed assessment of any missing packages and dependencies.

Opportunity 4: Refactoring

The final pattern to modernize your application is all about refactoring the code. Refactoring can be broken down into the following two sub-routines:

- **Going the microservices way**: In this routine, the application is decoupled into smaller, more manageable pieces of code that can ultimately be packaged as individual containers and made to run on any container platform, including Amazon ECS and Amazon EKS. By actually decoupling the monolith into microservices, your application can achieve higher levels of scalability and agility, as now, each component of the application can be managed as an individual resource with its own CICD pipelines, development lifecycle, and so on.

- **Going the serverless way**: This routine also requires decoupling your application code into smaller, more manageable pieces. However, instead of running all application components such as message queues and API gateways as code inside a container, you now substitute these services with AWS-managed serverless services, such as Amazon **Simple Queue Service (SQS)** or Amazon API Gateway.

Figure 20.3 is a representational workflow that depicts the path that a .NET application can undertake for refactoring using AWS native services:

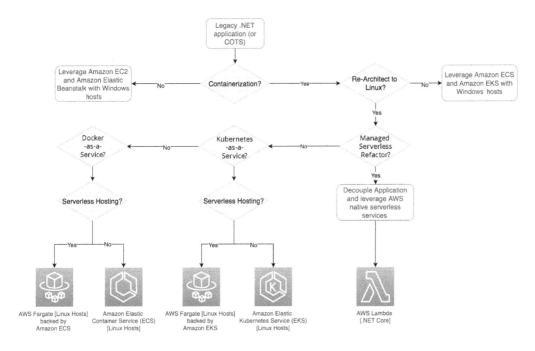

Figure 20.3: Refactoring pattern

Both the aforementioned routines provide a lot of long-term benefits to both the organization and the developers in terms of cost savings, scalability, agility, and security. However, refactoring your application code into true microservices or even serverless services requires effort in terms of costs and time, and in some cases, can involve a steep learning curve for developers and operational teams as well.In the next section of this chapter, we shall dive a bit deeper into these two routines and review the right choices for working with containers and serverless solutions.

Identifying Opportunities for Containers and Serverless Solutions

As you saw during the refactoring stage of our sample .NET application, there are in essence two application re-design choices that you can select based on your requirements: microservices using containers or going true serverless with AWS-managed serverless services. In this section, you shall explore both these options in a bit more detail as well as reviewing some interesting application design architectures and patterns along the way.

First, take a moment to understand what decoupling an application actually means. The decoupling of an application essentially means breaking up complex monolith code into smaller, more manageable chunks, with each chunk of code performing a single, specific task. Each chunk of code or microservice is designed to operate independently from the other microservices so that a change made in one microservice should not require changes to be made in others. Therefore, decoupled applications such as these can provide way better scale, agility, and resiliency compared to traditionally developed monoliths.

Once an application is decoupled, users need to address the hosting choices that come either as containers or as serverless cloud-native services, such as Amazon SQS and Amazon API Gateway. *Table 20.1* outlines some common considerations and guidelines for selecting the right serverless hosting option when it comes to microservices:

	Containers	**Serverless Services**
Architecture	Designed for virtually any workload that can be packaged and made to run as a container, such as AWS Fargate.	Designed for event-driven architectures, such as AWS Lambda.
Duration	Ideal for long-lived tasks.	Best suited for short-lived tasks (max. 15 minutes) that are generated based on certain events.
Costs	You do pay for idle resources, that is, even though the application is not performing any tasks, you still pay for the underlying compute, and storage. Charges are calculated and rounded off to the nearest second.	You do not pay for idle resources. Charges are calculated and rounded off to the nearest millisecond.
Performance	Min: 0.25 vCPU and 512 MiB RAM Max: 16 vCPU and 120 GB RAM	Min: 1 vCPU and 128 MB RAM Max: 6 vCPU and 10 GB RAM
Startup	Takes time for the container image to be pulled and deployed on a Fargate container, resulting in slow startup times.	Almost instantaneous startup times.

	Containers	Serverless Services
Operational overheads	Although the underlying hosts that run the containers are managed by AWS, you still need to package your application as a container image and deploy it, as well as manage the application's runtime.	Minimalistic overheads in terms of packaging the application, and the application runtime is managed by the cloud provider itself.
Scaling	Based on configurable rules.	Automated scaling based on incoming requests.
Application packaging and deployments	Relies on standard container/Docker packaging.	Supports application code as ZIP files as well as container/Docker images.
Integration	Limited integration capabilities with other AWS native services, except core services such as storage (Amazon EFS, Amazon FSx for Lustre), monitoring (Amazon CloudWatch, AWS X-Ray), and so on.	A lot more integration capabilities with AWS native services, including application development services (Amazon API Gateway, Amazon SNS, Amazon SQS), databases (Amazon DynamoDB), storage (Amazon S3, Amazon Kinesis), and so on.
Orchestration	Relies on external services such as Amazon ECS and Amazon EKS for container orchestration.	Relies on external services such as AWS Step Functions and Amazon Event Bridge for function orchestration.

Table 20.1: Comparing container and serverless hosting on AWS

Now, the common question that solutions architects often get asked is "Can we combine the benefits of both worlds, containers and serverless, in order to create loosely coupled systems?" The answer to this is yes, of course! There are lots of application development architectures and patterns out there, such as the hexagonal architecture, that focus on solving a particular problem statement by decomposing it into specific *domains* based on the concept of ports and adapters.

How does it work? The hexagonal architecture essentially divides a system into several loosely coupled domains. These domains represent the system's inner workings, such as the application code, its user interface, the backend database, the APIs, and so on. Each of these domains interacts with the others using one or more "ports," which can be implemented based on the application's requirements; for example; a port here could be anything from an API gateway to a web service, to a **Remote Procedure Call** (**RPC**). Ports are used to interface with the outside world using one or more adapters, as depicted in *Figure 20.4*:

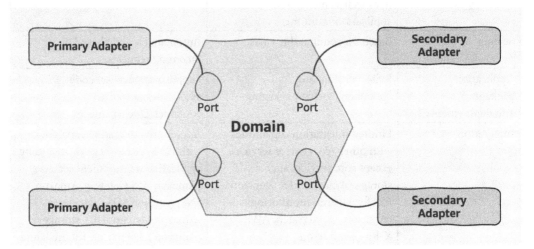

Figure 20.4: Anatomy of hexagonal architecture

Now, if we were to take this architecture and apply it to our sample web application (assuming it's already decoupled and broken down into microservices), *Figure 20.5* is what we would initially start out with. Assume the static portions of the web application are now re-architected for Amazon S3 and have a simple Amazon API Gateway and AWS Lambda interface in front of it for querying, as depicted in *Figure 20.5*:

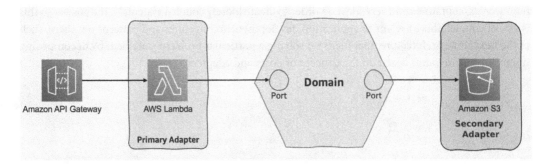

Figure 20.5: Starting out with a hexagonal architecture implementation for our sample application

Now if we were to apply the hexagonal architecture principles here, you can add more functionality to the sample web application by introducing different clients such as Amazon API Gateway, each client interfacing with the domain (application code) using a particular adapter (either a Lambda function for short-running task or a Fargate container for a long-running task). *Figure 20.6* depicts what the application looks like after more adapters are added to it:

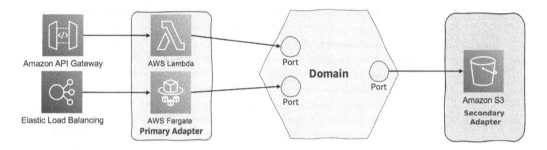

Figure 20.6: Adding more clients and adapters

As the domain's requirements increase, so does the number of changes needed in the adapters and ports as well. The best part of this type of architecture is that the domain always remains abstracted from the adapters using one or more ports. Since it is the adapter's responsibility to implement its corresponding port, the domain remains unchanged even if the adapters were to be swapped or replaced completely by a new set of adapters. With this design, it's much easier to add more functionality to your application as well since, in this case, all you need to do is write a new adapter without affecting the domain.

For example, suppose you are adding a NoSQL database in the form of Amazon DynamoDB to our sample application. Since the application is loosely coupled and relies on the hexagonal architecture, you only need to create a new DynamoDB adapter with its corresponding port and attach it to our domain!

Figure 20.7 depicts a system architecture involving various AWS components and their interactions with a central hexagonal domain that has four ports for communication.

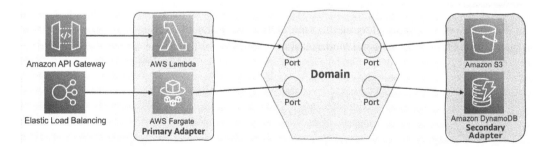

Figure 20.7: Adding new adapters and ports to extend the application's functionality

Similarly, you can now have multiple domains, each working on a set business objective, that are interconnected with each other using various adapters. Using this architectural best practice, you can now scale each of the domains and adapters individually based on your requirements, as depicted in *Figure 20.8*:

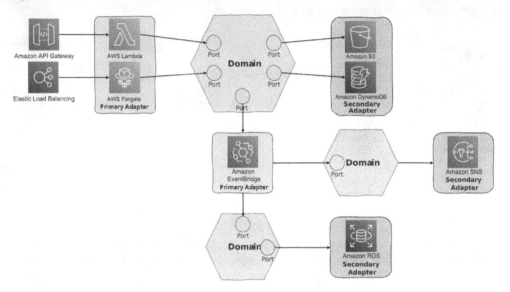

Figure 20.8: Expanding domains as per application requirements

With this, we come to the end of this section. In the next section, you will examine how to identify opportunities for purpose-built databases and how to leverage them for application modernizations.

Identifying Opportunities for Purpose-Built Databases

When it comes to purpose-built databases, AWS offers a select variety of databases that customers can choose from for their application workloads, including NoSQL, graph, time series, in-memory, search, and so on. In this section of the chapter, you will also explore these few select categories of database offerings and learn about identifying the right service for the right workloads.

For this section as well, we shall leverage the same SQL Server monolith database as an example and map it against each of the migration and modernization opportunities listed here, starting with rehosting.

Opportunity 1: Rehosting

Rehosting is the preferred choice for migrating even database workloads if you wish to have a faster mechanism of moving workloads to the cloud. Keeping in mind that our sample application is a SQL server, the options for rehosting are once again limited to leveraging Amazon EC2 instances itself. AWS provides licensed SQL Server AMIs that customers can choose to leverage if they do not wish to bring their own licenses to the cloud, or alternatively, spin up dedicated EC2 hosts as well for BYOL compatibility, as demonstrated in the following flowchart:

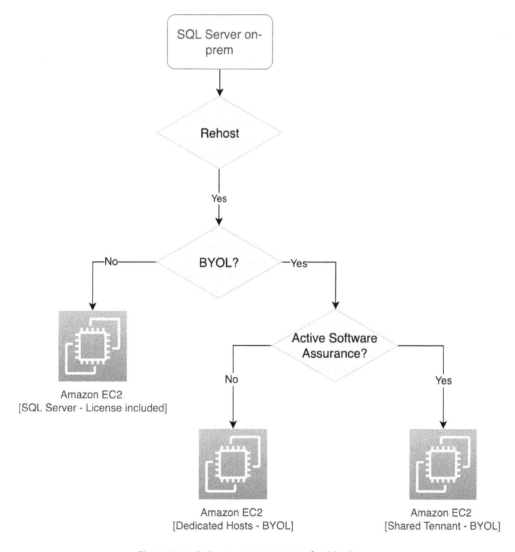

Figure 20.9: Rehosting opportunity for SQL Server

Opportunity 2: Replatforming

When it comes to replatforming commercial-grade databases such as Microsoft SQL Server, AWS provides you with two options:

- **Replatforming from a Windows OS to Linux:** This option is suitable for SQL Server workloads that are leveraging Windows as their underlying operating system that a customer wishes to migrate away from. AWS provides a Windows to Linux replatforming assistant that customers can leverage to assess their current SQL database, identify incompatibilities, export the backup to Amazon S3, and then upload it to a SQL server running on Ubuntu Linux as the operating system.**Replatforming to a managed service:** Unlike the previous instance where the customer is still responsible for the management of the underlying infrastructure, its patching, security, and so on, customers can choose to replatform their SQL Server database to other AWS managed services, including Amazon RDS for SQL Server to name one. *Figure 20.11* presents a representational workflow for replatforming a SQL Server database.

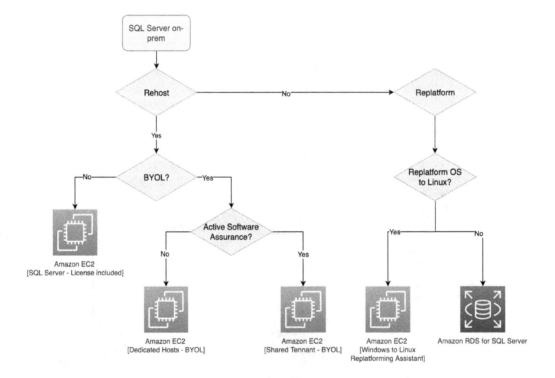

Figure 20.10: Replatforming opportunity for SQL Server

Opportunity 3: Refactoring

Refactoring is a heterogeneous database migration where customers take advantage of either the open source community-backed databases or cloud-native, purpose-built databases in order to modernize database workloads.

Both approaches are explained here:

- **Refactoring to Open Source**: Customers can choose to migrate away from SQL Server to open source databases such as MySQL, PostgreSQL, and MariaDB as well as using AWS **Database Migration Service** (**DMS**) coupled with the **Schema Conversion Tool** (**SCT**). Customers can also leverage Amazon RDS as a managed database offering to run MySQL, MariaDB, and PostgreSQL workloads with relative ease.

- **Refactoring to Cloud Native**: Depending on the workloads and customer requirements, sometimes open source databases are not performant enough compared to proprietary database offerings, including SQL Server and Oracle. In this case, AWS offers a purpose-built database-as-a-service called Amazon Aurora for customers who are specifically looking to move away from complex licensing and the costs of proprietary databases but still wish to have the same performance and features provided by them. Amazon Aurora is MySQL and PostgreSQL compatible and also comes in a totally serverless variant. In this case, customers do not have to manage any of the database's underlying capacity. The database scales dynamically based on the workload's requirements.

Figure 20.12 is a representational workflow depicting the options for refactoring a SQL Server database on AWS:

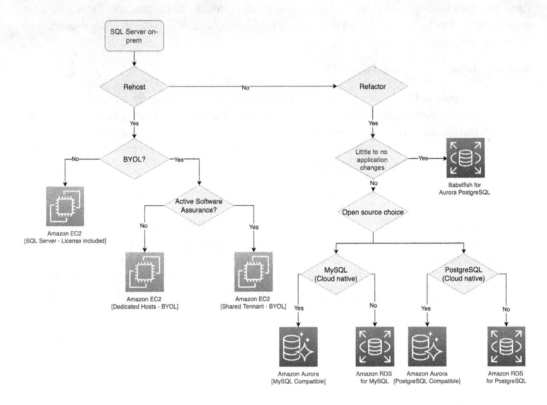

Figure 20.11: Refactoring opportunities for SQL Server on AWS

Opportunity 4: Rearchitecting

Rearchitecting here refers to moving away from the SQL Server database while leveraging specific, purpose-built databases provided by AWS to achieve the desired outcomes. These database options could include migrating away from SQL Server to NoSQL options such as Amazon DynamoDB or even Amazon DocumentDB to migrate the database to even more specific options such as Amazon Redshift for data warehousing.

Table 20.2 summarizes the migration methodology as well as the common use cases for moving workloads to purpose-built AWS databases:

	SQL using Amazon Aurora – PostgreSQL	NoSQL using Amazon DynamoDB	NoSQL using Amazon DocumentDB	Data warehouse using Amazon Redshift
Use case for migration	Refactoring exercise. Usually involves sticking with RDMS but moving away from proprietary database technology to something a bit more open source for cost (license) savings.	Rearchitecting exercise. Will require mapping of tables to key-value collections as well as modifying the application's data access layer to perform CRUD operations.	Rearchitecting exercise. Will require mapping of tables to JSON documents as well as modifying the application's data access layer to perform CRUD operations.	Rearchitecting exercise. Will require mapping of tables to columnar structures.
Migration execution	Migration using AWS services such as Babelfish for Aurora PostgreSQL	Migration using AWS DMS	Migration using AWS DMS	Migration using AWS DMS, as well as leveraging data extractors provided by the SCT
Effort required	No or minimal code changes to the application	Requires code overhaul to work with NoSQL	Requires code overhaul to work with NoSQL	Minimal to many code changes may be required depending on the application

Table 20.2: Rearchitecting SQL Server database to purpose-built options

With this, we come to the end of yet another section in the migration and modernization journey with AWS. The last section of this chapter will talk about how to identify and select the right integration service for your modernized workload and also take you through some basic application integration strategies.

Selecting the Appropriate Application Integration Service

When it comes to creating microservice-based applications and decoupling them, it becomes crucial to select the right set of integration services and patterns for your workloads. In this section, you will explore a few essential integration services provided by AWS as well as some interesting implementation patterns for microservices and serverless applications.

Table 20.3 summarizes the various AWS integration services that are available for customers to select from when it comes to serverless and microservices-based applications:

Category	Use Case	AWS Services Available
API Management	Create secure and scalable RESTful and WebSocket-based APIs for web applications	Amazon API Gateway
	Create secure GraphQL and pub/sub APIs to query and update data across multiple SaaS vendors, databases, microservices, and serverless applications	AWS AppSync
Messaging	Create and fan out notifications using mobile push, SMS, and email with high throughput	Amazon Simple Notification Service (SNS)
	Manage and store messages from various applications using a serverless queue	Amazon SQS
	Leverage Apache ActiveMQ and RabbitMQ as a managed service	Amazon MQ
Event Bus	Ingest, filter, transform, and deliver events to loosely coupled services, architectures	Amazon EventBridge
Workflows	Orchestrate serverless applications at scale	AWS Step Functions
	Coordinate complex workflows using a managed Apache Airflow	Amazon Managed Workflows for Apache Airflow (MWAA)
No code	Integrate data flows between most SaaS applications and AWS services with ease	Amazon AppFlow

Table 20.3: Application integration services and use cases

As you can see from the table, there are different mechanisms for integrating serverless and microservice-based applications using AWS native services. The use cases mentioned in the table represent a fraction of what can be achieved, and there is much more that you as a solutions architect can do with a combination of one or more integration services. The following subsections discuss some of these combinations and patterns and highlight the aspects to focus on when selecting the right service for the right job.

Pattern 1: API Gateway Pattern

As the name suggests, the API Gateway pattern provides an easy integration of serverless as well as microservices-based applications using RESTful APIs. The API Gateway pattern is ideal for web-based workloads that contain multiple clients interfacing across multiple services.

The pattern can also be used with applications that require a long connection poll to be maintained between clients and the application using WebSockets.

The following is a representational view of the API Gateway pattern being leveraged for a simple web application. You will notice that the pattern leverages a single Amazon API gateway for the login function (/login). However, you could also have dedicated API gateways for each of the other microservices as well if required. The added advantage of having multiple API gateways (/products, /processing APIs) designed in such a pattern is that it provides fine-grained control over the selected APIs' requests, throttle, monitoring, and chargeback. The obvious downside to this fan-out pattern is the added costs that each API gateway brings along with it.

Figure 20.13 presents a pattern for microservices architecture where different applications communicate with the backend services through one or more API gateways. The figure shows two scenarios: one with a single API gateway for both mobile and web applications, and another with multiple API gateways, each dedicated to a specific application.

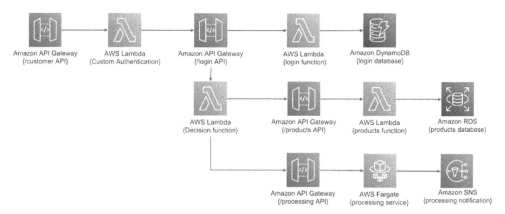

Figure 20.12: API Gateway pattern with single and multiple API
gateways implemented within the same architecture

Pattern 2: Messaging Pattern

A key integration method for highly decoupled applications is to leverage messaging queues as an asynchronous messaging mechanism. The clients publish and store messages in queues that are then asynchronously polled by a backend system. If a message is present, the backend system processes the message and if successful on completion, provides a form of acknowledgment back to the clients. If the processing fails, the message is either retried or sent to a special dead letter queue for further analysis. Such types of loosely coupled architectures allow the clients and backend systems to work completely independently of each other and thus reduce the impact on the overall system even if either of the components should fail.

Figure 20.13 is a representational diagram depicting Amazon SQS as a decoupling mechanism for processing images asynchronously using AWS Lambda functions:

Figure 20.13: Messaging pattern implementation using Amazon SQS

Pattern 3: Publish/Subscribe Pattern

The Publish/Subscribe pattern or pub/sub allows decoupled systems to communicate with each other without having to create any interdependencies on each other. This pattern can also encompass the messaging pattern wherein a client publishes a message to a particular channel that is subscribed to by one or more subscribers. Each subscriber gets a copy of the message and then in turn processes it as per their requirements without having to rely on the clients. The most common services used in this pattern are either Amazon SNS, through which a message is essentially broadcasted to various subscribers, or Amazon EventBridge, which relies on event rules to parse messages to their respective subscribers, as demonstrated in *Figure 20.14*:

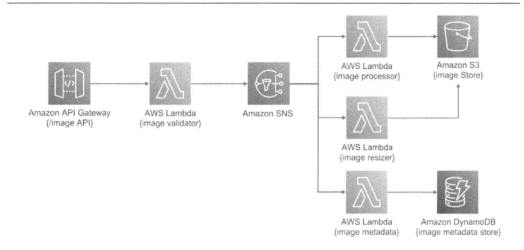

Figure 20.14: Pub/Sub pattern implementation using Amazon SNS

Summary

Throughout the chapters dealing with migration planning, you explored and understood the need for migrating applications and workloads to the cloud as well as the various mechanisms and tools available for undertaking such complex programs. You also examined various AWS services that aid the migration process and dived deep into the compute, storage, and database options that you can select to run your newly migrated workloads on AWS. Finally, in this chapter, you briefly learned about the steps involved in identifying modernization opportunities and how you can leverage AWS services including compute, database, and integrational services to achieve them.

Although this is the last chapter, you are not at the end of your learning journey. Reading this book, you will have covered the key domains of the SAP-C02 exam. You will have seen how you can use AWS to design solutions for organizational complexity, create new solutions for a growing modern enterprise, and continuously improve existing solutions. You will have also learned how AWS helps you migrate your workloads and accelerate modernization. However, to prepare for the exam, you should also check out the additional resources available online, including practice questions. The whole package should help you pass the SAP-C02 with ease, enabling you to continue your journey as an AWS Certified Solutions Architect Professional.

Further Reading

- Understand the hexagonal architecture pattern: `https://packt.link/N5gsI`

- Learn how to decouple large applications with Amazon EventBridge: `https://packt.link/FTThU`

- Explore how to modernize a SQL Server database to Amazon DynamoDB: `https://packt.link/Kndfi`

21
Accessing the Online Practice Resources

Your copy of *AWS Certified Solutions Architect – Professional Exam Guide (SAP-C02)* comes with free online practice resources. Use these to hone your exam readiness even further by attempting practice questions on the companion website. The website is user-friendly and can be accessed from mobile, desktop, and tablet devices. It also includes interactive timers for an exam-like experience.

How to Access These Resources

Here's how you can start accessing these resources depending on your source of purchase.

Purchased from Packt Store (packtpub.com)

If you've bought the book from the Packt store (`packtpub.com`) eBook or Print, head to `https://packt.link/sapc02practice`. There, log in using the same Packt account you created or used to purchase the book.

Packt+ Subscription

If you're a *Packt+ subscriber*, you can head over to the same link (`https://packt.link/sapc02practice`), log in with your `Packt ID`, and start using the resources. You will have access to them as long as your subscription is active.

If you face any issues accessing your free resources, contact us at `customercare@packt.com`.

Purchased from Amazon and Other Sources

If you've purchased from sources other than the ones mentioned above (like *Amazon*), you'll need to unlock the resources first by entering your unique sign-up code provided in this section. **Unlocking takes less than 10 minutes, can be done from any device, and needs to be done only once**. Follow these five easy steps to complete the process:

STEP 1

Open the link `https://packt.link/sapc02unlock` OR scan the following **QR code** (*Figure 21.1*):

Figure 21.1 – QR code for the page that lets you unlock this book's free online content.

Either of those links will lead to the following page as shown in *Figure 21.2*:

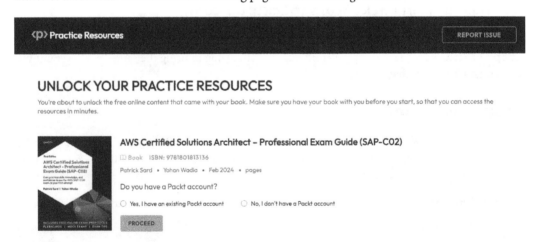

Figure 21.2 – Unlock page for the online practice resources

STEP 2

If you already have a Packt account, select the option `Yes, I have an existing Packt account`. If not, select the option `No, I don't have a Packt account`.

If you don't have a Packt account, you'll be prompted to create a new account on the next page. It's free and only takes a minute to create.

Click `Proceed` after selecting one of those options.

STEP 3

After you've created your account or logged in to an existing one, you'll be directed to the following page as shown in *Figure 21.3*.

Make a note of your unique unlock code:

`TDF5722`

Type in or copy this code into the text box labeled 'Enter Unique Code':

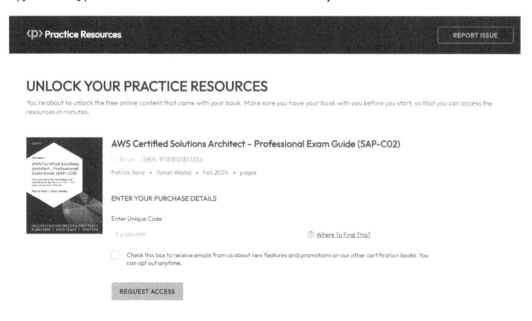

Figure 21.3 – Enter your unique sign-up code to unlock the resources

> **Troubleshooting Tip**
>
> After creating an account, if your connection drops off or you accidentally close the page, you can reopen the page shown in *Figure 21.2* and select `Yes, I have an existing account`. Then, sign in with the account you had created before you closed the page. You'll be redirected to the screen shown in *Figure 21.3*.

STEP 4

> **Note**
>
> You may choose to opt into emails regarding feature updates and offers on our other certification books. We don't spam, and it's easy to opt out at any time.

Click `Request Access`.

STEP 5

If the code you entered is correct, you'll see a button that says, OPEN PRACTICE RESOURCES, as shown in *Figure 21.4*:

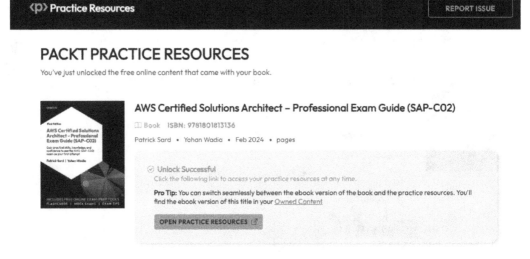

Figure 21.4 – Page that shows up after a successful unlock

Click the OPEN PRACTICE RESOURCES link to start using your free online content. You'll be redirected to the Dashboard shown in *Figure 21.5*:

Figure 21.5 – Dashboard page for AWS (SAP-C02) practice resources

Bookmark this link

Now that you've unlocked the resources, you can come back to them anytime by visiting https://packt.link/sapc02practice or scanning the following QR code provided in *Figure 21.6*:

Figure 21.6 – QR code to bookmark practice resources website

Troubleshooting Tips

If you're facing issues unlocking, here are three things you can do:

- Double-check your unique code. All unique codes in our books are case-sensitive and your code needs to match exactly as it is shown in *STEP 3*.

- If that doesn't work, use the `Report Issue` button located at the top-right corner of the page.

- If you're not able to open the unlock page at all, write to `customercare@packt.com` and mention the name of the book.

Practice Resources – A Quick Tour

This book will equip you with all the knowledge necessary to clear the exam. As important as learning the key concepts is, your chances of passing the exam are much higher if you apply and practice what you learn in the book. This is where the online practice resources come in. With interactive mock exams, flashcards, and exam tips, you can practice everything you learned in the book on the go. Here's a quick walkthrough of what you get.

A Clean, Simple Cert Practice Experience

You get a clean, simple user interface that works on all modern devices, including your phone and tablet. All the features work on all devices, provided you have a working internet connection. From the `Dashboard` (*Figure 21.7*), you can access all the practice resources that come with this book with just a click. If you want to jump back to the book, you can do that from here as well:

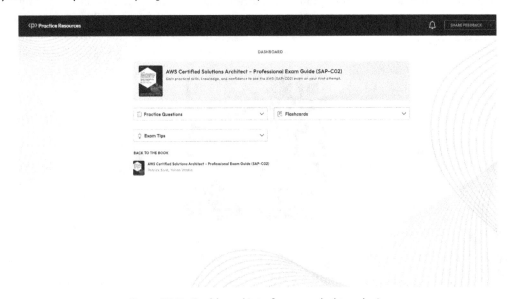

Figure 21.7 – Dashboard interface on a desktop device

Practice Questions

The Quiz Interface (*Figure 21.8*) is designed to help you focus on the question without any clutter.

You can navigate between multiple questions quickly and skip a question if you don't know the answer.

The interface also includes a live timer that auto-submits your quiz if you run out of time.

Click End Quiz if you want to jump straight to the results page to reveal all the solutions.

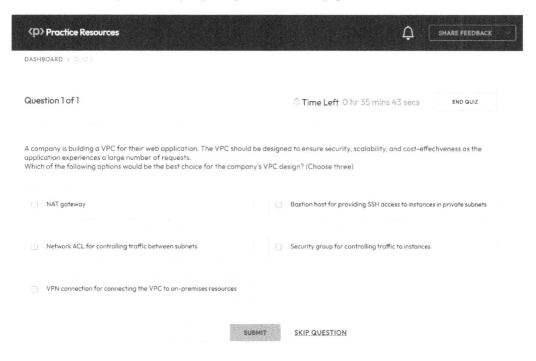

Figure 21.8: Practice questions interface on a desktop device

Be it a long train ride to work with just your phone or a lazy Sunday afternoon on the couch with your tablet, the quiz interface works just as well on all your devices as long as they're connected to the internet.

Figure 21.9 shows a screenshot of how the interface looks on mobile devices:

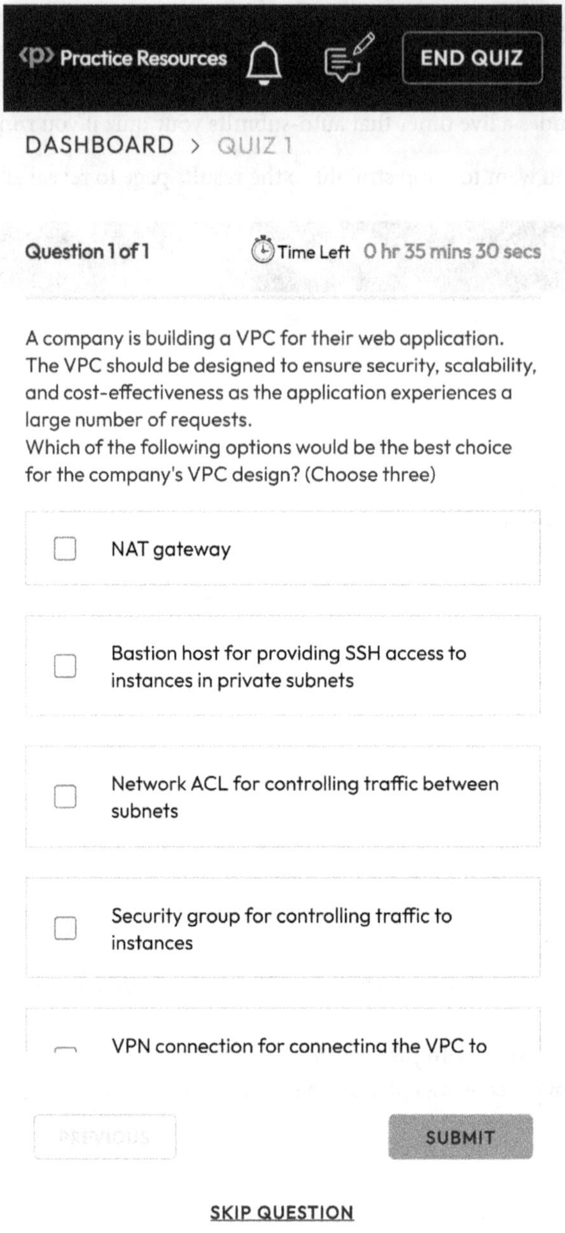

‹p› **Practice Resources**   🔔 ✎   **END QUIZ**

DASHBOARD > QUIZ 1

Question 1 of 1   ⏱ Time Left   0 hr 35 mins 30 secs

A company is building a VPC for their web application.
The VPC should be designed to ensure security, scalability,
and cost-effectiveness as the application experiences a
large number of requests.
Which of the following options would be the best choice
for the company's VPC design? (Choose three)

☐   NAT gateway

☐   Bastion host for providing SSH access to
      instances in private subnets

☐   Network ACL for controlling traffic between
      subnets

☐   Security group for controlling traffic to
      instances

☐   VPN connection for connecting the VPC to

PREVIOUS   SUBMIT

SKIP QUESTION

Figure 21.9: Quiz interface on a mobile device

Flashcards

Flashcards are designed to help you memorize key concepts. Here's how to make the most of them:

- We've organized all the flashcards into stacks. Think of these like an actual stack of cards in
- your hand.
- You start with a full stack of cards.
- When you open a card, take a few minutes to recall the answer.
- Click anywhere on the card to reveal the answer (*Figure 21.10*).
- Flip the card back and forth multiple times and memorize the card completely.
- Once you feel you've memorized it, click the Mark as memorized button on the top-right corner of the card. Move on to the next card by clicking Next.
- Repeat this process as you move to other cards in the stack.

You may not be able to memorize all the cards in one go. That's why, when you open the stack the next time, you'll only see the cards you're yet to memorize.

Your goal is to get to an empty stack, having memorized each flashcards in that stack.

Figure 21.10: Flashcards interface

Exam Tips

Exam Tips (see *Figure 21.11*) are designed to help you get exam-ready. From the start of your preparation journey to your exam day, these tips are organized such that you can review all of them in one go. If an exam tip comes in handy in your preparation, make sure to mark it as helpful so that other readers can benefit from your insights and experiences.

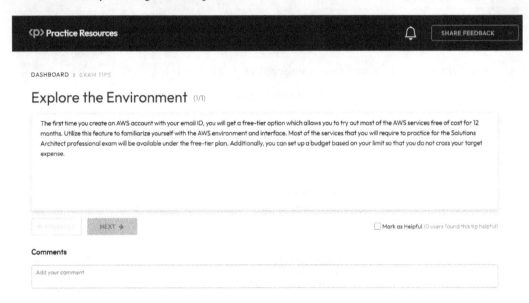

Figure 21.11 – Exam Tips Interface

Index

A

abnormal: 219, 230, 276

abstracted: 93, 369

abstracts: 181, 298

abundance: 128

accelerate: 58, 167, 226, 229, 313, 318, 333, 337, 379

acceptable: 36, 117, 202, 221, 249, 289

access-key: 75

account-id: 75, 339

accounts: 4, 8, 10, 13-17, 21, 38, 44-45, 49-51, 54-64, 67-72, 75, 84, 87, 97-98, 104, 106, 109, 113, 117, 128, 136, 154, 169, 189-190, 213, 215, 226, 233, 235, 238, 242, 266, 272, 276, 280, 284-286, 289-290, 305, 312, 340

actors: 106, 271, 283

adaptation: 253

adapted: 36, 95, 166, 273

addressed: 142, 270, 325

admins: 4

adopted: 150, 156, 175, 225, 307

advocate: 228, 293

affinity: 105

agility: 121, 142, 175, 248, 302, 308-309, 315, 364-366

agreement: 241

aligned: 141, 219

alignment: 219, 221, 317

aligns: 278

amazonvpc: 239

anonymized: 58

anycast: 115

appconfig: 299

apprunner: 196

appspec: 184-185

appsync: 97, 282

architect: 1, 72-73, 94, 114-115, 119, 121, 155, 174, 187, 256, 258, 260-262, 284, 301, 309, 319, 338, 343-345, 362, 377, 379

arm-based: 156, 302

associated: 3-4, 39, 41, 46, 55, 61, 75, 87-88, 90, 95, 102, 119, 130, 135, 141, 150, 182, 185, 203-205, 221, 270, 307

atomic: 133

attack: 97-98, 278, 282

attacker: 276, 289

attacks: 94, 97, 113, 275-277, 282

attribute: 6, 8, 11, 74, 286

attributes: 11-12, 16, 75, 201, 286

auditing: 250, 270, 280, 282

aurora: 131, 145-146, 148-149, 166-168,

D

dashboard: 2, 69, 77, 80, 129, 150, 173, 208-209, 211-212, 214, 256, 267, 280, 356

dashboards: 213, 230, 259

datagram: 97

dataset: 36, 168, 247, 331

datasets: 202, 334, 336, 356

debugging: 130

decouple: 345, 347, 361-362, 380

defense: 98

denied: 5, 10, 64, 67

deploy: 17, 31, 33-34, 36-37, 44, 46, 49-51, 62, 69, 93-94, 97, 112, 115-116, 118-119, 123, 133, 136-137, 142, 154, 158-159, 167, 171-172, 175-177, 179-186, 190, 192-196, 225, 232-233, 242, 248, 251-252, 256, 270, 276, 278, 283, 287-288, 291-299, 309, 326, 346

dnssec: 277

docker: 178, 192, 347, 350

document: 90-91, 109, 132, 140, 166, 235, 251, 262, 266, 287-288, 355

dummyuser: 75

duplicate: 8

duplicated: 45

durability: 36, 161-162, 165-166, 351

durable: 35-36, 143, 352

duration: 16, 23, 88, 146, 160, 200, 202, 230, 308, 313, 350

during: 9, 58, 97, 132, 164, 173, 176, 180, 184, 200-202, 206, 252, 261, 283, 289, 292, 295, 311, 314, 324-325, 348, 365

dynamic: 24, 46, 95, 131-132, 141, 161, 166, 263

dynamodb: 11, 42-43, 127, 131, 136, 143, 145-146, 148-149, 154, 166, 168, 191, 195, 204, 248, 262, 355-356, 358, 369, 374, 380

E

e-commerce: 172-173, 221, 258, 260

economic: 97

ecosystem: 158, 192, 194

elasticity: 113, 157

encrypt: 100-102, 104, 143, 280-281, 287

evolution: 8, 172, 174, 195, 220, 228, 266, 299

executive: 228

F

failing: 101, 126, 133

failover: 23-25, 116-117, 146, 149, 241

failovers: 325

families: 156-157, 167, 214

family: 33, 156-157, 164-165, 177, 182, 185, 201-202, 213-214, 259, 263, 298, 302, 332-334, 351

federal: 101, 280

federated: 9, 12, 16

file-level: 161, 290

filesystem: 33-34, 161-162, 164-165

finops: 73-75, 82

firewalls: 95-96, 180, 277

flowchart: 362, 371

folders: 34

footprint: 36, 99, 144, 258

forecast: 73, 85, 206, 208-209, 244

G

gained: 235, 275, 297

gaming: 170, 356

generating: 100, 126, 160

generation: 128, 156, 171, 201, 256, 280, 302

gitlab: 232

gitops: 291

www.packtpub.com

Subscribe to our online digital library for full access to over 7,000 books and videos, as well as industry leading tools to help you plan your personal development and advance your career. For more information, please visit our website.

Why subscribe?

- Spend less time learning and more time coding with practical eBooks and Videos from over 4,000 industry professionals

- Improve your learning with Skill Plans built especially for you

- Get a free eBook or video every month

- Fully searchable for easy access to vital information

- Copy and paste, print, and bookmark content

At www.packtpub.com, you can also read a collection of free technical articles, sign up for a range of free newsletters, and receive exclusive discounts and offers on Packt books and eBooks.

Other Books You May Enjoy

If you enjoyed this book, you may be interested in these other books by Packt:

AWS Certified Machine Learning - Specialty (MLS-C01) Certification Guide, Second Edition

Somanath Nanda and Weslley Moura

ISBN: 978-1-83508-220-1

- Identify ML frameworks for specific tasks
- Apply CRISP-DM to build ML pipelines
- Combine AWS services to build AI/ML solutions
- Apply various techniques to transform your data, such as one-hot encoding, binary encoder, ordinal encoding, binning, and text transformations
- Visualize relationships, comparisons, compositions, and distributions in the data
- Use data preparation techniques and AWS services for batch and real-time data processing
- Create training and inference ML pipelines with Sage Maker
- Deploy ML models in a production environment efficiently

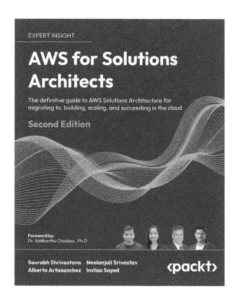

AWS for Solutions Architects, Second Edition

Saurabh Shrivastava, Neelanjali Srivastav, Alberto Artasanchez, and Imtiaz Sayed

ISBN: 978-1-80323-895-1

- Optimize your Cloud Workload using the AWS Well-Architected Framework
- Learn methods to migrate your workload using the AWS Cloud Adoption Framework
- Apply cloud automation at various layers of application workload to increase efficiency
- Build a landing zone in AWS and hybrid cloud setups with deep networking techniques
- Select reference architectures for business scenarios, like data lakes, containers, and serverless apps
- Apply emerging technologies in your architecture, including AI/ML, IoT and blockchain

Share Your Thoughts

Now you've finished *AWS Certified Solutions Architect – Professional Exam Guide (SAP-C02)*, we'd love to hear your thoughts! Scan the QR code below to go straight to the Amazon review page for this book and share your feedback or leave a review on the site that you purchased it from.

https://packt.link/r/1801813132

Your review is important to us and the tech community and will help us make sure we're delivering excellent quality content.

Download a Free PDF Copy of This Book

Thanks for purchasing this book!

Do you like to read on the go but are unable to carry your print books everywhere?

Is your eBook purchase not compatible with the device of your choice?

Don't worry, now with every Packt book you get a DRM-free PDF version of that book at no cost.

Read anywhere, any place, on any device. Search, copy, and paste code from your favorite technical books directly into your application.

The perks don't stop there, you can get exclusive access to discounts, newsletters, and great free content in your inbox daily.

Follow these simple steps to get the benefits:

1. Scan the QR code or visit the link below:

https://packt.link/free-ebook/9781801813136

2. Submit your proof of purchase.
3. That's it! We'll send your free PDF and other benefits to your email directly.